Springer Handbook of Auditory Research

Volume 65

AF801330

Series Editor

Richard R. Fay, Ph.D., Loyola University Chicag
Arthur N. Popper, Ph.D., University of Marylanc

Editorial Board

Karen Avraham, Ph.D., Tel Aviv University, Israel
Andrew Bass, Ph.D., Cornell University
Lisa Cunningham, Ph.D., National Institutes of Health
Bernd Fritzsch, Ph.D., University of Iowa
Andrew Groves, Ph.D., Baylor University
Ronna Hertzano, M.D., Ph.D., School of Medicine, University of Maryland
Colleen Le Prell, Ph.D., University of Texas, Dallas
Ruth Litovsky, Ph.D., University of Wisconsin
Paul Manis, Ph.D., University of North Carolina
Geoffrey Manley, Ph.D., University of Oldenburg, Germany
Brian Moore, Ph.D., Cambridge University, UK
Andrea Simmons, Ph.D., Brown University
William Yost, Ph.D., Arizona State University

More information about this series at http://www.springer.com/series/2506

The ASA Press

The ASA Press imprint represents a collaboration between the Acoustical Society of America and Springer dedicated to encouraging the publication of important new books in acoustics. Published titles are intended to reflect the full range of research in acoustics. ASA Press books can include all types of books published by Springer and may appear in any appropriate Springer book series.

Editorial Board

Mark F. Hamilton (Chair), University of Texas at Austin
James Cottingham, Coe College
Diana Deutsch, University of California, San Diego
Timothy F. Duda, Woods Hole Oceanographic Institution
Robin Glosemeyer Petrone, Threshold Acoustics
William M. Hartmann, Michigan State University
James F. Lynch, Woods Hole Oceanographic Institution
Philip L. Marston, Washington State University
Arthur N. Popper, University of Maryland
Martin Siderius, Portland State University
Andrea M. Simmons, Brown University
Ning Xiang, Rensselaer Polytechnic Institute
William Yost, Arizona State University.

ASA Press

Douglas L. Oliver · Nell B. Cant
Richard R. Fay · Arthur N. Popper
Editors

The Mammalian Auditory Pathways

Synaptic Organization and Microcircuits

ASA Press

Springer

Editors
Douglas L. Oliver
Department of Neuroscience
UConn Health, University of Connecticut
Farmington, CT, USA

Nell B. Cant
Department of Neurobiology
Duke University Medical Center
Durham, NC, USA

Richard R. Fay
Loyola University Chicago
Chicago, IL, USA

Arthur N. Popper
Department of Biology
University of Maryland
College Park, MD, USA

ISSN 0947-2657 ISSN 2197-1897 (electronic)
Springer Handbook of Auditory Research
ISBN 978-3-030-10111-4 ISBN 978-3-319-71798-2 (eBook)
https://doi.org/10.1007/978-3-319-71798-2

© Springer International Publishing AG 2018
Softcover re-print of the Hardcover 1st edition 2018
This work is subject to copyright. All rights are reserved by the Publisher, whether the whole or part of the material is concerned, specifically the rights of translation, reprinting, reuse of illustrations, recitation, broadcasting, reproduction on microfilms or in any other physical way, and transmission or information storage and retrieval, electronic adaptation, computer software, or by similar or dissimilar methodology now known or hereafter developed.
The use of general descriptive names, registered names, trademarks, service marks, etc. in this publication does not imply, even in the absence of a specific statement, that such names are exempt from the relevant protective laws and regulations and therefore free for general use.
The publisher, the authors and the editors are safe to assume that the advice and information in this book are believed to be true and accurate at the date of publication. Neither the publisher nor the authors or the editors give a warranty, express or implied, with respect to the material contained herein or for any errors or omissions that may have been made. The publisher remains neutral with regard to jurisdictional claims in published maps and institutional affiliations.

Printed on acid-free paper

This Springer imprint is published by Springer Nature
The registered company is Springer International Publishing AG
The registered company address is: Gewerbestrasse 11, 6330 Cham, Switzerland

Dr. Kent Morest (March, 2007).
(Photo by Frank Barton/UConn Health)

This volume is dedicated to D. Kent Morest, one of the premier neuroanatomists of the twentieth century. In a career spanning almost 60 years, Kent's research focused on the synaptic organization of the auditory system and on relationships among the neurons that make up its microcircuitry. Kent's mentorship and dedication to understanding the development and synaptic organization of specific neuron types motivated and guided several generations of students and colleagues. Thus, it is particularly appropriate to dedicate this work on the structure and function of the microcircuitry of the auditory system to him.

Acoustical Society of America

The purpose of the Acoustical Society of America (www.acousticalsociety.org) is to generate, disseminate, and promote the knowledge of acoustics. The Acoustical Society of America (ASA) is recognized as the world's premier international scientific society in acoustics, and counts among its more than 7000 members, professionals in the fields of bioacoustics, engineering, architecture, speech, music, oceanography, signal processing, sound and vibration, and noise control.

Since its first meeting in 1929, the ASA has enjoyed a healthy growth in membership and in stature. The present membership of approximately 7000 includes leaders in acoustics in the United States of America and around the world. The ASA has attracted members from various fields related to sound including engineering, physics, oceanography, life sciences, noise and noise control, architectural acoustics; psychological and physiological acoustics; applied acoustics; music and musical instruments; speech communication; ultrasonics, radiation, and scattering; mechanical vibrations and shock; underwater sound; aeroacoustics; macrosonics; acoustical signal processing; bioacoustics; and many more topics.

To assure adequate attention to these separate fields and to new ones that may develop, the Society establishes technical committees and technical groups charged with keeping abreast of developments and needs of the membership in their specialized fields. This diversity and the opportunity it provides for interchange of knowledge and points of view has become one of the strengths of the Society.

The ASA's publishing program has historically included *The Journal of the Acoustical Society of America, JASA-Express Letters, Proceedings of Meetings on Acoustics*, the magazine *Acoustics Today*, and various books authored by its members across the many topical areas of acoustics. In addition, ASA members are involved in the development of acoustical standards concerned with terminology, measurement procedures, and criteria for determining the effects of noise and vibration.

Series Preface

Springer Handbook of Auditory Research

The following preface is the one that we published in volume 1 of the Springer Handbook of Auditory Research back in 1992. As anyone reading the original preface, or the many users of the series, will note, we have far exceeded our original expectation of eight volumes. Indeed, with books published to date and those in the pipeline, we are now set for over 65 volumes in SHAR, and we are still open to new and exciting ideas for additional books.

We are very proud that there seems to be consensus, at least among our friends and colleagues, that SHAR has become an important and influential part of the auditory literature. While we have worked hard to develop and maintain the quality and value of SHAR, the real value of the books is very much because of the numerous authors who have given their time to write outstanding chapters and our many co-editors who have provided the intellectual leadership for the individual volumes. We have worked with a remarkable and wonderful group of people, many of whom have become great personal friends of both of us. We also continue to work with a spectacular group of editors at Springer. Indeed, several of our past editors have moved on in the publishing world to become senior executives. To our delight, this includes the current president of Springer US, Dr. William Curtis.

But the truth is that the series would and could not be possible without the support of our families, and we want to take this opportunity to dedicate all of the SHAR books, past and future, to them. Our wives, Catherine Fay and Helen Popper, and our children, Michelle Popper Levit, Melissa Popper Levinsohn, Christian Fay, and Amanda Fay Sierra, have been immensely patient as we developed and worked on this series. We thank them and state, without doubt, that this series could not have happened without them. We also dedicate the future of SHAR to our next generation of (potential) auditory researchers—our grandchildren—Ethan and Sophie Levinsohn, Emma Levit, and Nathaniel, Evan, and Stella Fay.

Preface 1992

The Springer Handbook of Auditory Research presents a series of comprehensive and synthetic reviews of the fundamental topics in modern auditory research. The volumes are aimed at all individuals with interests in hearing research including advanced graduate students, postdoctoral researchers, and clinical investigators. The volumes are intended to introduce new investigators to important aspects of hearing science and to help established investigators to better understand the fundamental theories and data in fields of hearing that they may not normally follow closely.

Each volume presents a particular topic comprehensively, and each serves as a synthetic overview and guide to the literature. As such, the chapters present neither exhaustive data reviews nor original research that has not yet appeared in peer-reviewed journals. The volumes focus on topics that have developed a solid data and conceptual foundation rather than on those for which a literature is only beginning to develop. New research areas will be covered on a timely basis in the series as they begin to mature.

Each volume in the series consists of a few substantial chapters on a particular topic. In some cases, the topics will be ones of traditional interest for which there is a substantial body of data and theory, such as auditory neuroanatomy (Vol. 1) and neurophysiology (Vol. 2). Other volumes in the series deal with topics that have begun to mature more recently, such as development, plasticity, and computational models of neural processing. In many cases, the series editors are joined by a co-editor having special expertise in the topic of the volume.

Richard R. Fay, Chicago, IL, USA
Arthur N. Popper, College Park, MD, USA

SHAR logo by Mark B. Weinberg, Potomac, Maryland, used with permission.

Preface 1992

Volume Preface

In 1992, volume 1 of the Springer Handbook of Auditory Research series, *The Mammalian Auditory Pathway: Neuroanatomy*,[1] presented the results of many decades of work that produced a basic understanding of the important structures in the auditory system and their major interconnections. Each structure was presented in terms of its major subdivisions and, when possible, its component cell types, its major sources of inputs and projection targets, and, in some cases, the fine structure of its neurons and synapses. Since then, largely due to the development of new methodologies, the ability to discuss different types of neurons and the microcircuits they form has expanded considerably. As a consequence, ways of studying the anatomy of the auditory system have changed dramatically, and these changes in approach to the subject are the focus of this "update" of volume 1 in the series.

In order to set the stage for the discussion of synaptic organization and microcircuits in the mammalian auditory pathway, Nell B. Cant and Douglas L. Oliver summarize the known afferent projections to major regions of the auditory pathway in Chap. 2. These inputs represent the potential sources of extrinsic control of activity in auditory nuclei. The discussion of auditory microcircuits begins in Chap. 3, by Maria E. Rubio, with the functional organization of specific, well-defined neuronal populations in the ventral cochlear nucleus and an emphasis on molecular specializations among their synaptic inputs.

In Chap. 4, Laurence O. Trussell and Donata Oertel discuss complex synaptic arrangements in the dorsal cochlear nucleus and the role of its circuitry in multisensory integration and synaptic plasticity. In Chap. 5, Conny Kopp-Scheinpflug and Ian D. Forsythe discuss synaptic integration in the superior olivary complex, a group of nuclei that are interconnected by a rich network of mainly inhibitory intrinsic circuits.

In Chap. 6, Tetsufumi Ito and Manuel S. Malmierca describe the inferior colliculus at a cellular level and present recent evidence that has shed new light on the neuron types in that structure. Following this, the medial geniculate body and the

[1] Webster, D. B., Popper, A. N., & Fay, R. R. (Eds.). (1992). *The mammalian auditory pathways: Neuroanatomy*. New York: Springer-Verlag.

thalamocortical auditory system are described by Heather Read and Alex D. Reyes in Chap. 7. They discuss new concepts based on the results of powerful new imaging techniques for monitoring the activity of large regions of the brain simultaneously.

In Chap. 8, Eike Budinger and Patrick Kanold discuss new recording techniques that have been developed over the past few decades that have allowed investigators to begin to understand how the cell types and microcircuitry of the auditory cortex perform the integration of information that results in perceptual representation of the auditory scene. Finally, in Chap. 9, Brett R. Schofield and Laura Hurley discuss the organization of four modulatory systems that have a widespread influence on all parts of the brain, including the auditory system.

Douglas L. Oliver, Farmington, CT, USA
Nell B. Cant, Durham, NC, USA
Richard R. Fay, Chicago, IL, USA
Arthur N. Popper, College Park, MD, USA

Contents

Contributors

Eike Budinger Department of Systems Physiology of Learning, Leibniz Institute for Neurobiology, Magdeburg, Germany

Nell B. Cant Department of Neurobiology, Duke University Medical Center, Durham, NC, USA

Ian D. Forsythe Department of Neuroscience, Psychology and Behaviour, University of Leicester, Leicester, UK

Laura Hurley Department of Biology, Indiana University, Bloomington, IN, USA

Tetsufumi Ito Department of Anatomy, School of Medicine, Kanazawa Medical University, Uchinada, Ishikawa, Japan

Patrick O. Kanold Department of Biology and Institute for Systems Research, University of Maryland, College Park, MD, USA

Conny Kopp-Scheinpflug Division of Neurobiology, Department of Biology II, Ludwig-Maximilians-University Munich, Planegg-Martinsried, Germany

Manuel S. Malmierca Auditory Neuroscience Laboratory, Institute for Neuroscience of Castilla y León, University of Salamanca, Salamanca, Spain

Department of Cell Biology and Pathology, Faculty of Medicine, University of Salamanca, Salamanca, Spain

The Salamanca Institute for Biomedical Research (IBSAL), Salamanca, Spain

Donata Oertel Department of Neuroscience, University of Wisconsin, Madison, WI, USA

Douglas L. Oliver Department of Neuroscience, UConn Health, University of Connecticut, Farmington, CT, USA

Heather L. Read Department of Psychological Sciences and Department of Biomedical Engineering, University of Connecticut, Storrs, CT, USA

Kavli Institute for Theoretical Physics, University of California, Santa Barbara, CA, USA

Alex D. Reyes Center for Neural Science, New York University, New York, NY, USA

Maria E. Rubio Department of Neurobiology and Otolaryngology, University of Pittsburgh Medical School, Pittsburgh, PA, USA

Brett R. Schofield Department of Anatomy and Neurobiology, Northeast Ohio Medical University, Rootstown, OH, USA

Laurence O. Trussell Oregon Hearing Research Center and Vollum Institute, Oregon Health and Science University, Portland, OR, USA

Chapter 1
Introduction to Mammalian Auditory Pathways

Douglas L. Oliver and Nell B. Cant

Abstract This volume in the Springer Handbook of Auditory Research series discusses the anatomy of the auditory system with an emphasis on how the local connections of specific neurons within auditory structures control the processing of sound. The volume is dedicated to D. Kent Morest, a leader of neuroanatomical studies of the auditory system in the twentieth century who stressed the importance of the circuits connecting specific types of neurons in auditory processing. In this introductory chapter, a brief history of the field of functional neuroanatomy is provided and the contents of each chapter are summarized.

Keywords Auditory cortex · Auditory system · Neuroanatomy · Neurophysiology

1.1 Introduction

Twenty-five years ago, volume one of the Springer Handbook of Auditory Research (SHAR) series (*The Mammalian Auditory Pathway: Neuroanatomy*) began with an overview of the anatomy of the mammalian auditory system (Webster 1992). The system was illustrated by "wiring diagrams" emphasizing the neural pathways leading from the ear to the cerebral cortex. Basic research since that time has resulted in significant increases in knowledge about the components of these fiber pathways and the synaptic organization of their terminations within auditory nuclei. In this new volume, the emphasis shifts from an analysis of connections *between* structures

D. L. Oliver (✉)
Department of Neuroscience, UConn Health, University of Connecticut, Farmington, CT, USA
e-mail: doliver@uchc.edu

N. B. Cant
Department of Neurobiology, Duke University Medical Center, Durham, NC, USA
e-mail: nellcant@neuro.duke.edu

© Springer International Publishing AG 2018

D. L. Oliver et al. (eds.), *The Mammalian Auditory Pathways*,
Springer Handbook of Auditory Research 65,
https://doi.org/10.1007/978-3-319-71798-2_1

to an analysis of the patterns of connectivity *within* those structures, that is, an analysis of the microcircuitry of the auditory system.

The ability to discuss different types of neurons and the microcircuits they form has been enhanced significantly by the advent of new methods and approaches. The first definitions of neuron types were based on their morphologies. Ramón y Cajal was an early master of this approach, using the Golgi method to describe specializations in dendritic and axonal structures (e.g., Ramón y Cajal 1911). Comprehensive studies in the later twentieth century provided catalogs of neuronal types based on morphological specializations of dendritic and axonal anatomy (reviewed in Cant 1992; Winer 1992). About the same time, physiologists began to define neuron types based on their responses to relatively simple sound stimuli (reviewed in Rhode and Greenberg 1992). Twenty-five years ago, those approaches were primarily employed independently as evidenced by the separation of anatomical and physiological descriptions into two SHAR volumes (Popper and Fay 1992; Webster et al. 1992). Even at that time, however, new methods were emerging that allowed combined studies of the structures and functions of the auditory system. For example, intracellular recordings in the cochlear nucleus were combined with intracellular injections of markers that allowed visualization of the morphology of the recorded cells. Such studies were further extended by the examination of the synaptic organization of terminals arising from known sources (reviewed in Rhode and Greenberg 1992). The chapters in this new volume serve to illustrate how combinations of new anatomical, cellular, molecular, and physiological techniques have transformed the discussion of neuronal types and microcircuitry.

It is now possible to classify neurons based on a wide variety of features, such as the characteristics of their ion channels, the somatic and dendritic arrangements of their inputs from specific sources, the specific proteins involved in neurotransmitter synthesis and trafficking, and the targeting of their axons. Further, powerful new methods, including viral transfection, have been developed for the genetic labeling of specific cell types and for visualization of specific gene products. Advances in optical imaging and manipulation of membrane channels through photostimulation have been combined with genetic approaches to allow systematic examination of the physiological interactions of known cell populations.

As discussed in this volume, these and other approaches have been employed to identify the cellular and synaptic components of specific auditory circuits. In each part of the auditory system, it is now possible to visualize synaptic inputs to defined neuronal populations from specific extrinsic sources as well as those from local axons. Instead of intracellular recording and labeling in single neurons, it is now possible to study the function(s) of two or more interconnected neurons at the same time. The new methods allow an integration of synaptic structure and function that is far more powerful than possible previously. Ultimately, the goals are to identify the local microcircuits that incorporate each neuronal type and to understand the interactions within the microcircuitry that create the output that is transmitted to the other parts of the auditory system. This volume tracks the progress in identification of the microcircuits that define distinct regions of the auditory system.

1.2 Overview of Chapters

The major nuclei and cortical areas and their connections to each other are reviewed in Chap. 2 by Nell B. Cant and Douglas L. Oliver as an introduction to the potential components of microcircuitry at each level of the auditory pathways. In summarizing the sources of afferent projections to each region, it becomes apparent that there is a unique pattern of long connections for each major nucleus and cortical subdivision. New information about basic auditory connectivity continues to accumulate as evidenced by the fact that well over half of the references cited in this chapter were published after 1992. Although the emphasis in the subsequent chapters in this volume is on the use of new methods and approaches, many of the studies cited in Chap. 2 were conducted using traditional neuroanatomical techniques, which serves as a reminder that these methods are still appropriate in some instances. Information about potential sources of extrinsic input to each region provides background for the subsequent chapters. In each chapter, experts in auditory neuroanatomy and physiology discuss new insights into the structural basis for functions in the auditory nuclei at every level of the central nervous system from the cochlear nucleus to the auditory cortex. At each level, combinations of approaches are described that have been chosen to address questions specific to that level.

The ventral cochlear nucleus has a rich history of anatomical analysis, and the characterization of unique neuronal types in that structure serves as a model for other areas of the auditory system. As discussed in Chap. 3 by Maria E. Rubio, a variety of molecular techniques combined with tracing and recording techniques have allowed known cell populations to be tagged and studied to provide detailed information about the synaptic organization of each cell type. Rubio also discusses some of the key molecular components of specialized synapses in the ventral cochlear nucleus and describes ways in which these synapses may be altered by experience-dependent plasticity and after hearing loss.

In Chap. 4, Laurence O. Trussell and Donata Oertel address synaptic interactions in the dorsal cochlear nucleus, which is a highly organized structure containing a number of inhibitory and excitatory interneurons. This neural circuitry integrates information from both auditory and nonauditory sources, including the somatosensory system. Activity in the circuits drastically alters incoming information from the cochlea, and the output of the dorsal cochlear nucleus to the midbrain bears little resemblance to its input. Physiological analyses of specific components of these circuits have allowed insights into the role of synaptic plasticity even at this very early stage of auditory processing.

Conny Kopp-Scheinpflug and Ian D. Forsythe, the authors of Chap. 5, describe interactions within the superior olivary complex, a region that contains neurons that may be among the most specialized in the brain in terms of their synaptic, molecular, and circuit properties. The segregation of different neuron populations into separate areas in the complex allows detailed analysis of the cellular and electrical properties of neurons with well-defined sources of input and with known projection targets. Kopp-Scheinpflug and Forsythe discuss mechanisms of local inhibition in a functional context, especially with respect to the role of the superior olivary complex in binaural integration. The importance of specializations in intrinsic ionic conductances is emphasized; together with morphological and synaptic specializations, they shape the responses to acoustic stimuli.

Information from the dorsal and ventral cochlear nuclei and from the nuclei of the superior olivary complex converges in the inferior colliculus. Until recently, it has been difficult to define cell populations in this structure unambiguously, but new methods have revealed molecular and cellular properties that allow precise definitions of neural types. In Chap. 6, Tetsufumi Ito and Manuel S. Malmierca discuss molecular techniques used to identify the large GABAergic neurons (neurotransmitter: gamma-aminobutyric acid) that differ from other collicular neurons in terms of their synaptic organization, inputs, and axonal targeting. Ito and Malmierca also explore the evidence that inputs to the inferior colliculus from different sources are actually segregated within the boundaries of the larger anatomical regions, which may reflect a substrate for parallel channels processing different types of auditory information. Thus, Chap. 6 shows how the neuron types, microcircuits, and subregions of the inferior colliculus constitute an interface between the auditory brainstem and the auditory forebrain.

The medial geniculate body and the thalamocortical auditory pathways are described by Heather L. Read and Alex D. Reyes in Chap. 7. Even today, most summaries of the forebrain auditory pathways represent the projection from the ventral division of the medial geniculate nucleus to the core auditory cortex as unitary, but this chapter emphasizes that there are multiple parallel pathways connecting these forebrain areas. Optical imaging and high density electrophysiological recordings have revealed functionally distinct fields within the core auditory cortex, and companion anatomical tracing studies have shown how specific regions of the medial geniculate convey information to these separate cortical fields. The studies described represent a major advance in thinking about forebrain pathways with insights into thalamocortical architecture and termination patterns that should help to explain how the areas in core auditory cortex differ in the processing of information about sound.

Within the auditory cortex, basic cell types are well-established; however, the intrinsic circuitry in auditory cortex is less understood than the circuitry in other areas. Indeed, most of the synapses on cortical neurons arise intracortically, making it difficult or impossible to study their specific sources of input with traditional neuroanatomical tracing techniques. In Chap. 8, Eike Budinger and Patrick Kanold discuss the use of new in vivo and in vitro imaging and recording methods that are transforming our understanding of the structure and function of the microcircuits in the auditory cortical areas. These approaches will help to shed light on how the cell types and microcircuitry of the cortex perform the integration of information that results in perceptual representation of the auditory scene.

The final chapter by Brett R. Schofield and Laura Hurley (Chap. 9) is a review of the nonauditory inputs to auditory nuclei whose function in auditory circuitry has been relatively unclear. With the advent of techniques for visualization and mapping of these inputs, this is changing. Schofield and Hurley discuss four major modulatory neurochemical systems that have widespread influence on almost all auditory nuclei. The sources of the modulators are described along with the issue of the co-release of neuromodulators. Schofield and Hurley present potential pathways for global modification of the behavior of microcircuits in the auditory system in response to changes in both internal brain states and the external environment.

1.3 Perspective

In 1975, D. Kent Morest, to whom this volume is dedicated, noted that, "It is necessary to define circuits in terms of the connections of individual neurons or, at least, specific types of neurons" (Morest 1975). The chapters in this volume illustrate progress toward this goal, but this is not the end of the road. There remains much to learn about the synaptic organization and microcircuits of the mammalian auditory pathway. Techniques from molecular biology and genetics will make it possible to identify and mark specific types of neurons. Multicellular recording techniques will show how these neurons are interconnected to function in local microcircuits and to contribute to the auditory network as a whole. The ability to identify specific neuronal populations, combined with a knowledge of their synaptic connections, opens a new world of possible manipulations of the central auditory system and offers many new opportunities for developing therapies to treat auditory processing disorders of many types, including diseases such as tinnitus. A continued emphasis on the neurons and microcircuits of the auditory system will be necessary to understand how neural processing of sound stimuli results in hearing and perception. As the microcircuitry within the auditory system is further defined, future investigations can address the mechanisms that control how information about sound is processed and transformed as it passes through the auditory pathway.

Compliance with Ethics Requirements Douglas L. Oliver is a consultant for Pfizer, Inc.

Nell Cant declares that she has no conflict of interest.

References

Cant, N. B. (1992). The cochlear nucleus: Neuronal types and their synaptic organization. In D. B. Webster, A. N. Popper, & R. R. Fay (Eds.), *The mammalian auditory pathway: Neuroanatomy* (pp. 66–116). New York: Springer.

Morest, D. K. (1975). Structural organization of the auditory pathways. In D. B. Tower (Ed.), *The nervous system: Vol 3. Human communication and its disorders* (pp. 19–29). New York: Raven Press.

Popper, A. N., & Fay, R. R. (Eds.). (1992). *The mammalian auditory pathway: Neurophysiology*. New York: Springer.

Ramón y Cajal, S. (1911). *Histology of the nervous system of man and vertebrates* (Vols. 1, 2). (N. Swanson & L. W. Swanson, Trans. 1995). New York: Oxford University Press.

Rhode, W. S., & Greenberg, S. (1992). Physiology of the cochlear nuclei. In A. N. Popper & R. R. Fay (Eds.), *The mammalian auditory pathway: Neurophysiology* (pp. 94–152). New York: Springer.

Webster, D. B. (1992). An overview of mammalian auditory pathways with an emphasis on humans. In D. B. Webster, A. N. Popper, & R. R. Fay (Eds.), *The mammalian auditory pathway: Neuroanatomy* (pp. 1–22). New York: Springer.

Webster, D. B., Popper, A. N., & Fay, R. R. (Eds.). (1992). *The mammalian auditory pathway: Neuroanatomy*. New York: Springer.

Winer, J. A. (1992). The functional architecture of the medial geniculate body and the primary auditory cortex. In D. B. Webster, A. N. Popper, & R. R. Fay (Eds.), *The mammalian auditory pathway: Neuroanatomy* (pp. 222–409). New York: Springer.

Chapter 2
Overview of Auditory Projection Pathways and Intrinsic Microcircuits

Nell B. Cant and Douglas L. Oliver

Abstract The major structures in the auditory brainstem and forebrain are described in terms of the complement of inputs from extrinsic sources that, along with local collaterals and interneurons, make up the microcircuitry that is characteristic of each structure.

Keywords Auditory cortex · Auditory neuroanatomy · Cochlear nucleus · Inferior colliculus · Lateral lemniscus · Medial geniculate · Superior olivary complex

2.1 Introduction

The neuroanatomy of the mammalian auditory system is complex, and the details are often a source of confusion and frustration for both students and auditory researchers alike. The ascending pathways consist of a maze of interconnected brainstem and forebrain structures that culminate in projections to the auditory cortex (AC). In turn, cortical neurons project to the thalamus and brainstem. Both the brainstem and forebrain contain numerous auditory nuclei, each with its own unique neural organization and physiological response properties. In general, each nucleus contains multiple neuronal populations that are distinguished by differences in morphology, physiology, intrinsic membrane properties, neurotransmitter profiles, sources of inputs, and projection targets, among other variables. The functional role of each auditory structure is ultimately determined by the patterns of synaptic

---

N. B. Cant (✉)
Department of Neurobiology, Duke University Medical Center, Durham, NC, USA
e-mail: nellcant@neuro.duke.edu

D. L. Oliver
Department of Neuroscience, UConn Health, University of Connecticut, Farmington, CT, USA
e-mail: doliver@uchc.edu

© Springer International Publishing AG 2018
D. L. Oliver et al. (eds.), *The Mammalian Auditory Pathways*,
Springer Handbook of Auditory Research 65,
https://doi.org/10.1007/978-3-319-71798-2_2

organization of each neuronal type present. Both extrinsic and intrinsic connections contribute to the organization.

Studies of microcircuitry are almost by definition studies of the interdependence of structure and function. Neuroanatomical techniques and methods of analysis allow physiological studies to be targeted at well-defined neuronal populations. Questions about mechanisms underlying specific functional properties help to focus and guide neuroanatomical investigations. The interrelationship is evident at all levels of the central auditory system and is a major theme of this volume.

Structures from the cochlea to the cortex are connected by long pathways made up of myelinated and, less often, unmyelinated fibers. Although most long-range projections terminate in excitatory synapses, a significant number are inhibitory. Every auditory structure receives input from a variety of sources, and every structure, in turn, is connected to a variety of targets. In some cases, the sources of input and the targets of projections can be identified as distinct cell types, but in no case is it currently possible to trace with certainty the flow of information from one well-characterized neuronal population to the next along the entire neuraxis. Continued research in fundamental neuroanatomy and neurophysiology is necessary to supply this information.

The auditory pathways are often divided into ascending and descending components, but at the neuronal level, this is a somewhat arbitrary distinction. The terminal synaptic fields formed by the axons of these two components often overlap and *together* they form the basic circuitry that determines function. The neurons in each structure receive a combination of excitatory and inhibitory inputs arranged in specific patterns on the cell surface, and their activity patterns are shaped by all of these ascending and descending inputs. In some cases, the descending inputs appear to give rise to as many (or more) synaptic terminations as do the ascending inputs, and the functions of many neurons are dependent on the interactions among both types of inputs (e.g., Nakamoto et al. 2008; Bajo and King 2013).

A large number of reviews are available that summarize the overall connectivity of the auditory system (e.g., Smith and Spirou 2002; Malmierca 2003). This chapter focuses on the components that make up the microcircuitry at each level. Knowledge of the synaptic organization characteristic of each neuronal population serves to constrain the interpretation of physiological data and helps to refine computational models of circuit behavior. The chapters in this volume are devoted to discussions of the functional significance of the microcircuitry elaborated at each level of the auditory pathway.

2.2 The Cochlear Nucleus

The cochlear nucleus (CN) is made up of two distinct components, the laminated dorsal cochlear nucleus (DCN) and the ventral cochlear nucleus (VCN) (see Table 2.1 for a list of all abbreviations). Both subdivisions contain several types of large projection neurons that send axons to a variety of brainstem nuclei (reviewed

Table 2.1 Abbreviations

AC	Auditory cortex
AVCN	Anteroventral cochlear nucleus
BIC	Brachium of the inferior colliculus
c (used as a subscript)	Contralateral
CN	Cochlear nucleus
DCN	Dorsal cochlear nucleus
DNLL	Dorsal nucleus of the lateral lemniscus
GABA	Gamma-aminobutyric acid
GCD	Granule cell domain (of the CN)
i (used as a subscript)	Ipsilateral
IC	Inferior colliculus
INLL	Intermediate nucleus of the lateral lemniscus
LNTB	Lateral nucleus of the trapezoid body
LSO	Lateral superior olivary nucleus
MG	Medial geniculate nucleus
MNTB	Medial nucleus of the trapezoid body
MOC	Medial olivocochlear system
MSO	Medial superior olivary nucleus
NLL	Nuclei of the lateral lemniscus
PO	Periolivary
PVCN	Posteroventral cochlear nucleus
SOC	Superior olivary complex
SPN	Superior paraolivary nucleus
SR	Spontaneous rate
TRN	Thalamic reticular nucleus
VCN	Ventral cochlear nucleus
VNLL	Ventral nucleus of the lateral lemniscus
VNTB	Ventral nucleus of the trapezoid body

by Cant and Benson 2003; Oertel et al. 2011). The DCN also contains several distinct types of interneurons that make up a complex internal circuitry (Diño and Mugnaini 2008). The VCN is often divided into the anteroventral cochlear nucleus (AVCN) and the posteroventral cochlear nucleus (PVCN), which are separated by the fibers of the cochlear nerve root. To a certain extent, these two regions contain different neuronal populations, but the segregation is not absolute, as some types of neurons may be located in both subdivisions. For the purposes of the present discussion, the most important distinction is between the magnocellular core of the VCN, where the large projection neurons are located, and the granule cell domain (GCD), which surrounds the core in a thin layer and contains granule cells and several types of small cells (Mugnaini et al. 1980). Axons (parallel fibers) arising from granule cells located throughout the GCD project into the DCN and form an integral part of its internal circuitry. Inputs to the cochlear nuclei are summarized in Fig. 2.1.

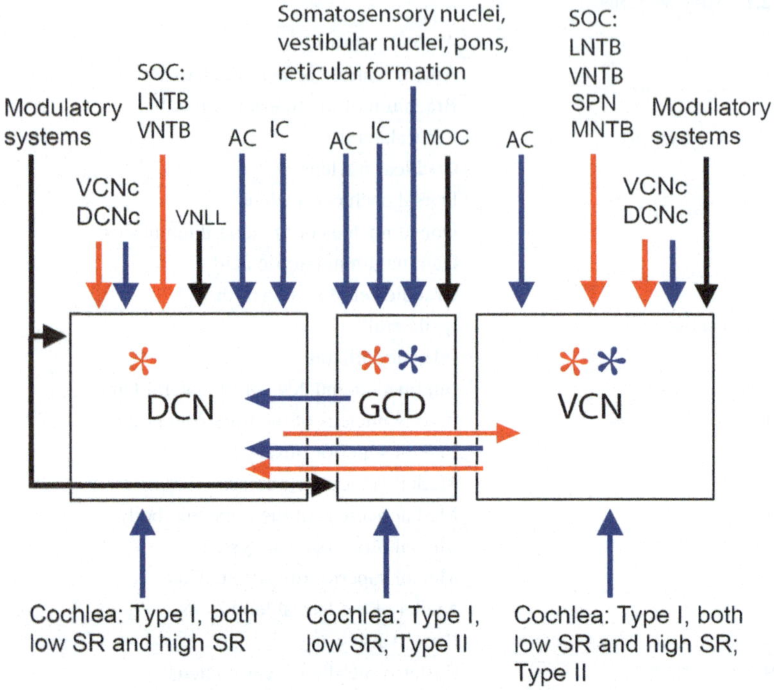

Fig. 2.1 Schematic diagram that summarizes the major sources of input to the cochlear nuclei (Dorsal, *DCN*; Ventral, *VCN*) and the granule cell domain (*GCD*). Known excitatory inputs are indicated in *blue*; known inhibitory inputs are indicated in *red*. Inputs for which the synaptic effects have not been fully characterized are indicated in *black*. *Asterisks* represent excitatory (*blue*) and inhibitory (*red*) synaptic inputs arising from neurons intrinsic to the structures illustrated. (*AC*, auditory cortex; *c*, contralateral; *IC*, inferior colliculus; *LNTB*, lateral nucleus of the trapezoid body; *MNTB*, medial nucleus of the trapezoid body; *MOC*, medial olivocochlear system; *SOC*, superior olivary complex; *SPN*, superior paraolivary nucleus; *SR*, spontaneous rate; *VNLL*, ventral nucleus of the lateral lemniscus; *VNTB*, ventral nucleus of the trapezoid body)

2.2.1 Projections to the Cochlear Nucleus from the Cochlea

The auditory pathways originate in the cochlea. Information from the cochlea to the brain is carried by the auditory nerve, which supplies the majority of excitatory input to the large cells of the VCN and part of the excitatory input to the large cells of the DCN. Compared to the central auditory pathways, the auditory nerve is relatively simple with three basic components (reviewed by Ryugo and Parks 2003). Large, myelinated Type I fibers form approximately 95% of the nerve, and they are postsynaptic to inner hair cells. They can be further subdivided into two groups distinguished by their rates of spontaneous (SR) activity (high SR and low SR). Although the responses of high and low SR fibers to simple sound stimuli are similar, there are numerous subtle differences in their terminations in both the cochlea (Liberman 1980) and the cochlear nucleus (Liberman 1991, 1993). Thin, unmyelinated Type II fibers make up the remaining auditory nerve fibers, and they are postsynaptic to outer hair cells.

All cochlear nerve fibers enter the brain at the ventral surface of the VCN. Type I fibers branch extensively, forming boutons of various sizes and, in the rostral part of the AVCN, the distinctive, large endbulbs of Held. At the caudal pole of the PVCN, the axons turn dorsally to enter the DCN, where they branch and form small terminal boutons. The branches in both the VCN and DCN are organized topographically such that the axons from the apical cochlea terminate ventrally, and axons from the basal cochlea terminate dorsally. This segregation of axons responding to specific frequencies forms the basis for the tonotopic organization that is maintained throughout the auditory pathways.

Terminals arising from Type I fibers are distributed almost exclusively to the magnocellular core of the VCN, where they form synapses with each of the large cell types (Liberman 1991, 1993), and to deep layers of the DCN, where they form synapses with DCN projection neurons (Smith and Rhode 1985). The main target of Type II fibers is the GCD and an adjacent layer of small cells (Brown et al. 1988; Morgan et al. 1994). The small cells also receive input from Type I fibers with low SR but not from Type I fibers with high SR (Ryugo 2008). Type II fibers rarely enter the DCN proper (Brown and Ledwith 1990); like the Type I fibers, they are presumed to be excitatory (Berglund et al. 1996).

2.2.2 *Projections to the Cochlear Nucleus from Noncochlear Sources*

Although the cochlea is the major source of excitatory input to the large cells of the cochlear nucleus, there are numerous additional sources of input, both excitatory and inhibitory. These include the contralateral cochlear nucleus, the superior olivary complex, the inferior colliculus, the auditory cortex, and a variety of structures not considered part of the auditory system per se.

2.2.2.1 Projections to the Cochlear Nucleus from the Superior Olivary Complex

A number of distinct neuronal populations in the superior olivary complex (SOC) project to both the VCN and DCN (see Sect. 2.4). Most of the projections arise in the ipsilateral lateral nucleus of the trapezoid body (LNTB) and the contralateral ventral nucleus of the trapezoid body (VNTB) (Shore et al. 1991). At least 80% of the neurons that project to the cochlear nucleus from the SOC are positive for markers of gamma-aminobutyric acid (GABA), glycine, or both, indicating that they are inhibitory (Saint Marie et al. 1993; Ostapoff et al. 1997). Most of the glycinergic projections appear to arise ipsilaterally with only a few arising in the contralateral LNTB and VNTB (Benson and Potashner 1990). Therefore, much of the substantial projection from the contralateral VNTB is presumably GABAergic.

Other olivary nuclei with projections to the cochlear nuclei include the ipsilateral superior paraolivary nucleus (SPN) and the medial nucleus of the trapezoid body (MNTB) (reviewed in Thompson and Schofield 2000). Although SOC projections to the cochlea and cochlear nucleus are largely separate (Adams 1983; Winter et al. 1989), a small percentage of the neurons of the medial olivocochlear system give rise to collateral branches to the cochlear nucleus that terminate mainly in the GCD (Brown 1993). Cholinergic input to a specific neuronal population located in the cochlear nerve root (cochlear root neurons) arises from cells in the VNTB (Gómez-Nieto et al. 2008a), but whether these are collaterals of medial olivocochlear axons is not known.

2.2.2.2 Projections to the Cochlear Nucleus from the Ventral Nucleus of the Lateral Lemniscus

Inputs to the DCN from the ventral nucleus of the lateral lemniscus (VNLL) have been reported (Winter et al. 1989; Benson and Potashner 1990), but these projections have not been described in any detail.

2.2.2.3 Projections to the Cochlear Nucleus from the Inferior Colliculus

Projections from the inferior colliculus (IC) terminate mainly in the ipsilateral DCN (Caicedo and Herbert 1993; Weedman and Ryugo 1996). Most of the projections arise from neurons located in the central and external nuclei (Schofield 2001). The terminations are topographically organized (Malmierca et al. 1996) and arise from neurons separate from those that project to the medial geniculate nucleus, contralateral IC, and SOC (Coomes and Schofield 2004a; Okoyama et al. 2006).

The main target of the IC projections is the granule cell layer in the DCN (Malmierca et al. 1996), where the synapses appear to be excitatory (Milinkeviciute et al. 2016). The cells in the IC that project to the DCN may receive inputs from the auditory cortex (Schofield and Coomes 2006). Together with inputs directly from the auditory cortex itself (Sect. 2.2.2.4), these connections form pathways for higher order control of neuronal responses to primary input. No collicular inputs to the VCN have been reported and in some cases have been specifically denied (e.g., Caicedo and Herbert 1993).

2.2.2.4 Projections to the Cochlear Nucleus from the Auditory Cortex

Bilateral (although predominantly ipsilateral) projections to the DCN and the GCD arise in pyramidal cells of layer V of the auditory cortex (Feliciano et al. 1995). At least some cortical fibers terminate as mossy fibers and form synapses with the dendrites of granule cells and several other types of small cells (Weedman et al. 1996; Meltzer and Ryugo 2006). Although most of the cortical fibers terminate in the DCN and the GCD, a few terminals are also scattered in the VCN (Schofield and Coomes 2005b). At least some of the cells in both the DCN and the VCN that receive cortical input project, in turn, to the IC (Schofield and Coomes 2005a; Schofield and Coomes 2006).

2.2.2.5 Cochlear Nucleus to Cochlear Nucleus: Reciprocal, Commissural, and Intrinsic Connections

Reciprocal pathways interconnect the DCN and VCN on the same side. Vertical cells in the DCN give rise to the tuberculoventral pathway, which projects topographically into the magnocellular areas of the VCN (with the exception of the octopus cell area) (Wickesberg and Oertel 1990; Wickesberg et al. 1991). These projections are inhibitory and, most likely, glycinergic.

Both glycinergic and nonglycinergic projections arising in the VCN project into the DCN. The glycinergic cells have radiate dendritic fields and form narrow (topographic) terminal fields in the DCN; the nonglycinergic cells have flatter (or planar) dendritic fields and form widespread terminal fields in the DCN (Doucet and Ryugo 1997). Cells that project to the DCN may also give rise to extensive collaterals in the VCN itself (Arnott et al. 2004).

Multipolar cells scattered throughout the VCN and a few cells in the DCN project to the opposite cochlear nuclear complex, where they terminate throughout much of the VCN and in the DCN (Schofield and Cant 1996; Needham and Paolini 2007). At least some of the commissural neurons in both the DCN and VCN are glycinergic (Wenthold 1987; Alibardi 2000).

In addition to the reciprocal and commissural interconnections, local intrinsic circuitry adds to the complexity of the networks formed in the cochlear nuclei. Many (although not all) of the large projection neurons in both the VCN and DCN have collateral branches that terminate locally within the parent subdivision (e.g., Smith and Rhode 1989; Zhang and Oertel 1993). Both bushy cells (that project to the SOC) and multipolar cells (that project to the IC) may be locally inhibited or excited (Campagnola and Manis 2014). The DCN contains a variety of small interneurons that form complex inhibitory circuits with the large projection cells (reviewed by Young and Oertel 2003).

2.2.2.6 Projections to the Cochlear Nucleus from Nonauditory Areas of the Brain

Projections from a wide variety of nonauditory areas terminate preferentially in the GCD and/or DCN. These include somatosensory inputs via the dorsal column and trigeminal nuclei (e.g., Wright and Ryugo 1996; Zhou and Shore 2004) as well as direct primary input via the second cervical dorsal root (Zhan et al. 2006).

Other sources are: the vestibular system (e.g., Zhao et al. 1995; Newlands and Perachio 2003); areas of the brainstem reticular formation that are involved in the control of respiration, the cardiovascular system, and other similar brainstem functions (Kamiya et al. 1988; Zhan and Ryugo 2007); and areas that appear to play roles in global modulation of neuronal activity (Behrens et al. 2002; Gómez-Nieto et al. 2008b). Inputs to the GCD also arise in the pontine nuclei (Ohlrogge et al. 2001). The inputs from many (if not all) of these sources terminate as mossy fibers, forming excitatory synapses with granule cell dendrites and small cells (Ryugo et al. 2003; Haenggeli et al. 2005). Since the granule cells project to the molecular layer of the DCN, where they form parallel fibers that synapse with the fusiform cells (Mugnaini et al. 1980), the integrated information from all of these sources would be expected to have an important impact on fusiform cell activity.

2.3 The Superior Olivary Complex

Like the cochlear nuclei, the superior olivary complex (SOC) is made up of a large number of distinct neuronal populations, and each population has its own pattern of synaptic organization and projection targets. Because most of the major cell types are segregated from one another, the nuclei of the SOC have been particularly attractive subjects for research into physiological mechanisms of auditory coding (e.g., Yin 2002) and developmental principles (e.g., Sanes and Friauf 2000; Kandler 2004). A major function of the SOC is the encoding of binaural information critical for localization of sounds. Other functions include modulation of cochlear transduction through the olivocochlear system (Warr 1992) and initiation of the inhibitory pathways that feed forward to the midbrain (Sect. 2.5.2) and back to the cochlear nucleus (Sect. 2.2.2.1).

Neurons in the SOC may be either excitatory or inhibitory. The projections from the two major nuclei in the SOC (the medial and lateral superior olivary nuclei; MSO and LSO, respectively) are predominantly excitatory (e.g., Oliver et al. 1995), although the LSO also contains a population of inhibitory (glycinergic) projection neurons (Saint Marie et al. 1989). Surrounding these two nuclei are a number of cell groups in which most of the neurons are inhibitory (glycinergic, GABAergic, or both) (e.g., Adams and Mugnaini 1990; Kulesza and Berrebi 2000). These include the LNTB, MNTB, and VNTB and, in rodents, the SPN. These latter nuclei and other less well-defined groups of neurons are sometimes referred to collectively as the periolivary (PO) nuclei, although the inputs and outputs of the included neurons are quite diverse. The organization of the SOC in various species and its intrinsic and extrinsic connections were reviewed comprehensively by Thompson and Schofield (2000). This section presents an overview of the major inputs to the SOC. Kopp-Scheinpflug and Forsythe explain and illustrate the connectivity of specific cell types in Chap. 5 (Figs. 5.1, 5.2, 5.3, 5.4 and 5.5).

2.3.1 Projections to the Superior Olivary Complex from the Cochlear Nuclei

Most of the large cell types in the VCN project to the SOC (reviewed by Cant and Benson 2003). Spherical bushy cells in the VCN project to the principal cells of the MSO and LSO, forming excitatory synapses. One axon sends branches to the MSO on both sides of the brainstem. Consequently, MSO neurons are in a position to compare the timing of excitatory inputs from both ears. In contrast, the LSO receives excitatory input mainly from the ipsilateral ear. Projections into the LSO from other olivary nuclei convey inhibitory information from the opposite ear (Sect. 2.3.2); therefore, the neurons of the LSO are in a position to compare the intensity of sounds at the two ears. These two nuclei give rise to pathways that carry binaural information to the midbrain (Sect. 2.5.2).

The LSO, particularly the high frequency region, also receives inputs from planar multipolar cells in the VCN (Doucet and Ryugo 2003). Whether they terminate on the principal cells or the cells of the lateral olivocochlear system (or both) is not known. Planar multipolar cells also terminate in the SPN (Schofield 1995) and in the VNTB, where they form synapses on medial olivocochlear neurons (Darrow et al. 2012).

Globular bushy cells project to the LNTB on the ipsilateral side of the brain and to the MNTB on the contralateral side (Tolbert et al. 1982). In the MNTB, the axons terminate in the large calyces of Held that form multiple synapses on the cell bodies of principal cells (reviewed by Borst and van Hoeve 2012). The globular bushy cells also project into periolivary areas and may supply a few collaterals to the LSO (Friauf and Ostwald 1988), although they do not appear to be a significant source of its input (Gómez-Álverez and Saldaña 2015). Octopus cells in the PVCN form terminations in the SPN and some other periolivary areas (Schofield 1995).

Projections from the DCN to the SOC are not usually reported. In the North American opossum, the DCN appears to provide a significant proportion of the input to the SPN (Willard and Martin 1983), but this was not observed in the guinea pig (Schofield 1995).

2.3.2 Intra-Olivary Connections

The cell groups of the SOC have extensive interconnections on the ipsilateral side. The only reported commissural connections arise in the VNTB, which projects to the contralateral LSO (Warr and Beck 1996). The most studied intrinsic connections within the SOC are those made by neurons of the MNTB. Neurons in both the LSO and MSO receive inhibitory inputs from the glycinergic principal cells of the ipsilateral MNTB (Bledsoe et al. 1990). In both cases, the inhibition plays an important role in shaping the responses of the neurons to binaural inputs (reviewed by

Grothe et al. 2010). The principal cells of the MNTB also send collateral branches to the LNTB (Sommer et al. 1993), SPN (Gómez-Álverez and Saldaña 2015), and VNTB (Banks and Smith 1992). The VNTB, in turn, projects back to the MNTB (Albrecht et al. 2014). Other olivary cells also form collateral branches within the SOC. The MSO receives inputs from the ipsilateral LNTB that are presumably inhibitory (Cant and Hyson 1992; Spirou and Berrebi 1997), and the SPN receives inputs from the MSO that are presumably excitatory (Kuwabara and Zook 1999). These examples serve to illustrate the rich circuitry that underlies complex inhibitory and disinhibitory interactions within the SOC (Adams and Mugnaini 1990).

2.3.3 Descending Projections to the Superior Olivary Complex

Projections to the SOC arise in the auditory cortex, thalamus, inferior colliculus, nuclei of the lateral lemniscus, and several other midbrain sources (e.g., AC: Feliciano et al. 1995; thalamus: Yasui et al. 1992; IC: Faye-Lund 1986; ventral nucleus of the lateral lemniscus: Whitley and Henkel 1984; dorsal nucleus of the lateral lemniscus: Bajo et al. 1993). The ipsilateral VNTB is consistently identified as the major recipient of the descending inputs, although there are terminations in other nuclei as well (e.g., AC: Feliciano et al. 1995; Coomes and Schofield 2004b; IC: Caicedo and Herbert 1993; Malmierca et al. 1996). The projections from the IC arise mainly from neurons in the central and external nuclei that are distinct from those that project to the medial geniculate body or to the contralateral IC (Okoyama et al. 2006). Known targets of descending projections from the central nucleus are the medial olivocochlear neurons that lie within the VNTB (Thompson and Thompson 1993). The IC may contact nonolivocochlear cells as well (Vetter et al. 1993).

Cortical projections arise from both primary and secondary areas of the auditory cortex. The projecting cells are not the same as those that project to the cochlear nucleus (Sect. 2.2.2.4) (Doucet et al. 2002). In addition to inputs to the VNTB, the cortex also projects to the LSO; however, whether these inputs target the principal cells of the LSO or the lateral olivocochlear cells (found both within and surrounding the LSO) is not known. Additional projections from the forebrain arise in the subparafasicular nucleus of the thalamus and terminate mainly in the SPN (Yasui et al. 1992).

2.3.4 Projections to the Superior Olivary Complex from Nonauditory Sources

Inputs from the trigeminal ganglion terminate laterally in the SOC, mainly around the margins of the LSO (Shore et al. 2000). The inputs may play a role in circuits that involve the lateral olivocochlear system. Modulatory inputs from serotonergic and noradrenergic brainstem nuclei are abundant in the SOC, especially in the periolivary regions (Schofield and Hurley, Chap. 9).

2.4 The Nuclei of the Lateral Lemniscus

The nuclei of the lateral lemniscus (NLL) are generally divided into a dorsal nucleus (DNLL), an intermediate nucleus (INLL), and a ventral nucleus (VNLL), although in many studies the INLL and VNLL are grouped together (reviewed by Merchán et al. 1997; Oertel and Wickesberg 2002). The nuclei are distinguished by their neurochemistry (Ito et al. 2011). Typically, the DNLL neurons are GABAergic (Adams and Mugnaini 1984), the VNLL neurons are glycinergic (Riquelme et al. 2001), and the INLL neurons are glutamatergic (Saint Marie et al. 1997a; Ito et al. 2011).

A majority of the inputs to the VNLL and INLL (about 90%) are from the contralateral cochlear nucleus (Glendenning et al. 1981). The remainder of the inputs arise from the MNTB and periolivary nuclei. Little is known about the intrinsic microcircuitry of the INLL and VNLL, but evidence from physiological experiments (e.g., Nayagam et al. 2005) suggests an important role for interneuronal connectivity. Neurons in the VNLL give rise to elaborate collateral branches (Zhao and Wu 2001), providing an anatomical basis for these results.

The DNLL consists of a compact group of cells lying ventral to the inferior colliculus. Its input arises from a variety of sources, including the cochlear nuclei (Glendenning et al. 1981), the LSO and MSO (Shneiderman et al. 1988; Yang et al. 1996), and commissural inputs from the contralateral DNLL (Oliver and Shneiderman 1989). As with the VNLL, little is known about its intrinsic microcircuitry. Both the VNLL and DNLL are important sources of inhibitory input to the inferior colliculus (Sect. 2.5.3).

2.5 The Inferior Colliculus

The IC forms the interface between hindbrain and forebrain auditory systems. It is generally subdivided into a tonotopically organized central nucleus that is surrounded by lateral, dorsal, and rostral cortices. Almost all of the brainstem nuclei discussed in the preceding sections contain one or several different cell populations that project to the IC (Fig. 2.2). Descending projections from the forebrain and commissural connections add to the complexity of the midbrain auditory areas. From the beginning of modern studies of IC connections, it has been clear that the termination zones of the myriad inputs from the brainstem and forebrain do not necessarily follow subdivision boundaries and may be overlapping, partially overlapping, or segregated to different parts of one or more of the subdivisions (summarized in Aitkin 1989). The implication is that the number of functionally different areas may be far greater than is suggested by the number of traditionally defined subdivisions.

Comprehensive reviews of the sources and terminal distributions of inputs to the IC are available (e.g., Winer and Schreiner 2005). For the purpose of this brief overview, a simple and useful way to conceptualize the organization of the IC is to divide it into a lemniscal core (central nucleus) and nonlemniscal (cortical) regions. The central nucleus receives the majority of the ascending inputs from the cochlear nuclei and the superior olivary complex (Sects. 2.5.1 and 2.5.2) (Oliver 2000).

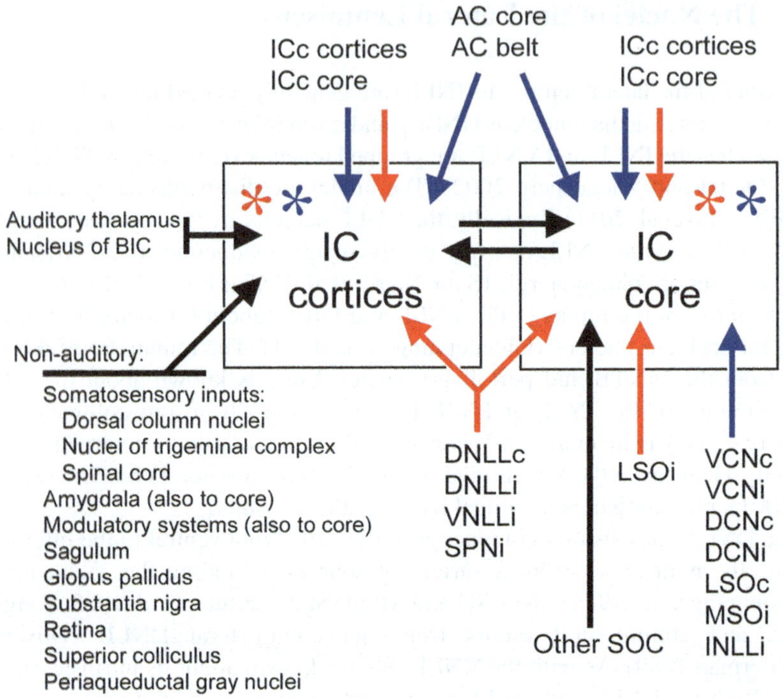

Fig. 2.2 Schematic diagram that summarizes the major sources of input to the inferior colliculus (*IC*). Known excitatory inputs are indicated in *blue*; known inhibitory inputs are indicated in *red*. Inputs for which the synaptic effects have not been fully characterized are indicated in *black*. *Asterisks* represent excitatory (*blue*) and inhibitory (*red*) synaptic inputs arising from neurons intrinsic to the structures illustrated. (*AC,* auditory cortex; *BIC,* brachium of the inferior colliculus; *c,* contralateral; *DCN,* dorsal cochlear nucleus; *DNLL,* dorsal nucleus of the lateral lemniscus; *i,* ipsilateral; *INLL,* intermediate nucleus of the lateral lemniscus; *LSO,* lateral superior olive; *MSO,* medial superior olive; *SOC,* superior olivary complex; *SPN,* superior paraolivary nucleus; *VCN,* ventral cochlear nucleus; *VNLL,* ventral nucleus of the lateral lemniscus)

In contrast, the nonlemniscal cortices receive substantial inputs from the contralateral IC and the auditory cortex (Sects. 2.5.3 and 2.5.4). The cortices also receive a number of nonauditory inputs (Sect. 2.5.5).

2.5.1 *Projections to the Inferior Colliculus from the Cochlear Nuclei*

Inputs from the DCN arise from fusiform and giant cells and inputs from the VCN arise from multipolar cells (reviewed by Cant and Benson 2003). All of the inputs from the cochlear nuclei are organized topographically with respect to frequency representation, preserving the tonotopy established in the cochlea (Osen 1972). The terminations have the anatomical characteristics of excitatory synapses

(Oliver 1985, 1987) and are associated with markers of glutamatergic neurotransmission (Ito and Oliver 2010). The terminal zones of the DCN and VCN overlap throughout most, if not all, of the central core of the IC, although terminals from the DCN may extend further into the surrounding cortical areas (Oliver et al. 1997). Although it is quite possible that terminations from the VCN and DCN target different cell populations (Sect. 2.5.2), there is little evidence on this matter.

2.5.2 Projections to the Inferior Colliculus from the Superior Olivary Complex

Major projections from the SOC arise from the ipsilateral MSO (Henkel and Spangler 1983), both the ipsilateral and contralateral LSOs (Shneiderman and Henkel 1987), and the SPN (Kulesza and Berrebi 2000). Like those from the cochlear nuclei, the projections from the MSO and LSO terminate in the core of the IC but are less widespread, occupying only part of the central nucleus (Cant and Benson 2006; Ito and Oliver 2010). For example, in the cat the LSO and DCN inputs overlap in the ventrolateral central nucleus, but only DCN input is present more dorsomedially (Oliver et al. 1997; Shneiderman and Henkel 1987). The inputs from the SPN, in contrast, terminate throughout both core and cortical areas (Saldaña et al. 2009). Synaptic terminals formed by axons from the ipsilateral MSO and the contralateral LSO are glutamatergic (Ito and Oliver 2010) and have the fine structural characteristics of excitatory synapses (Oliver et al. 1995). Inputs from the ipsilateral LSO are glycinergic (Saint Marie and Baker 1990), whereas projections from the SPN are GABAergic (Kulesza and Berrebi 2000). The remaining periolivary nuclei send both excitatory and inhibitory projections to the IC (Ito et al. 2011), but for the most part, these inputs appear to be considerably sparser than those from the MSO, LSO, and SPN.

Terminations from the LSO and the MSO do not completely overlap within a given isofrequency lamina (Shneiderman and Henkel 1987; Loftus et al. 2004). Therefore, small regions within a lamina could be dominated by inputs from only a subset of the potential sources of ascending inputs. Loftus and colleagues (2010) confirmed that neurons with different potential sets of input sources tend to have different physiological response properties. Indeed, a consistent theme in studies of IC connectivity is that the layers that make up the tonotopic map appear to be composed of multiple functional regions, each with a distinct pattern of inputs (Ross et al. 1988; Oliver et al. 1997).

2.5.3 Projections to the Inferior Colliculus from the Nuclei of the Lateral Lemniscus

Projections to the IC from the DNLL are bilateral and GABAergic (Shneiderman and Oliver 1989). They are distributed widely throughout the central nucleus and collicular cortices on both sides (Shneiderman et al. 1988). In the central core of the IC,

DNLL terminals form dense bands contralaterally and more diffuse bands ipsilaterally (Kudo 1981; Henkel et al. 2003). The banded projections from the contralateral DNLL overlap with those from the ipsilateral LSO (Sect. 2.5.2), but they appear to be segregated from projections from the contralateral DCN and LSO (Fathke and Gabriele 2009), providing further evidence for regional specializations within a given isofrequency lamina.

In the cat the projections from both the VNLL and the INLL are ipsilateral (Glendenning et al. 1981), but in mice the projection from the INLL is bilateral (Frisina et al. 1998). The axons make widespread terminations throughout the central core and extend into cortical areas (Kudo 1981; Whitley and Henkel 1984). Most of the projections from the VNLL are glycinergic, GABAergic, or both (Saint Marie and Baker 1990; González-Hernández et al. 1996), whereas those from the INLL appear to be glutamatergic and, therefore, presumably excitatory (Ito et al. 2011).

In general, the VNLL itself receives excitatory inputs from the contralateral ventral cochlear nucleus, whereas the DNLL receives bilateral inputs from the cochlear nuclei and SOC (Sect. 2.4). Therefore, the inhibitory effects of the VNLL in the IC are driven mainly by the contralateral ear and those of the DNLL are driven bilaterally. Inputs from both of these sources, along with those from the SPN (Sect. 2.5.2), play important roles in shaping neuronal responses in the IC (Kidd and Kelly 1996; Pollak et al. 2011).

2.5.4 Commissural and Intrinsic Connections

Several different groups of neurons (defined morphologically) project from one IC to the other (González-Hernández et al. 1986), forming a substantial proportion of collicular input (Moore 1988). The projections are topographic and appear to arise from both GABAergic and glutamatergic cell populations in both the medial cortical regions and the core of the IC (Saldaña and Merchán 2005). To a certain extent, the projections from one side to the other appear to be point-to-point with respect to position within a putative isofrequency lamina, but crossing axons also diverge to terminate throughout a layer (Malmierca et al. 2009). The densest terminations appear to be in the dorsal and rostral cortices, although axons also extend into the central core of the IC (Malmierca et al. 1995).

Many neurons in the IC give rise to highly branched axons that form extensive intrinsic plexuses (González-Hernández et al. 1989; Oliver et al. 1991). Both the commissural and intrinsic terminal arbors are arranged in laminar patterns (that reflect the frequency organization of the central nucleus), and they also extend into both the medial and lateral cortices of the IC (Saldaña and Merchán 2005). One specific target of both commissural and intrinsic axons is a well-defined population of large GABAergic tectothalamic cells (Ito and Oliver 2014).

2.5.5 Projections to the Inferior Colliculus from the Auditory Cortex and Auditory Thalamus

The IC is a major target of descending projections from the auditory cortex. The densest projections are to the dorsal cortex of the IC (e.g., Andersen et al. 1980; Faye-Lund 1985), but terminals are also found in the central nucleus (Saldaña et al. 1996). The cortical terminals have the fine structure of excitatory synapses (Saldaña et al. 1996) and are glutamatergic (Feliciano and Potashner 1995). The projections are topographic and arise from many different cortical fields (Herbert et al. 1991; Winer et al. 1998). Each field has different patterns of projections to specific parts of the IC (Bajo et al. 2007). Commissural neurons throughout the IC are a direct target of at least some of the cortical inputs. Most of the postsynaptic commissural neurons are excitatory, but a small proportion are GABAergic (Nakamoto et al. 2013). The auditory cortex also projects to a number of targets, such as the SOC (Sect. 2.3.3) and midbrain cholinergic nuclei, that in turn project to the IC (Peterson and Schofield 2007; Schofield 2010). Forebrain projections to the IC also arise in nonlemniscal nuclei of the auditory thalamus (Sect. 2.6.1) (Patel et al. 2016). Most, if not all, of these descending projections terminate in the nonlemniscal divisions of the IC. The GABAergic neurons in the nucleus of the brachium of the IC, which lies between the IC and the medial geniculate body (Sect. 2.6), also terminate in the nonlemniscal IC.

2.5.6 Projections to the Inferior Colliculus from Nonauditory Areas

Several sites not associated specifically with the auditory system project into the inferior colliculus. These include the dorsal column nuclei and nuclei of the trigeminal complex (Robards 1979), the somatosensory cortex (Robards 1979), and the spinal cord (Künzle 1993). These somatosensory inputs terminate mainly in the ventrolateral part of the external cortex of the IC where they overlap to some extent with inputs from the cochlear nuclei (Zhou and Shore 2006). Neurons in this region may be responsive to somatic or auditory stimuli or to both (Aitkin et al. 1981).

Other inputs to the IC originate in the amygdala (Marsh et al. 2002), the globus pallidus (Moriizumi and Hattori 1991), the substantia nigra (Olazábal and Moore 1989), the retina (Itaya and Van Hoesen 1982), nucleus sagulum (Henkel and Shneiderman 1988), the deep layers of the superior colliculus (Coleman and Clerici 1987), periaqueductal gray nuclei (Adams 1980), and nuclei surrounding the medial geniculate nucleus (Senatorov and Hu 2002). This is not an exhaustive list but serves to indicate the great variety of inputs that may influence activity in the IC. The inputs from these nonauditory sources vary widely in the density and extent of their terminal fields. For example, the inputs from the globus pallidus and retina are quite sparse and appear to be limited to the most superficial surface of the midbrain, whereas the inputs from the amygdala appear to be more widespread (Marsh et al. 2002).

In general, little is known about which cell populations constitute the targets of the nonauditory inputs. Inputs from cholinergic, dopaminergic, serotonergic, and other modulatory systems are discussed in Chap. 9 by Schofield and Hurley.

2.6 The Medial Geniculate

The medial geniculate (MG) is the principle thalamic structure providing sensory input to the auditory cortex. It is generally divided into ventral, dorsal, and medial divisions (Winer 1992) and is surrounded by a number of additional cell groups associated with the auditory pathways (Márquez-Legorreta et al. 2016). Both the ventral and dorsal divisions can be further subdivided into regions that receive unique combinations of inputs and project to different neocortical targets. Such further subdivision reveals significant species differences in organization and connectivity (Wenstrup 2005; Winer et al. 1988)

Comprehensive reviews of the neuroanatomy of the MG emphasize the details of its connectivity (Wenstrup 2005; Winer 1992). For the purposes of this overview, the MG can be divided into lemniscal and nonlemniscal regions as was done for the IC. The ventral division of the MG receives its main ascending inputs from the central nucleus of the IC and, therefore, is considered to represent the continuation of the lemniscal pathways into the forebrain. The tonotopic organization of the MG (reviewed by Clarey et al. 1992) is reflected anatomically in fibrodendritic layers made up of neurons with oriented dendrites and associated axons (McMullen et al. 2005; Morest 1965a). The nonlemniscal portions of the auditory thalamus (the dorsal and medial divisions of the MG and surrounding cell groups) contain broadly tuned neurons (Clarey et al. 1992) and receive most of their ascending inputs from the nonlemniscal regions of the IC, although the segregation of the pathways is not absolute (Mellott et al. 2014). Input to the MG is dominated by projections from the inferior colliculus and the auditory cortex, but several sources of sparser input have been identified (Fig. 2.3).

2.6.1 Projections to the Medial Geniculate from the Inferior Colliculus

Almost all of the ascending projections to the MG arise in the ipsilateral IC (Andersen et al. 1980) with a small contribution from the contralateral IC (Kudo and Niimi 1978). Several morphologically distinct cell populations participate in the lemniscal projection from the central nucleus of the IC to the ventral division of the MG (Oliver 1984). Areas of the central nucleus that receive ascending inputs from both the superior olivary complex and the cochlear nuclei (Sect. 2.5.2) project to a different part of the ventral division than do those areas that receive ascending inputs mainly from the cochlear nuclei (Cant and Benson 2007). This suggests that

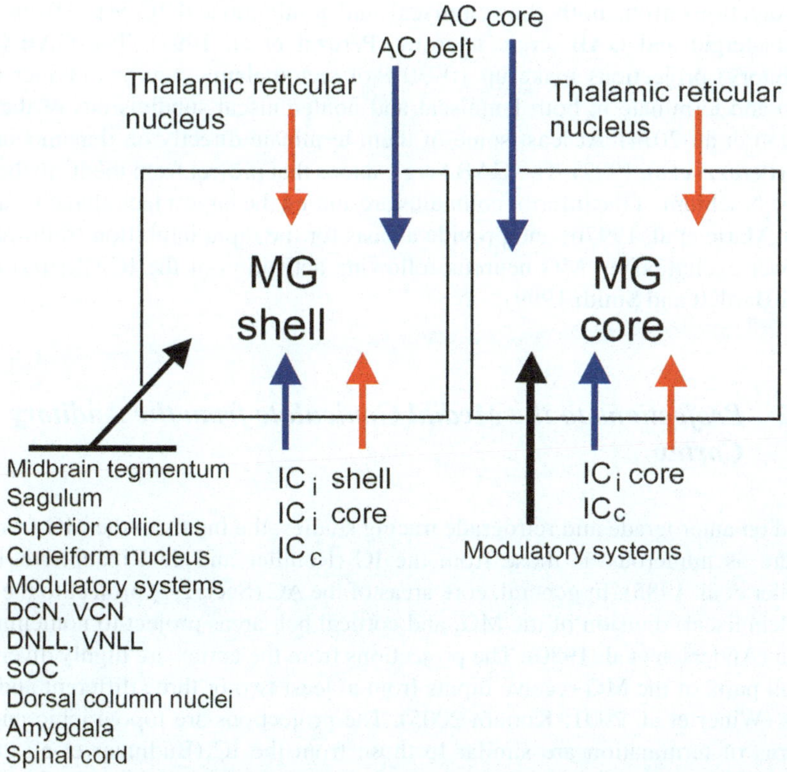

Fig. 2.3 Schematic diagram that summarizes the major sources of input to the medial geniculate (*MG*). Known excitatory inputs are indicated in *blue*; known inhibitory inputs are indicated in *red*. Inputs for which the synaptic effects have not been fully characterized are indicated in *black*. (*AC*, auditory cortex; *c*, contralateral; *DCN*, dorsal cochlear nucleus; *DNLL*, dorsal nucleus of the lateral lemniscus; *i*, ipsilateral; *IC*, inferior colliculus; *MG*, medial geniculate; *SOC*, superior olivary complex; *VCN*, ventral cochlear nucleus; *VNLL*, ventral nucleus of the lateral lemniscus)

the lemniscal component is made up of *at least* two functionally distinct pathways (see also Straka et al. 2014). Differences in physiological response properties in rostral versus caudal parts of the ventral division (Rodrigues-Dagaeff et al. 1989) may reflect the separate pathways.

Projections from cortical regions of the IC terminate primarily in the dorsal and medial divisions of the MG and in the associated nuclei lying near its borders (reviewed by Wenstrup 2005). The nonlemniscal portions of the auditory thalamus receive subsets of inputs from specific combinations of midbrain tectal and tegmental sources (Oliver and Hall 1978; Calford and Aitkin 1983). The nonlemniscal regions are interconnected with many areas of the brain in addition to the tectal connections (Sect. 2.6.4) and may play a role in a variety of modulatory and basic physiological functions (reviewed by Lee 2015).

Projections from both the lemniscal and nonlemniscal IC arise from both glutamatergic and GABAergic neurons (Peruzzi et al. 1997). The GABAergic (inhibitory) projections make up 10–30% of tectothalamic neurons (Winer et al. 1996) and terminate in both lemniscal and nonlemniscal subdivisions of the MG (Mellott et al. 2014). At least some of them terminate directly on thalamocortical cells (Peruzzi et al. 1997). The GABAergic axons that project from the IC to the MG via the brachium of the inferior colliculus are among the largest tectothalamic axons (Saint Marie et al. 1997b) and provide a basis for the rapid inhibition (followed by a slower excitation) in MG neurons following activation of the IC (Peruzzi et al. 1997; Bartlett and Smith 1999).

2.6.2 Projections to the Medial Geniculate from the Auditory Cortex

Based on anterograde and retrograde tracing studies, the inputs to the MG from the AC are as numerous as those from the IC (Rouiller and de Ribaupierre 1985; Rouiller et al. 1985). In general, core areas of the AC (Sect. 2.7) project to the ventral (lemniscal) division of the MG, and cortical belt areas project to nonlemniscal nuclei (Andersen et al. 1980). The projections from the cortex are highly divergent, and all parts of the MG receive inputs from at least two or three different auditory fields (Winer et al. 2001; Kimura 2005). The projections are topographic and the patterns of termination are similar to those from the IC (Budinger et al. 2000). Cortical synaptic terminals are of two types: small boutons are found throughout the MG and "giant" terminals are found in the dorsal division of the nucleus (Rouiller and Welker 2000; Bajo et al. 1995). The projections giving rise to the two types of terminals may participate in different types of corticothalamic modulation (He 2003).

2.6.3 Intrinsic Connections

Very little is known about intrinsic connectivity in the MG. In rodents there are few GABAergic intrinsic neurons (Winer and Larue 1996), and the extent to which the neurons that project out of the MG give rise to intrinsic collaterals is not known. Most of the extrinsic inputs to the MG, including the large terminals, terminate on dendrites (Bartlett et al. 2000; Ojima and Murakami 2010). Elaborate dendrodendritic synaptic configurations, which are characteristic of the MG (Morest 1975), could allow for complex intrageniculate interactions. In contrast to structures at other levels of the auditory system, there appear to be no commissural connections between the thalamic nuclei on the two sides of the brain.

2.6.4 *Projections to the Medial Geniculate from Other Sources*

Although most projections to the dorsal and ventral divisions of the MG arise in the IC and AC (Sects. 2.6.1 and 2.6.2), there are several other notable sources of input to one or more of its subdivisions. The thalamic reticular nucleus (TRN) is a GABAergic nucleus that sends inhibitory feedback to many thalamic nuclei (reviewed by Guillery et al. 1998). The auditory sector of the TRN receives input from the MG and, in turn, projects back to all subdivisions of the MG (Rouiller et al. 1985; Crabtree 1998). This portion of the TRN receives excitatory inputs from the AC (Zhang et al. 2008) and from nonprimary visual and somatosensory cortices (Kimura et al. 2012).

Just as in the IC, the nonlemniscal portions of the MG appear to receive a greater variety of inputs than do the lemniscal portions. The dorsal division receives input from the midbrain tegmentum ventral to the IC and medial to the lateral lemniscus via the lateral tegmental system (Morest 1965b). Inputs to the dorsal division also include the sagulum and the cuneiform nuclei of the midbrain reticular formation and the deep layers of the superior colliculus (Oliver and Hall 1978). Both cholinergic and noncholinergic cells from midbrain tegmental nuclei project to the MG (Motts and Schofield 2010).

The medial division receives the widest variety of inputs with projections from both auditory and nonauditory sources, including the dorsal and ventral cochlear nuclei (Malmierca et al. 2002; Schofield et al. 2014), the dorsal and ventral nuclei of the lateral lemniscus (Bajo et al. 1993), the SOC (Henkel 1983), and the spinal cord and dorsal column nuclei (LeDoux et al. 1987). The terminations of these projections are generally described as diffuse and nontopographic. The medial division also receives projections from the amygdala and, in turn, an important target of the medial division is the lateral division of the amygdala (LeDoux and Farb 1991). The interconnections with the amygdala point to a central role of the medial division in auditory fear conditioning (Weinberger 2010).

2.7 The Auditory Cortex

The ascending auditory pathways terminate in the AC, but the information flow does not stop there. The AC has extensive interconnections with other cortical areas (reviewed by Kraus and Canlon 2012), and descending cortical projections target the thalamus and brainstem. In all species, the AC is organized into multiple structurally and functionally distinct fields (reviewed by Hackett 2011). Species-specific nomenclature makes generalizations difficult, but to a first approximation, the AC in all species can be divided into core and shell (or belt) regions (reviewed by Winer 1992). The core areas exhibit complete topographic representations of the cochlea, whereas tonotopic representations are not so evident in the surrounding fields.

Core and belt areas differ in their cytoarchitecture, microcircuitry, and physiology (Hackett et al. 2014), and they play different roles in auditory processing (Lomber and Malhotra 2008; Higgins et al. 2010).

Almost all ascending auditory input to the AC is funneled through the thalamus, but thalamic synapses make up only a small fraction of the total synaptic contacts on cortical cells (Hackett et al. 2014; Bizley et al. 2015). Based on numbers of projecting neurons, Lee and Winer (2011) estimated that thalamic input accounts for about 15% of the total input to core AC. Most input to cortical neurons arises intracortically from commissural connections and both long-range and short-range ipsilateral connections (Sect. 2.7.2) (Budinger and Kanold, Chap. 8).

2.7.1 Projections to the Auditory Cortex from the Medial Geniculate

All regions of auditory cortex receive inputs from the MG (reviewed by Imig and Morel 1983; de Ribaupierre 1997). Ascending inputs to core areas arise in the ventral (lemniscal) division of the MG. These inputs are topographic and terminate in layer four, conferring a basic tonotopic organization on the core cortex, although layer four cells also receive intracortical connections from areas that represent a wider frequency spectrum (Metherate et al. 2005; Kratz and Manis 2015). In addition, upper and lower layers of core areas receive inputs from nonlemniscal nuclei, including the medial division of the MG and the suprageniculate nucleus (reviewed by Winer 1992). Ascending inputs from the dorsal (nonlemniscal) division of the MG terminate in layer 4 in belt and parabelt auditory cortical areas.

Thalamocortical projections are highly specific. The multiple thalamic subdivisions have characteristic patterns of regional and laminar distribution (e.g., Huang and Winer 2000; Storace et al. 2010). Further, different cell populations in the ventral division project to different subdivisions of the core cortex (Lee et al. 2004; Read et al. 2011). Even *within* a core area, patchy projections (McMullen and de Venecia 1993) confer different physiological response properties on clustered groups of neurons (Velenovsky et al. 2003). Projections from the thalamus are not positive for GABAergic markers (Smith et al. 2012) and presumably are excitatory.

2.7.2 Corticocortical Connections: Commissural and Intrinsic

Like other areas of cortex, the AC receives topographically organized inputs from its counterpart on the opposite side of the brain (reviewed by Hackett and Phillips 2011), but most intracortical connections arise on the ipsilateral side (Lee and Winer 2011). Core and belt areas are highly interconnected through both short-range and long-range projections (Wallace et al. 2002; Hackett et al. 2014). Intrinsic interconnectivity has been difficult to study with traditional neuroanatomical methods, but the advent of new techniques has made the synaptic organization of the cortex more amenable to detailed study. Recent studies confirm that cortical circuits are as specific and highly organized as the rest of the auditory system. These new results and insights are discussed in detail in Chap. 8 by Budinger and Kanold.

2.7.3 *Nonauditory Cortical Inputs to Auditory Cortex*

Although most intracortical connections arise within the auditory fields, both core and belt areas also receive inputs from nonauditory cortical areas, including visual, somatosensory, and olfactory cortices (reviewed by Bizley and King 2009; Budinger and Scheich 2009). Like thalamic inputs and intrinsic connections, axons from nonauditory areas terminate in specific patterns. For example, two different extrastriate visual areas terminate in distinctly different laminar patterns in the AC (Smith et al. 2010). Areas of the forebrain implicated in global modulation of cortical states, such as the cholinergic basal complex, also project to the AC, as do a number of brainstem modulatory pathways (Schofield and Hurley, Chap. 9). The interconnections with other cortices and with modulatory centers serve to integrate auditory activity with overall behavioral and cognitive states (Zatorre 2007).

2.8 Summary

Multiple synaptic interruptions lie between the cochlea and the auditory cortex. At one time, auditory structures in the brainstem and diencephalon were referred to as "relay nuclei," but this characterization fails to acknowledge the information processing that occurs at each level of the auditory pathway. The activity of individual neurons is probably never a reflection of single inputs but, rather, is a reflection of integration of activity from convergent inputs often arising from a large number of diverse sources. Much remains to be learned about the synaptic organization of the structures that make up the auditory system, but the available evidence allows some generalizations. All auditory structures, from the cochlear nuclei to the auditory cortex, consist of a number of neuronal populations with different structural and functional properties. All of these populations appear to receive both excitatory and inhibitory inputs, which may arise from a few or many sources. Each area of the brainstem and forebrain receives both ascending and descending projections, and almost all of these connections are organized topographically with respect to the cochlear frequency map. Most of the areas are interconnected bilaterally through commissural pathways.

Most auditory areas also receive inputs from areas associated with other sensory modalities and with modulatory systems that provide information about the overall state of the brain. The synaptic inputs are arranged on the dendritic and somatic surfaces of individual neurons in ways characteristic of each cell population.

The sources and arrangements of the inputs, together with the intrinsic membrane properties of the postsynaptic cells, determine the responses of the neurons to acoustic stimuli in any given brain state. One of the most important challenges in neuroanatomical research is the continued search for ways to identify and define neuronal populations and to characterize the specific sources of inputs to each population.

Compliance with Ethics Requirements Nell B. Cant declares that she has no conflict of interest.

Douglas L. Oliver is a consultant for Pfizer, Inc.

References

Adams, J. C. (1980). Crossed and descending projections to the inferior colliculus. *Neuroscience Letters, 19*, 1–5.

Adams, J. C. (1983). Cytology of periolivary cells and the organization of their projections in the cat. *The Journal of Comparative Neurology, 215*, 275–289.

Adams, J. C., & Mugnaini, E. (1984). Dorsal nucleus of the lateral lemniscus: A nucleus of GABAergic projection neurons. *Brain Research Bulletin, 13*, 585–590.

Adams, J. C., & Mugnaini, E. (1990). Immunocytochemical evidence for inhibitory and disinhibitory circuits in the superior olive. *Hearing Research, 49*, 281–298.

Aitkin, L. M. (1989). The auditory system. In A. Björklund, T. Hökfelt, & L. W. Swanson (Eds.), *Handbook of chemical neuroanatomy* (Vol. 7, pp. 165–218). Amsterdam: Elsevier.

Aitkin, L. M., Kenyon, C. E., & Philpott, P. (1981). The representation of the auditory and somatosensory systems in the external nucleus of the cat inferior colliculus. *The Journal of Comparative Neurology, 196*, 25–40.

Albrecht, O., Dondzillo, A., Mayer, F., Thompson, J. A., & Klug, A. (2014). Inhibitory projections from the ventral nucleus of the trapezoid body to the medial nucleus of the trapezoid body in the mouse. *Frontiers in Neural Circuits, 8*, 83. https://doi.org/10.3389/fncir.2014.00083.

Alibardi, L. (2000). Cytology, synaptology and immunocytochemistry of commissural neurons and their putative axonal terminals in the dorsal cochlear nucleus of the rat. *Annals of Anatomy, 182*, 207–220.

Andersen, R. A., Knight, P. L., & Merzenich, M. M. (1980). The thalamocortical and corticothalamic connections of AI, AII, and the anterior auditory field (AAF) in the cat: Evidence for two largely segregated systems of connections. *The Journal of Comparative Neurology, 194*, 663–701.

Arnott, R. H., Wallace, M. N., Shackleton, T. M., & Palmer, A. R. (2004). Onset neurones in the anteroventral cochlear nucleus project to the dorsal cochlear nucleus. *Journal of the Association for Research in Otolaryngology, 5*, 153–170.

Bajo, V. M., & King, A. J. (2013). Cortical modulation of auditory processing in the midbrain. *Frontiers in Neural Circuits, 6*, 114. https://doi.org/10.3389/fncir.2012.00114.

Bajo, V. M., Merchán, M. A., López, D. E., & Rouiller, E. M. (1993). Neuronal morphology and efferent projections of the dorsal nucleus of the lateral lemniscus in the rat. *The Journal of Comparative Neurology, 334*, 241–262.

Bajo, V. M., Nodal, F. R., Bizley, J. K., Moore, D. R., & King, A. J. (2007). The ferret auditory cortex: Descending projections to the inferior colliculus. *Cerebral Cortex, 17*, 475–491.

Bajo, V. M., Rouiller, E. M., Welker, E., Clarke, S., Villa, A. E., de Ribaupierre, Y., & de Ribaupierre, F. (1995). Morphology and spatial distribution of corticothalamic terminals originating from cat auditory cortex. *Hearing Research, 83*, 161–174.

Banks, M. I., & Smith, P. H. (1992). Intracellular recordings from neurobiotin-labeled cells in brain slices of the rat medial nucleus of the trapezoid body. *The Journal of Neuroscience, 12*, 2819–2837.

Bartlett, E. L., & Smith, P. H. (1999). Anatomic, intrinsic, and synaptic properties of dorsal and ventral division neurons in rat medial geniculate body. *Journal of Neurophysiology, 81*, 1999–2016.

Bartlett, E. L., Stark, J. M., Guillery, R. W., & Smith, P. H. (2000). Comparison of the fine structure of cortical and collicular terminals in the rat medial geniculate body. *Neuroscience, 100*, 811–828.

Behrens, E. G., Schofield, B. R., & Thompson, A. M. (2002). Aminergic projections to cochlear nucleus via descending auditory pathways. *Brain Research, 955*, 34–44.

Benson, C. G., & Potashner, S. J. (1990). Retrograde transport of [^{3}H]glycine from the cochlear nucleus to the superior olive in the guinea pig. *The Journal of Comparative Neurology, 296*, 415–426.

Berglund, A. M., Benson, T. E., & Brown, M. C. (1996). Synapses from labeled type II axons in the mouse cochlear nucleus. *Hearing Research, 94*, 31–46.

Bizley, J. K., Bajo, V. M., Nodal, F. R., & King, A. J. (2015). Cortico-cortical connectivity within ferret auditory cortex. *The Journal of Comparative Neurology, 523*(15), 2187–2210. https://doi.org/10.1002/cne.23784.

Bizley, J. K., & King, A. J. (2009). Visual influences on ferret auditory cortex. *Hearing Research, 258*, 55–63.

Bledsoe, S. C., Snead, C. R., Helfert, R. H., Prasad, V., Wenthold, R. J., & Altschuler, R. A. (1990). Immunocytochemical and lesion studies support the hypothesis that the projection from the medial nucleus of the trapezoid body to the lateral superior olive is glycinergic. *Brain Research, 517*, 189–194.

Borst, J. G. G., & van Hoeve, J. S. (2012). The calyx of Held synapse: From model synapse to auditory relay. *Annual Reviews of Physiology, 74*, 199–224.

Brown, M. C. (1993). Fiber pathways and branching patterns of biocytin-labeled olivocochlear neurons in the mouse brainstem. *The Journal of Comparative Neurology, 337*, 600–613.

Brown, M. C., & Ledwith, J. V., III. (1990). Projections of thin (type-II) and thick (type-I) auditory-nerve fibers into the cochlear nucleus of the mouse. *Hearing Research, 49*, 105–118.

Brown, M. C., Liberman, M. C., Benson, T. E., & Ryugo, D. K. (1988). Brainstem branches from olivocochlear axons in cats and rodents. *The Journal of Comparative Neurology, 278*, 591–603.

Budinger, E., Heil, P., & Scheich, H. (2000). Functional organization of auditory cortex in the Mongolian gerbil (*Meriones unguiculatus).* IV. Connections with anatomically characterized subcortical structures. *European Journal of Neuroscience, 12*, 2452–2472.

Budinger, E., & Scheich, H. (2009). Anatomical connections suitable for direct processing of neuronal information of different modalities via the rodent primary auditory cortex. *Hearing Research, 258*, 16–27.

Caicedo, A., & Herbert, H. (1993). Topography of descending projections from inferior colliculus to auditory brainstem nuclei in the rat. *The Journal of Comparative Neurology, 328*, 377–392.

Calford, M. B., & Aitkin, L. M. (1983). Ascending projections to the medial geniculate body of the cat: Evidence for multiple, parallel auditory pathways through thalamus. *The Journal of Neuroscience, 3*, 2365–2380.

Campagnola, L., & Manis, P. B. (2014). A map of functional synaptic connectivity in the mouse anteroventral cochlear nucleus. *The Journal of Neuroscience, 34*, 2214–2230.

Cant, N. B., & Benson, C. G. (2003). Parallel auditory pathways: Projection patterns of different neuronal populations in the dorsal and ventral cochlear nuclei. *Brain Research Bulletin, 60*, 457–474.

Cant, N. B., & Benson, C. G. (2006). Organization of the inferior colliculus of the gerbil (*Meriones unguiculatus*): Differences in distribution of projections from the cochlear nuclei and the superior olivary complex. *The Journal of Comparative Neurology, 495*, 511–528.

Cant, N. B., & Benson, C. G. (2007). Multiple topographically organized projections connect the central nucleus of the inferior colliculus to the ventral division of the medial geniculate nucleus in the gerbil, *Meriones unguiculatus. The Journal of Comparative Neurology, 503*, 432–453.

Cant, N. B., & Hyson, R. L. (1992). Projections from the lateral nucleus of the trapezoid body to the medial superior olivary nucleus in the gerbil. *Hearing Research, 58*, 26–34.

Clarey, J. C., Barone, P., & Imig, T. J. (1992). Physiology of thalamus and cortex. In A. N. Popper & R. R. Fay (Eds.), *The mammalian auditory pathway: Neurophysiology* (pp. 232–334). New York: Springer.

Coleman, J. R., & Clerici, W. J. (1987). Sources of projections to subdivisions of the inferior colliculus in the rat. *The Journal of Comparative Neurology, 262*, 215–226.

Coomes, D. L., & Schofield, B. R. (2004a). Separate projections from the inferior colliculus to the cochlear nucleus and thalamus in guinea pigs. *Hearing Research, 191*, 67–78.

Coomes, D. L., & Schofield, B. R. (2004b). Projections from the auditory cortex to the superior olivary complex in guinea pigs. *European Journal of Neuroscience, 19*, 2188–2200.

Crabtree, J. W. (1998). Organization in the auditory sector of the cat's thalamic reticular nucleus. *The Journal of Comparative Neurology, 390*, 167–182.

Darrow, K. N., Benson, T. E., & Brown, M. C. (2012). Planar multipolar cells in the cochlear nucleus project to medial olivocochlear neurons in mouse. *The Journal of Comparative Neurology, 520*, 1365–1375.

De Ribaupierre, F. (1997). Acoustical information processing in the auditory thalamus and cerebral cortex. In G. Ehret & R. Romand (Eds.), *The central auditory system* (pp. 317–404). New York: Oxford University Press.

Diño, M. R., & Mugnaini, E. (2008). Distribution and phenotypes of unipolar brush cells in relation to the granule cell system of the rat cochlear nucleus. *Neuroscience, 154*, 29–50.

Doucet, J. R., Rose, L., & Ryugo, D. K. (2002). The cellular origin of corticofugal projections to the superior olivary complex in the rat. *Brain Research, 925*, 28–41.

Doucet, J. R., & Ryugo, D. K. (1997). Projections from the ventral cochlear nucleus to the dorsal cochlear nucleus in rats. *The Journal of Comparative Neurology, 385*, 245–264.

Doucet, J. R., & Ryugo, D. K. (2003). Axonal pathways to the lateral superior olive labeled with biotinylated dextran amine injections in the dorsal cochlear nucleus of rats. *The Journal of Comparative Neurology, 461*, 452–465.

Fathke, R. L., & Gabriele, M. L. (2009). Patterning of multiple layered projections to the auditory midbrain prior to experience. *Hearing Research, 249*, 36–43.

Faye-Lund, H. (1985). The neocortical projection to the inferior colliculus in the albino rat. *Anatomy and Embryology, 173*, 53–70.

Faye-Lund, H. (1986). Projection from the inferior colliculus to the superior olivary complex in the albino rat. *Anatomy and Embryology, 175*, 35–52.

Feliciano, M., & Potashner, S. J. (1995). Evidence for a glutamatergic pathway from the guinea pig auditory cortex to the inferior colliculus. *Journal of Neurochemistry, 65*, 1348–1357.

Feliciano, M., Saldaña, E., & Mugnaini, E. (1995). Direct projections from the rat primary auditory neocortex to nucleus sagulum, paralemniscal regions, superior olivary complex and cochlear nuclei. *Auditory Neuroscience, 1*, 287–308.

Friauf, E., & Ostwald, J. (1988). Divergent projections of physiologically characterized rat ventral cochlear neurons as shown by intra-axonal injection of horseradish peroxidase. *Experimental Brain Research, 73*, 263–285.

Frisina, R. D., Walton, J. P., Lynch-Armour, M. A., & Byrd, J. D. (1998). Inputs to a physiologically characterized region of the inferior colliculus of the young adult CBA mouse. *Hearing Research, 115*, 61–81.

Glendenning, K. K., Brunso-Bechtold, J. K., Thompson, G. C., & Masterton, R. B. (1981). Ascending auditory afferents to the nuclei of the lateral lemniscus. *The Journal of Comparative Neurology, 197*, 673–703.

Gómez-Álverez, M., & Saldaña, E. (2015). Different tonotopic regions of the lateral superior olive receive a similar combination of afferent inputs. *The Journal of Comparative Neurology, 524*(11), 2230–2250. https://doi.org/10.1002/cne.23942.

Gómez-Nieto, R., Horta-Junior, J. A. C., Castellano, O., Herrero-Turrión, M. J., Rubio, M. E., & López, D. E. (2008). Neurochemistry of the afferents to the rat cochlear root nucleus: Possible synaptic modulation of acoustic startle. *Neuroscience, 154*, 51–64.

Gómez-Nieto, R., Rubio, M. E., & López, D. E. (2008). Cholinergic input from the ventral nucleus of the trapezoid body to cochlear root neurons in rats. *The Journal of Comparative Neurology, 506*, 452–468.

González-Hernández, T., Mantolán-Sarmiento, B., González-González, B., & Pérez-González, H. (1996). Sources of GABAergic input to the inferior colliculus of the rat. *The Journal of Comparative Neurology, 372*, 309–326.

González-Hernández, T. H., Meyer, G., & Ferres-Torres, R. (1986). The commissural interconnections of the inferior colliculus in the albino mouse. *Brain Research, 368*, 268–276.

González-Hernández, T. H., Meyer, G., & Ferres-Torres, R. (1989). Development of neuronal types and laminar organization in the central nucleus of the inferior colliculus in the cat. *Neuroscience, 30*, 127–141.

Grothe, B., Pecka, M., & McAlpine, D. (2010). Mechanisms of sound localization in mammals. *Physiological Reviews, 90*, 983–1012.

Guillery, R. W., Feig, S. L., & Lozsádi, D. A. (1998). Paying attention to the thalamic reticular nucleus. *Trends in Neurosciences, 21*, 28–32.

Hackett, T. A. (2011). Information flow in the auditory cortical network. *Hearing Research, 271*, 133–146.

Hackett, T. A., de la Mothe, L. A., Camalier, C. R., Falchier, A., Lakatos, P., Kajikawa, Y., & Schroeder, C. E. (2014). Feedforward and feedback projections of caudal belt and parabelt areas of auditory cortex: Refining the hierarchical model. *Frontiers in Neuroscience, 8*, 72. https://doi.org/10.3389/fnins.2014.00072.

Hackett, T. A., & Phillips, D. P. (2011). The commissural auditory system. In J. A. Winer & C. E. Schreiner (Eds.), *The auditory cortex* (pp. 117–132). New York: Springer.

Haenggeli, C.-A., Pongstaporn, T., Doucet, J. R., & Ryugo, D. K. (2005). Projections from the spinal trigeminal nucleus to the cochlear nucleus in the rat. *The Journal of Comparative Neurology, 484*, 191–205.

He, J. (2003). Corticofugal modulation of the auditory thalamus. *Experimental Brain Research, 153*, 579–590.

Henkel, C. K. (1983). Evidence of sub-collicular auditory projections to the medial geniculate nucleus in the cat: An autoradiographic and horseradish peroxidase study. *Brain Research, 259*, 21–30.

Henkel, C. K., Fuentes-Santamaria, V., Alvarado, J. C., & Brunso-Bechtold, J. K. (2003). Quantitative measurement of afferent layers in the ferret inferior colliculus: DNLL projections to sublayers. *Hearing Research, 177*, 32–42.

Henkel, C. K., & Shneiderman, A. (1988). Nucleus sagulum: Projections of a lateral tegmental area to the inferior colliculus in the cat. *The Journal of Comparative Neurology, 271*, 577–588.

Henkel, C. K., & Spangler, K. M. (1983). Organization of the efferent projections of the medial superior olivary nucleus in the cat as revealed by HRP and autoradiographic tracing methods. *The Journal of Comparative Neurology, 221*, 416–428.

Herbert, H., Aschoff, A., & Ostwald, J. (1991). Topography of projections from the auditory cortex to the inferior colliculus in the rat. *The Journal of Comparative Neurology, 304*, 103–122.

Higgins, N. C., Storace, D. A., Escabí, M. A., & Read, H. L. (2010). Specialization of binaural responses in ventral auditory cortices. *The Journal of Neuroscience, 30*, 14522–14532.

Huang, C. L., & Winer, J. A. (2000). Auditory thalamocortical projections in the cat: Laminar and areal patterns of input. *The Journal of Comparative Neurology, 427*, 302–331.

Imig, T. J., & Morel, A. (1983). Organization of the thalamocortical auditory system in the cat. *Annual Reviews of Neuroscience, 6*, 95–120.

Itaya, S. K., & Van Hoesen, G. W. (1982). Retinal innervation of the inferior colliculus in rat and monkey. *Brain Research, 233*, 45–52.

Ito, T., Bishop, D. C., & Oliver, D. L. (2011). Expression of glutamate and inhibitory amino acid vesicular transporters in the rodent auditory brainstem. *The Journal of Comparative Neurology, 519*, 316–340.

Ito, T., & Oliver, D. L. (2010). Origins of glutamatergic terminals in the inferior colliculus identified by retrograde transport and expression of VGLUT1 and VGLUT2 genes. *Frontiers in Neuroanatomy, 4*, 135. https://doi.org/10.3389/fnana.2010.00135.

Ito, T., & Oliver, D. L. (2014). Local and commissural IC neurons make axosomatic inputs on large GABAergic tectothalamic neurons. *The Journal of Comparative Neurology, 522*, 3539–3554.

Kamiya, H., Itoh, K., Yasui, Y., Ino, T., & Mizuno, N. (1988). Somatosensory and auditory relay nucleus in the rostral part of the ventrolateral medulla: A morphological study in the cat. *The Journal of Comparative Neurology, 273*, 421–435.

Kandler, K. (2004). Activity-dependent organization of inhibitory circuits: Lessons from the auditory system. *Current Opinion in Neurobiology, 14*, 96–104.

Kidd, S. A., & Kelly, J. B. (1996). Contribution of the dorsal nucleus of the lateral lemniscus to binaural reponses in the inferior colliculus of the rat: Interaural time delays. *The Journal of Neuroscience, 16*, 7390–7397.

Kimura, A., Donishi, T., Okamoto, K., & Tamai, Y. (2005). Topography of projections from the primary and non-primary auditory cortical areas to the medial geniculate body and thalamic reticular nucleus in the rat. *Neuroscience, 135*, 1325–1342.

Kimura, A., Yokoi, I., Imbe, H., Donishi, T., & Kaneoke, Y. (2012). Auditory thalamic reticular nucleus of the rat: Anatomical nodes for modulation of auditory and cross-modal sensory processing in the loop connectivity between cortex and thalamus. *The Journal of Comparative Neurology, 520*, 1457–1480.

Kratz, M. B., & Manis, P. B. (2015). Spatial organization of excitatory synaptic inputs to layer 4 neurons in mouse primary auditory cortex. *Frontiers in Neural Circuits, 9*, 17. https://doi.org/10.3389/fncir.2015.00017.

Kraus, K. S., & Canlon, B. (2012). Neuronal connectivity and interactions between the auditory and limbic systems. Effects of noise and tinnitus. *Hearing Research, 288*, 34–46.

Kudo, M. (1981). Projections of the nuclei of the lateral lemniscus in the cat: An autoradiographic study. *Brain Research, 221*, 57–69.

Kudo, M., & Niimi, K. (1978). Ascending projections of the inferior colliculus onto the medial geniculate body in the cat studied by anterograde and retrograde tracing techniques. *Brain Research, 155*, 113–117.

Kulesza, R. J., & Berrebi, A. S. (2000). Superior paraolivary nucleus of the rat is a GABAergic nucleus. *Journal of the Association for Research in Otolaryngology, 1*, 255–269.

Künzle, H. (1993). Tectal and related target areas of spinal and dorsal column nuclear projections in hedgehog tenrec. *Somatosensory and Motor Research, 10*, 339–353.

Kuwabara, N., & Zook, J. M. (1999). Local collateral projections from the medial superior olive to the superior paraolivary nucleus in the gerbil. *Brain Research, 846*, 59–71.

LeDoux, J. E., & Farb, C. R. (1991). Neurons of the acoustic thalamus that project to the amygdala contain glutamate. *Neuroscience Letters, 134*, 145–149.

LeDoux, J. E., Ruggiero, D. A., Forest, E., Stornetta, R., & Reis, D. J. (1987). Topographic organization of convergent projections to the thalamus from the inferior colliculus and spinal cord in the rat. *The Journal of Comparative Neurology, 264*, 123–146.

Lee, C. C. (2015). Exploring functions for the non-lemniscal auditory thalamus. *Frontiers in Neural Circuits, 9*, 69. https://doi.org/10.3389/fncir.2015.00069.

Lee, C. C., Imaizumi, K., Schreiner, C. E., & Winer, J. A. (2004). Concurrent tonotopic processing streams in auditory cortex. *Cerebral Cortex, 14*, 441–451.

Lee, C. C., & Winer, J. A. (2011). Convergence of thalamic and cortical pathways in cat auditory cortex. *Hearing Research, 274*, 85–94.

Liberman, M. C. (1980). Morphological differences among radial afferent fibers in the cat cochlea: An electron microscopic study of serial sections. *Hearing Research, 3*, 45–63.

Liberman, M. C. (1991). Central projections of auditory-nerve fibers of differing spontaneous rate. I. Anteroventral cochlear nucleus. *The Journal of Comparative Neurology, 313*, 240–258.

Liberman, M. C. (1993). Central projections of auditory nerve fibers of differing spontaneous rate. II. Posteroventral and dorsal cochlear nuclei. *The Journal of Comparative Neurology, 327*, 17–36.

Loftus, W. C., Bishop, D. C., & Oliver, D. L. (2010). Differential patterns of inputs create functional zones in central nucleus of inferior colliculus. *The Journal of Neuroscience, 30*, 13396–13408.

Loftus, W. C., Bishop, D. C., Saint Marie, R. L., & Oliver, D. L. (2004). Organization of binaural excitatory and inhibitory inputs to the inferior colliculus from the superior olive. *The Journal of Comparative Neurology, 472*, 330–344.

Lomber, S. G., & Malhotra, S. (2008). Double dissociation of 'what' and 'where' processing in auditory cortex. *Nature Neuroscience, 11*, 609–616. https://doi.org/10.1038/nn.2108.

Malmierca, M. S. (2003). The structure and physiology of the rat auditory system: An overview. *International Reviews of Neurobiology, 56*, 147–211.

Malmierca, M. S., Hernández, O., Antunes, F. M., & Rees, A. (2009). Divergent and point-to-point connections in the commissural pathway between the inferior colliculi. *The Journal of Comparative Neurology, 514*, 226–239.

Malmierca, M. S., Le Beau, F. E. N., & Rees, A. (1996). The topographical organization of descending projections from the central nucleus of the inferior colliculus in guinea pig. *Hearing Research, 93*, 167–180.

Malmierca, M. S., Merchán, M. A., Henkel, C. K., & Oliver, D. L. (2002). Direct projections from cochlear nuclear complex to auditory thalamus in the rat. *The Journal of Neuroscience, 22*, 10891–10897.

Malmierca, M. S., Rees, A., Le Beau, F. E. N., & Bjaalie, J. G. (1995). Laminar organization of frequency-defined local axons within and between the inferior colliculi of the guinea pig. *The Journal of Comparative Neurology, 357*, 124–144.

Márquez-Legorreta, E., Horta-Júnior, J. A. C., Berrebi, A. S., & Saldaña, E. (2016). Organization of the zone of transition between the pretectum and the thalamus, with emphasis on the pretectothalamic lamina. *Frontiers in Neuroanatomy, 10*, 82. https://doi.org/10.3389/fnana.2016.00082.

Marsh, R. A., Fuzessery, Z. M., Grose, C. D., & Wenstrup, J. J. (2002). Projection to the inferior colliculus from the basal nucleus of the amygdala. *The Journal of Neuroscience, 22*, 10449–10460.

McMullen, N. T., & de Venecia, R. K. (1993). Thalamocortical patches in auditory neocortex. *Brain Research, 620*, 317–322.

McMullen, N. T., Velenovsky, D. S., & Holmes, M. G. (2005). Auditory thalamic organization: Cellular slabs, dendritic arbors and tectothalamic axons underlying the frequency map. *Neuroscience, 136*, 927–943.

Mellott, J. G., Foster, N. L., Ohl, A. P., & Schofield, B. R. (2014). Excitatory and inhibitory projections in parallel pathways from the inferior colliculus to the auditory thalamus. *Frontiers in Neuroanatomy, 8*, 124. https://doi.org/10.3389/fnana.2014.00124.

Meltzer, N. E., & Ryugo, D. K. (2006). Projections from auditory cortex to cochlear nucleus: A comparative analysis of rat and mouse. *Anatomical Record, 288A*, 397–408.

Merchán, M. A., Malmierca, M. S., Bajo, V. M., & Bjaalie, J. G. (1997). The nuclei of the lateral lemniscus. Old views and new perspectives. In J. Syka (Ed.), *Acoustical signal processing in the central auditory system* (pp. 211–226). New York: Plenum Press.

Metherate, R., Kaur, S., Kawai, H., Lazar, R., Liang, K., & Rose, H. J. (2005). Spectral integration in auditory cortex: Mechanisms and modulation. *Hearing Research, 206*, 146–158.

Milinkeviciute, G., Muniak, M. A., & Ryugo, D. K. (2016). Descending projections from the inferior colliculus to the dorsal cochlear nucleus are excitatory. *The Journal of Comparative Neurology, 525*(4), 773–793. https://doi.org/10.1002/cne.24095.

Moore, D. R. (1988). Auditory brainstem of the ferret: Sources of projections to the inferior colliculus in normal and neonatally cochlea-ablated gerbils. *The Journal of Comparative Neurology, 240*, 180–195.

Morest, D. K. (1965a). The laminar structure of the medial geniculate body of the cat. *Journal of Anatomy, 99*, 143–160.

Morest, D. K. (1965b). The lateral tegmental system of the midbrain and the medial geniculate body: Study with Golgi and Nauta methods in cat. *Journal of Anatomy, 99*, 611–634.

Morest, D. K. (1975). Synaptic relationships of Golgi type II cells in the medial geniculate body of the cat. *The Journal of Comparative Neurology, 162*, 157–194.

Morgan, Y. V., Ryugo, D. K., & Brown, M. C. (1994). Central trajectories of type II (thin) fibers of the auditory nerve in cats. *Hearing Research, 79*, 74–82.

Moriizumi, T., & Hattori, T. (1991). Pallidotectal projection to the inferior colliculus of the rat. *Experimental Brain Research, 87*, 223–226.

Motts, S. D., & Schofield, B. R. (2010). Cholinergic and non-cholinergic projections from the pedunculopontine and laterodorsal tegmental nuclei to the medial geniculate body in guinea pigs. *Frontiers in Neuroanatomy, 4*, 137. https://doi.org/10.3389/fnana.2010.00137.

Mugnaini, E., Warr, W. B., & Osen, K. K. (1980). Distribution and light microscopic features of granule cells in the cochlear nuclei of cat, rat, and mouse. *The Journal of Comparative Neurology, 191*, 581–606.

Nakamoto, K. T., Jones, S. J., & Palmer, A. R. (2008). Descending projections from auditory cortex modulate sensitivity in the midbrain to cues for spatial position. *Journal of Neurophysiology, 99*, 2347–2356.

Nakamoto, K. T., Sowick, C. S., & Schofield, B. R. (2013). Auditory cortical axons contact commissural cells throughout the guinea pig inferior colliculus. *Hearing Research, 306*, 131–144.

Nayagam, D. A. X., Clarey, J. C., & Paolini, A. G. (2005). Powerful, onset inhibition in the ventral nucleus of the lateral lemniscus. *Journal of Neurophysiology, 94*, 1651–1654.

Needham, K., & Paolini, A. G. (2007). The commissural pathway and cochlear nucleus bushy neurons: An in vivo intracellular investigation. *Brain Research, 1134*, 113–121.

Newlands, S. D., & Perachio, A. A. (2003). Central projections of the vestibular nerve: A review and single fiber study in the Mongolian gerbil. *Brain Research Bulletin, 60*, 475–495.

Oertel, D., & Wickesberg, R. E. (2002). Ascending pathways through ventral nuclei of the lateral lemniscus and their possible role in pattern recognition in natural sounds. In D. Oertel, R. R. Fay, & A. N. Popper (Eds.), *Integrative functions in the mammalian auditory pathway* (pp. 207–237). New York: Springer.

Oertel, D., Wright, S., Cao, X.-J., Ferragamo, M., & Bal, R. (2011). The multiple functions of T stellate/multipolar/chopper cells in the ventral cochlear nucleus. *Hearing Research, 276*, 61–69.

Ohlrogge, M., Doucet, J. R., & Ryugo, D. K. (2001). Projections of the pontine nuclei to the cochlear nucleus in rats. *The Journal of Comparative Neurology, 436*, 290–303.

Ojima, H., & Murakami, K. (2010). Triadic synaptic interactions of large corticothalamic terminals in non-lemniscal thalamic nuclei of the cat auditory system. *Hearing Research, 274*, 40–47.

Okoyama, S., Ohbayashi, M., Ito, M., & Harada, S. (2006). Neuronal organization of the rat inferior colliculus participating in four major auditory pathways. *Hearing Research, 218*, 72–80.

Olazábal, U. E., & Moore, J. K. (1989). Nigrotectal projection to the inferior colliculus: Horseradish peroxidase transport and tyrosine hydroxylase immunohistochemical studies in rats, cats, and bats. *The Journal of Comparative Neurology, 282*, 98–118.

Oliver, D. L. (1984). Neuron types in the central nucleus of the inferior colliculus that project to the medial geniculate body. *Neuroscience, 11*, 409–424.

Oliver, D. L. (1985). Quantitative analyses of axonal endings in the central nucleus of the inferior colliculus and distribution of 3H-labeling after injections in the dorsal cochlear nucleus. *The Journal of Comparative Neurology, 237*, 343–359.

Oliver, D. L. (1987). Projections to the inferior colliculus from the anteroventral cochlear nucleus in the cat: Possible substrates for binaural interaction. *The Journal of Comparative Neurology, 264*, 24–46.

Oliver, D. L. (2000). Ascending efferent projections of the superior olivary complex. *Microscopy Research Techniques, 51*, 355–363.

Oliver, D. L., Beckius, G. E., Bishop, D. C., & Kuwada, S. (1997). Simultaneous anterograde labeling of axonal layers from lateral superior olive and dorsal cochlear nucleus in the inferior colliculus of cat. *The Journal of Comparative Neurology, 382*, 215–229.

Oliver, D. L., Beckius, G. E., & Shneiderman, A. (1995). Axonal projections from the lateral and medial superior olive to the inferior colliculus of the cat: A study using electron microscopic autoradiography. *The Journal of Comparative Neurology, 360*, 17–32.

Oliver, D. L., & Hall, W. C. (1978). The medial geniculate body of the tree shrew, *Tupaia glis*. I. Cytoarchitecture and midbrain connections. *The Journal of Comparative Neurology, 182*, 423–458.

Oliver, D. L., Kuwada, S., Yin, T. C. T., Haberly, L. B., & Henkel, C. K. (1991). Dendritic and axonal morphology of HRP-injected neurons in the inferior colliculus of the cat. *The Journal of Comparative Neurology, 303*, 75–100.

Oliver, D. L., & Shneiderman, A. (1989). An EM study of the dorsal nucleus of the lateral lemniscus: Inhibitory, commissural, and synaptic connections between ascending auditory pathways. *The Journal of Neuroscience, 9*, 967–982.

Osen, K. K. (1972). Projections of the cochlear nuclei on the inferior colliculus in the cat. *The Journal of Comparative Neurology, 144*, 355–372.

Ostapoff, E.-M., Benson, C. G., & Saint Marie, R. L. (1997). GABA- and glycine-immunoreactive projections from the superior olivary complex to the cochlear nucleus in guinea pig. *The Journal of Comparative Neurology, 381*, 500–512.

Patel, M. B., Sons, S., Yudintsev, G., Lesicko, A. M. H., Yang, L., Taha, G. A., & Pierce, S. M. (2016). Anatomical characterization of subcortical descending projections to the inferior colliculus in mouse. *The Journal of Comparative Neurology, 525*(4), 885–900. https://doi.org/10.1002/cne.24106.

Peruzzi, D., Bartlett, E., Smith, P. H., & Oliver, D. L. (1997). A monosynaptic GABAergic input from the inferior colliculus to the medial geniculate body in rat. *The Journal of Neuroscience, 17*, 3766–3777.

Peterson, D. C., & Schofield, B. R. (2007). Projections from auditory cortex contact ascending pathways that originate in the superior olive and inferior colliculus. *Hearing Research, 232*, 67–77.

Pollak, G. D., Gittelman, J. X., Li, N., & Xie, R. (2011). Inhibitory projections from the ventral nucleus of the lateral lemniscus and superior paraolivary nucleus create directional selectivity of frequency modulations in the inferior colliculus: A comparison of bats with other mammals. *Hearing Research, 273*, 134–144.

Read, H. L., Nauen, D. W., Escabí, M., Miller, L. M., Schreiner, C. E., & Winer, J. A. (2011). Distinct core thalamocortical pathways to central and dorsal primary auditory cortex. *Hearing Research, 274*, 95–104.

Riquelme, R., Saldaña, E., Osen, K. K., Ottersen, O. P., & Merchán, M. A. (2001). Co-localization of GABA and glycine in the ventral nucleus of the lateral lemniscus in rat: An in-situ hybridization and semiquantitative immunocytochemical study. *The Journal of Comparative Neurology, 432*, 409–424.

Robards, M. J. (1979). Somatic neurons in the brainstem and neocortex projecting to the external nucleus of the inferior colliculus: An anatomical study in the opossum. *The Journal of Comparative Neurology, 184*, 547–566.

Rodrigues-Dagaeff, C., Simm, G., De Ribaupierre, Y., Villa, A., De Ribaupierre, F., & Rouiller, E. M. (1989). Functional organization of the ventral division of the medial geniculate body of the cat: Evidence for a rostro-caudal gradient of response properties and cortical projections. *Hearing Research, 39*, 103–126.

Ross, L. S., Pollak, G. D., & Zook, J. M. (1988). Origin of ascending projections to an isofrequency region of the mustache bat's inferior colliculus. *The Journal of Comparative Neurology, 270*, 488–505.

Rouiller, E. M., Colomb, E., Capt, M., & De Ribaupierre, F. (1985). Projections of the reticular complex of the thalamus onto physiologically characterized regions of the medial geniculate body. *Neuroscience Letters, 5*, 227–232.

Rouiller, E. M., & de Ribaupierre, F. (1985). Origin of afferents to physiologically defined regions of the medial geniculate body of the cat: Ventral and dorsal divisions. *Hearing Research, 19*, 97–114.

Rouiller, E. M., & Welker, E. (2000). A comparative analysis of the morphology of corticothalamic projections in mammals. *Brain Research Bulletin, 53*, 727–741.

Ryugo, D. K. (2008). Projections of low spontaneous rate, high threshold auditory nerve fibers to the small cell cap of the cochlear nucleus in cats. *Neuroscience, 154*, 114–126.

Ryugo, D. K., Haenggeli, C.-A., & Doucet, J. R. (2003). Multimodal inputs to the granule cell domain of the cochlear nucleus. *Experimental Brain Research, 153*, 477–485.

Ryugo, D. K., & Parks, T. N. (2003). Primary innervation of the avian and mammalian cochlear nucleus. *Brain Research Bulletin, 60*, 435–456.

Saint Marie, R. L., & Baker, R. A. (1990). Neurotransmitter-specific uptake and retrograde transport of [^{3}H]glycine from the inferior colliculus by ipsilateral projections of the superior olivary complex and nuclei of the lateral lemniscus. *Brain Research, 524*, 244–253.

Saint Marie, R. L., Morest, D. K., & Brandon, C. J. (1989). The form and distribution of GABAergic synapses on principal cell types of the ventral cochlear nucleus of the cat. *Hearing Research, 42*, 97–112.

Saint Marie, R. L., Ostapoff, E.-M., Benson, C. G., Morest, D. K., & Potashner, S. J. (1993). Non-cochlear projections to the ventral cochlear nucleus: Are they mainly inhibitory? In M. A. Merchán, J. M. Juiz, D. A. Godfrey, & E. Mugnaini (Eds.), *The mammalian cochlear nuclei. Organization and function* (pp. 121–131). New York: Plenum Press.

Saint Marie, R. L., Shneiderman, A., & Stanforth, D. A. (1997). Patterns of gamma-aminobutyric acid and glycine immunoreactivities reflect structural and functional differences of the cat lateral lemniscal nuclei. *The Journal of Comparative Neurology, 389*, 264–276.

Saint Marie, R. L., Stanforth, D. A., & Jubelier, E. M. (1997). Substrate for rapid feedforward inhibition of the auditory forebrain. *Brain Research, 765*, 173–176.

Saldaña, E., Aparicio, M. A., Fuentes-Santamaria, V., & Berrebi, A. S. (2009). Connections of the superior paraolivary nucleus of the rat: Projections to the inferior colliculus. *Neuroscience, 163*, 372–387.

Saldaña, E., Feliciano, M., & Mugnaini, E. (1996). Distribution of descending projections from primary auditory neocortex to inferior colliculus mimics the topography of intracollicular projections. *The Journal of Comparative Neurology, 371*, 15–40.

Saldaña, E., & Merchán, M. A. (2005). Intrinsic and commissural connections of the inferior colliculus. In J. A. Winer & C. E. Schreiner (Eds.), *The inferior colliculus* (pp. 155–181). New York: Springer.

Sanes, D. H., & Friauf, E. (2000). Development and influence of inhibition in the lateral superior olivary nucleus. *Hearing Research, 147*, 46–58.

Schofield, B. R. (1995). Projections from the cochlear nucleus to the superior paraolivary nucleus in guinea pigs. *The Journal of Comparative Neurology, 360*, 135–149.

Schofield, B. R. (2001). Origins of projections from the inferior colliculus to the cochlear nucleus in guinea pigs. *The Journal of Comparative Neurology, 429*, 206–220.

Schofield, B. R. (2010). Projections from auditory cortex to midbrain cholinergic neurons that project to the inferior colliculus. *Neuroscience, 166*, 231–240.

Schofield, B. R., & Cant, N. B. (1996). Origins and targets of commissural connections between the cochlear nuclei in guinea pigs. *The Journal of Comparative Neurology, 375*, 128–146.

Schofield, B. R., & Coomes, D. L. (2005a). Auditory cortical projections to the cochlear nucleus in guinea pigs. *Hearing Research, 199*, 89–102.

Schofield, B. R., & Coomes, D. L. (2005b). Projections from auditory cortex contact cells in the cochlear nucleus that project to the inferior colliculus. *Hearing Research, 206*, 3–11.

Schofield, B. R., & Coomes, D. L. (2006). Pathways from auditory cortex to the cochlear nucleus in guinea pigs. *Hearing Research, 216–217*, 81–89.

Schofield, B. R., Motts, S. D., Mellott, J. G., & Foster, N. L. (2014). Projections from the dorsal and ventral cochlear nuclei to the medial geniculate body. *Frontiers in Neuroanatomy, 8*, 10. https://doi.org/10.3389/fnana.2014.00010.

Senatorov, V. V., & Hu, B. (2002). Extracortical descending projections to rat inferior colliculus. *Neuroscience, 115*, 243–250.

Shneiderman, A., & Henkel, C. K. (1987). Banding of lateral superior olivary nucleus afferents in the inferior colliculus: A possible substrate for sensory integration. *The Journal of Comparative Neurology, 266*, 519–534.

Shneiderman, A., & Oliver, D. L. (1989). EM autoradiographic study of the projections from the dorsal nucleus of the lateral lemniscus: A possible source of inhibitory inputs to the inferior colliculus. *The Journal of Comparative Neurology, 286*, 28–47.

Shneiderman, A., Oliver, D. L., & Henkel, C. K. (1988). Connections of the dorsal nucleus of the lateral lemniscus: An inhibitory parallel pathway in the ascending auditory system? *The Journal of Comparative Neurology, 276*, 188–208.

Shore, S. E., Helfert, R. H., Bledsoe, S. C., Jr., Altschuler, R. A., & Godfrey, D. A. (1991). Descending projections to the dorsal and ventral divisions of the cochlear nucleus in guinea pig. *Hearing Research, 52*, 255–268.

Shore, S. E., Vass, Z., Wys, N. L., & Altschuler, R. A. (2000). Trigeminal ganglion innervates the auditory brainstem. *The Journal of Comparative Neurology, 419*, 271–285.

Smith, P. H., Manning, K. A., & Uhlrich, D. J. (2010). Evaluation of inputs to rat primary auditory cortex from the suprageniculate nucleus and extrastriate visual cortex. *The Journal of Comparative Neurology, 518*, 3679–3700.

Smith, P. H., & Rhode, W. S. (1985). Electron microscopic features of physiologically characterized, HRP-labeled fusiform cells in the cat dorsal cochlear nucleus. *The Journal of Comparative Neurology, 237*, 127–143.

Smith, P. H., & Rhode, W. S. (1989). Structural and functional properties distinguish two types of multipolar cells in the cat anteroventral cochlear nucleus. *The Journal of Comparative Neurology, 282*, 595–616.

Smith, P. H., & Spirou, G. A. (2002). From the cochlea to the cortex and back. In D. Oertel, R. R. Fay, & A. N. Popper (Eds.), *Integrative functions in the mammalian auditory pathway* (pp. 6–71). New York: Springer.

Smith, P. H., Uhlrich, D. J., Manning, K. A., & Banks, M. I. (2012). Thalamocortical projections to rat auditory cortex from the ventral and dorsal divisions of the medial geniculate nucleus. *The Journal of Comparative Neurology, 520*, 34–51.

Sommer, I., Lingenhöhl, K., & Friauf, E. (1993). Principal cells of the rat medial nucleus of the trapezoid body: An intracellular in vivo study of their physiology and morphology. *Experimental Brain Research, 95*, 223–239.

Spirou, G. A., & Berrebi, A. S. (1997). Glycine immunoreactivity in the lateral nucleus of the trapezoid body of the cat. *The Journal of Comparative Neurology, 383*, 473–488.

Storace, D. A., Higgins, N. C., & Read, H. L. (2010). Thalamic label patterns suggest primary and ventral auditory fields are distinct core regions. *The Journal of Comparative Neurology, 518*, 1630–1646.

Straka, M. M., Schmitz, S., & Lim, H. H. (2014). Response features across the auditory midbrain reveal an organization consistent with a dual lemniscal pathway. *The Journal of Neurophysiology, 112*, 981–998.

Thompson, A. M., & Schofield, B. R. (2000). Afferent projections of the superior olivary complex. *Microscopy Research Techniques, 51*, 330–354.

Thompson, A. M., & Thompson, G. C. (1993). Relationship of descending inferior colliculus projections to olivocochlear neurons. *The Journal of Comparative Neurology, 335*, 402–412.

Tolbert, L. P., Morest, D. K., & Yurgelun-Todd, D. A. (1982). The neuronal architecture of the anteroventral cochlear nucleus of the cat in the region of the cochlear nerve root: Horseradish peroxidase labeling of identified cell types. *Neuroscience, 7*, 3031–3052.

Velenovsky, D. S., Cetas, J. S., Price, R. O., Sinex, D. G., & McMullen, N. T. (2003). Functional subregions in primary auditory cortex defined by thalamocortical terminal arbors: An electrophysiological and anterograde labeling study. *Journal of Neuroscience, 23*, 308–316.

Vetter, D. E., Saldaña, E., & Mugnaini, E. (1993). Input from the inferior colliculus to medial olivocochlear neurons in the rat: A double label study with PHA-L and cholera toxin. *Hearing Research, 70*, 173–186.

Wallace, M. N., Rutkowski, R. G., & Palmer, A. R. (2002). Interconnections of auditory areas in the guinea pig neocortex. *Experimental Brain Research, 143*, 106–119.

Warr, W. B. (1992). Organization of olivocochlear efferent systems in mammals. In D. B. Webster, A. N. Popper, & R. R. Fay (Eds.), *The mammalian auditory pathway: Neuroanatomy* (pp. 410–448). New York: Springer.

Warr, W. B., & Beck, J. E. (1996). Multiple projections from the ventral nucleus of the trapezoid body in the rat. *Hearing Research, 93*, 83–101.

Weedman, D. L., Pongstaporn, T., & Ryugo, D. K. (1996). Ultrastructural study of the granule cell domain of the cochlear nucleus in rats: Mossy fiber endings and their targets. *The Journal of Comparative Neurology, 369*, 345–360.

Weedman, D. L., & Ryugo, D. K. (1996). Pyramidal cells in primary auditory cortex project to cochlear nucleus in rat. *Brain Research, 706*, 97–102.
Weinberger, N. M. (2010). The medial geniculate, not the amygdala, as the root of auditory fear conditioning. *Hearing Research, 274*, 61–74.
Wenstrup, J. J. (2005). The tectothalamic system. In J. A. Winer & C. E. Schreiner (Eds.), *The inferior colliculus* (pp. 200–230). New York: Springer.
Wenthold, R. J. (1987). Evidence for a glycinergic pathway connecting the two cochlear nuclei: An immunocytochemical and retrograde transport study. *Brain Research, 415*, 183–187.
Whitley, J. M., & Henkel, C. K. (1984). Topographical organization of the inferior collicular projection and other connections of the ventral nucleus of the lateral lemniscus in the cat. *The Journal of Comparative Neurology, 229*, 257–270.
Wickesberg, R. E., & Oertel, D. (1990). Delayed, frequency-specific inhibition in the cochlear nuclei of mice: A mechanism for monaural echo suppression. *The Journal of Neuroscience, 10*, 1762–1768.
Wickesberg, R. E., Whitlon, D., & Oertel, D. (1991). Tuberculoventral neurons project to the multipolar cell area but not to the octopus cell area of the posteroventral cochlear nucleus. *The Journal of Comparative Neurology, 313*, 457–468.
Willard, F. H., & Martin, G. F. (1983). The auditory brainstem nuclei and some of their projections to the inferior colliculus in the North American opossum. *Neuroscience, 10*, 1203–1232.
Winer, J. A. (1992). The functional architecture of the medial geniculate body and the primary auditory cortex. In D. B. Webster, A. N. Popper, & R. R. Fay (Eds.), *The mammalian auditory pathway: Neuroanatomy* (pp. 222–408). New York: Springer.
Winer, J. A., Diehl, J. J., & Larue, D. T. (2001). Projections of auditory cortex to the medial geniculate body of the cat. *The Journal of Comparative Neurology, 430*, 27–55.
Winer, J. A., & Larue, D. T. (1996). Evolution of GABAergic circuitry in the mammalian medial geniculate body. *Proceedings of the National Academy of Sciences of the United States of America, 93*, 3083–3087.
Winer, J. A., Larue, D. T., Diehl, J. J., & Hefti, B. J. (1998). Auditory cortical projections to cat inferior colliculus. *The. The Journal of Comparative Neurology, 400*, 147–174.
Winer, J. A., Morest, D. K., & Diamond, I. T. (1988). A cytoarchitectonic atlas of the medial geniculate body of the opossum, *Didelphys virginiana*, with a comment on the posterior intralaminar nuclei of the thalamus. *The Journal of Comparative Neurology, 274*, 422–448.
Winer, J. A., Saint Marie, R. L., Larue, D. T., & Oliver, D. L. (1996). GABAergic feedforward projections from the inferior colliculus to the medial geniculate body. *Proceedings of the National Academy of Sciences of the United States of America, 93*, 8005–8010.
Winer, J. A., & Schreiner, C. E. (Eds.). (2005). *The inferior colliculus*. New York: Springer.
Winter, I. M., Robertson, D., & Cole, K. S. (1989). Descending projections from auditory brainstem nuclei to the cochlea and cochlear nucleus of the guinea pig. *The Journal of Comparative Neurology, 280*, 143–157.
Wright, D. D., & Ryugo, D. K. (1996). Mossy fiber projections from the cuneate nucleus to the cochlear nucleus in the rat. *The Journal of Comparative Neurology, 365*, 159–172.
Yang, L., Liu, Q., & Pollak, G. D. (1996). Afferent connections to the dorsal nucleus of the lateral lemniscus of the mustache bat: Evidence for two functional subdivisions. *The Journal of Comparative Neurology, 373*, 575–592.
Yasui, Y., Nakano, K., & Mizuno, N. (1992). Descending projections from the subparafascicular thalamic nucleus to the lower brain stem in the rat. *Experimental Brain Research, 90*, 508–518.
Yin, T. C. T. (2002). Neural mechanisms of encoding binaural localization cues in the auditory brainstem. In D. Oertel, R. R. Fay, & A. N. Popper (Eds.), *Integrative functions in the mammalian auditory pathway* (pp. 99–159). New York: Springer.
Young, E. D., & Oertel, D. (2003). The cochlear nucleus. In G. M. Shepherd (Ed.), *Synaptic organization of the brain* (pp. 125–163). New York: Oxford University Press.
Zatorre, R. J. (2007). There's more to auditory cortex than meets the ear. *Hearing Research, 229*, 24–30.

Zhan, X., Pongstaporn, T., & Ryugo, D. K. (2006). Projections of the second cervical dorsal root ganglion to the cochlear nucleus in rats. *The Journal of Comparative Neurology, 496*, 335–348.
Zhan, X., & Ryugo, D. K. (2007). Projections of the lateral reticular nucleus to the cochlear nucleus in rats. *The Journal of Comparative Neurology, 504*, 583–598.
Zhang, S., & Oertel, D. (1993). Giant cells of the dorsal cochlear nucleus of mice: Intracellular recordings in slices. *Journal of Neurophysiology, 69*, 1398–1408.
Zhang, Z., Liu, C.-H., Yu, Y.-Q., Fujimoto, K., Chan, Y. S., & He, J. (2008). Corticofugal projection inhibits the auditory thalamus through the thalamic reticular nucleus. *Journal of Neurophysiology, 99*, 2938–2945.
Zhao, H. B., Parham, K., Ghoshal, S., & Kim, D. O. (1995). Small neurons in the vestibular nerve root project to the marginal shell of the anteroventral cochlear nucleus in the cat. *Brain Research, 700*, 295–298.
Zhao, M., & Wu, S. H. (2001). Morphology and physiology of neurons in the ventral nucleus of the lateral lemniscus in rat brain slices. *The Journal of Comparative Neurology, 433*, 255–271.
Zhou, J., & Shore, S. (2004). Projections from the trigeminal nuclear complex to the cochlear nuclei: A retrograde and anterograde tracing study in the guinea pig. *Journal of Neuroscience Research, 78*, 901–907.
Zhou, J., & Shore, S. (2006). Convergence of spinal trigeminal and cochlear nucleus projections in the inferior colliculus of the guinea pig. *The Journal of Comparative Neurology, 495*, 100–112.

Chapter 3
Microcircuits of the Ventral Cochlear Nucleus

Maria E. Rubio

Abstract The neurons of the cochlear nuclei are the first central processors of auditory information, and they provide inputs to all the major brainstem and midbrain auditory nuclei. The ventral region of the cochlear nucleus (the ventral cochlear nucleus) represents the beginning of the binaural pathway through its projections to the superior olivary complex. The synaptic circuitry of the ventral cochlear nucleus specializes in the precise and rapid representation of incoming signals from the cochlear afferents. The ventral cochlear nucleus has two main regions: the core or magnocellular region and the granular cell domain. The magnocellular region contains these main neuronal cell types: the bushy cells (spherical and globular) and the multipolar or stellate cells (T stellate and D stellate). Auditory nerve fibers are the major source of excitation to both bushy and stellate cells. The synaptic connectivity pattern of neural networks between the neurons in a brain region is an essential determinant of their role in information processing. This chapter concentrates on the connectivity and synaptic microcircuits, including the key molecular synaptic components that allow the principal cells (bushy and stellate cells) to accomplish their functions. In addition, putative aspects of experience-dependent plasticity and hearing loss are discussed.

Keywords Auditory anatomy · GABA receptors · Glutamate receptors · Glycine receptors · Hearing loss · Synaptic circuits · Synaptic plasticity

3.1 Introduction

The focus of this chapter is the circuitry formed by the principal cells in the anteroventral cochlear nucleus (AVCN) and the anterior portion of the posteroventral cochlear nucleus (PVCN) (see all abbreviations in Table 3.1). The discussion does

M. E. Rubio (✉)
Departments of Neurobiology and Otolaryngology,
University of Pittsburgh Medical School, Pittsburgh, PA, USA
e-mail: mer@pitt.edu

© Springer International Publishing AG 2018

D. L. Oliver et al. (eds.), *The Mammalian Auditory Pathways*,
Springer Handbook of Auditory Research 65,
https://doi.org/10.1007/978-3-319-71798-2_3

Table 3.1 Abbreviations

Abbreviation	Name
AC	Auditory cortex
AMPA	α-Amino-3-hydroxy-5-methyl-4-isoxazolepropionic acid
AN	Auditory nerve
AVCN	Anteroventral cochlear nucleus
BC	Bushy cell
DCN	Dorsal cochlear nucleus
D-s	D stellate cell
EPSCs	Excitatory postsynaptic currents
GABA	Gamma-aminobutyric acid
GCD	Granular cell domain
GluA1	Ionotropic AMPA glutamate receptor subunit 1
GluA2	Ionotropic AMPA glutamate receptor subunit 2
GluA3	Ionotorpic AMPA glutamate receptor subunit 3
GluA4	Ionotropic AMPA glutamate receptor subunit 4
GluN1	Ionotropic NMDA glutamate receptor subunit 1
GlyR	Glycine receptor
GlyRα1	Glycine receptor subunit alpha 1
IC	Inferior colliculus
LSO	Lateral superior olivary nucleus
mGluR	Metabotropic glutamate receptors
mGluR1	Metabotropic glutamate receptor subunit 1
MNTB	Medial nucleus of the trapezoid body
mRNA	Messenger ribonucleic acid
MSO	Medial superior olivary nucleus
NMDA	N-methyl-D-aspartate
OCA	Octopus cell area
P2XR	ATP ionotropic purine receptor
PVCN	Posteroventral cochlear nucleus
SOC	Superior olivary complex
T-s	T stellate cell
Tv	Tuberculoventral cell
VCN	Ventral cochlear nucleus
VGAT	Vesicular GABA transporter
VGLUT1	Vesicular glutamate transporter 1
VGLUT2	Vesicular glutamate transporter 2

not include the posterior part of the PVCN (the octopus cell area) as its circuitry seems largely separate from the rest of the nucleus (for information about octopus cells, see Oertel et al. 2000). For simplicity, the area discussed is referred to as the ventral cochlear nucleus (VCN). The anatomy and function of the main cell types is described briefly (for more information see Cant 1992; Young and Oertel 2003, 2010; Manis et al. 2012). The connectivity and synaptic microcircuits in the VCN are described in detail, including the key molecular synaptic components of the principal cells. Putative aspects of experience-dependent plasticity and hearing loss also are discussed.

3.2 General Anatomy and Function of Main Cell Types

The synaptic circuitry of the VCN specializes in the precise and rapid representation of incoming signals from the cochlear nerve (Young and Oertel 2003, 2010). Postsynaptic responses of the principal cells depend on several important factors, including specializations of the primary excitatory afferent terminals, the membrane properties of the neurons, and the distribution of inhibitory inputs. The VCN represents the beginning of the binaural pathway through its projections to the superior olivary complex. The functions of and the main neuronal types in the VCN are conserved in general across different animal species, including rhesus monkeys (*Macaca mulatta*) (Gómez-Nieto and Rubio 2011), although there are anatomical data indicating that one type of bushy cells (BC), the globular BC, is sparse in humans (Moore and Osen 1979; Richter et al. 1983; Adams 1986).

The VCN is made up of two main regions: the magnocellular (or core) region and the granular cell domain, also called the small cell cap in cats (Osen 1969) (Fig. 3.1). The magnocellular (or core) region contains the main neuronal cell types of the VCN: the BCs (spherical and globular) and the multipolar or stellate cells (T stellate and D stellate). Auditory nerve (AN) fibers are the major source of excitation to both bushy and stellate cells of the VCN (Figs. 3.1 and 3.2). However, these cell groups differ in the number of auditory nerve fibers that converge on them, which is important for determining the function of each cell type. The granular cell domain (or small cell cap) is located on the periphery of the VCN and contains granule cells and small cells, including Golgi cells, unipolar brush cells, and chestnut cells (Weedman et al. 1996). The main excitatory input to the granular cell domain is somatosensory (Zhan and Ryugo 2007; Zheng et al. 2011), although it also receives low spontaneous rate AN fibers (Ryugo 2008) and unmyelinated Type II AN fibers (Benson and Brown 2004).

In general, the organization of the main cell types within the nucleus appears to be similar across species. Spherical BCs are located more rostrally, whereas globular BCs, T stellate, and D stellate cells are located more caudally, closer to the AN root. The neuronal distribution and differentiation of these neuronal types is less clear in rats and mice (Gómez-Nieto and Rubio 2009; Lauer et al. 2013), which could be due to different auditory sensitivities as these two species hear low frequencies poorly (Heffner and Hefner 1985). Like almost all structures in the auditory pathways, the core of the VCN is tonotopically organized with low frequency AN fibers distributed more ventrally and rostrally, whereas higher frequency fibers are distributed more caudally and dorsally (Fig. 3.1) (Young and Oertel 2003). Whether there is tonotopy in the granular cell domain is unknown.

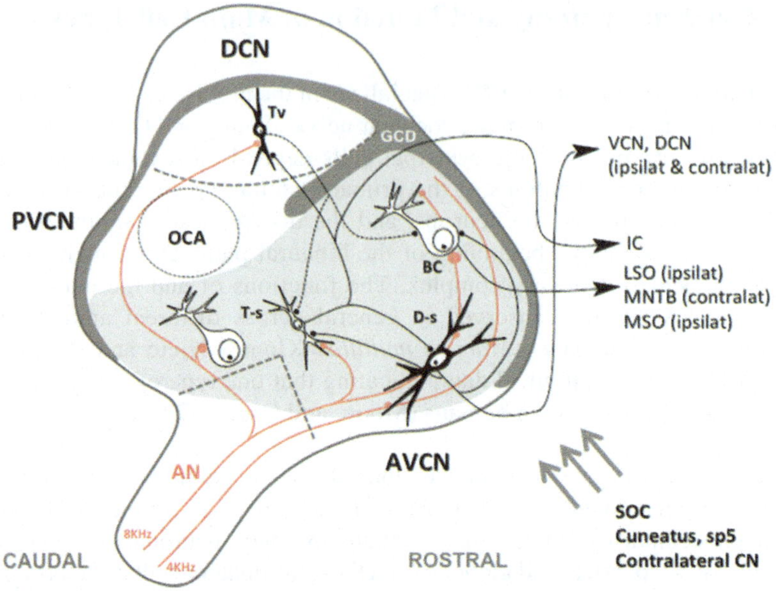

Fig. 3.1 Schematic of three main subdivisions of the cochlear nucleus (*CN*) and the intrinsic connections between the principal neurons [bushy cells (*BC*), T stellate cells (*T-s*), and D stellate cells (*D-s*)] within the anteroventral cochlear nucleus (*AVCN*) and the anterior posteroventral cochlear nucleus (*PVCN*). The *pale gray* background represents the areas of the ventral cochlear nucleus (*VCN*) described in this chapter: the AVCN and anterior PVCN. Intrinsic inhibitory (glycinergic) innervation arises from the tuberculoventral cells (*Tv*) of the dorsal cochlear nucleus (*DCN*). Axonal projections from the principal cells and their targets are also represented. The auditory nerve (*AN*) is represented entering the cochlear nucleus and dividing into descending and ascending branches. Middle frequency fibers (8 KHz) have a more caudal location within the AN, whereas lower frequency fibers (4 KHz) are more rostral. The extrinsic inputs to the AVCN from the superior olivary complex (*SOC*), cuneatus, spinal trigeminal nucleus (*sp5*), and contralateral cochlear nucleus (*CN*), are represented with arrows. (*GCD*, granular cell domain; *IC*, inferior colliculus; *LSO*, lateral superior olive; *MNTB*, medial nucleus of the trapezoid body; *MSO*, medial superior olive; *OCA*, octopus cell area)

3.2.1 Main Neuronal Cells in the Core Region

3.2.1.1 Bushy Cells

The two types of bushy cells (BCs) can be anatomically distinguished in the VCN based on cytological features. The spherical and globular bushy cells can be distinguished according to the shape and fine structure of their cell bodies (Cant and Morest 1979; Tolbert and Morest 1982) and their different locations within the VCN. Golgi and tracer studies have shown that BCs have one or two main dendritic branches (with an average length up to 54.5 ± 13.5 μm in rats). When there are two dendrites, they usually emerge from opposite poles of the cell body. Some BCs have thick primary dendritic processes forming a sparse bush; other BCs have thin dendrites that

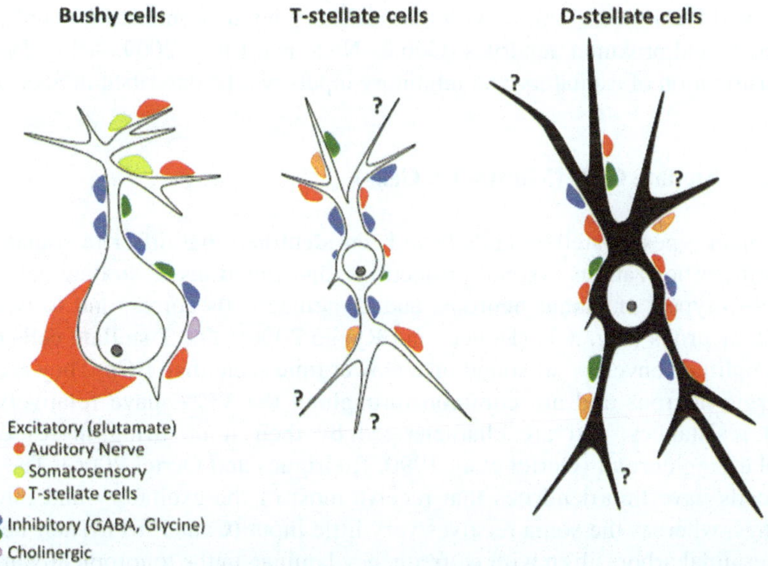

Fig. 3.2 Schematic of the excitatory (glutamate, cholinergic) and inhibitory (glycine, GABA) inputs on bushy cells and T and D stellate cells. Excitatory glutamatergic neurons (bushy and T stellate cells) are colored in *white*, whereas inhibitory glycinergic neurons (D stellate cells) are colored in *black*. *Question marks (?)* highlight the lack of information about synaptic inputs on dendrites of T and D stellate cells

form a dense bush with beaded enlargements more distally along the dendritic process. Bushy cell dendrites lack dendritic spines (Gómez-Nieto and Rubio 2009).

The globular and spherical BCs are glutamatergic neurons (Suneja et al. 1995; Alibardi 1998b) that can be described as the initiation of the pathways for binaural hearing (see reviews by Cant and Benson 2003; Young and Oertel 2003). They send their axons to the superior olivary complex (SOC) both contralaterally and ipsilaterally (Fig. 3.1). Globular BCs send their axons mainly to the contralateral medial nucleus of the trapezoid body and to the ipsilateral lateral nucleus of the trapezoid body (Tolbert et al. 1982; Kuwabara et al. 1991). Spherical BCs send their axons ipsilaterally to the medial superior olivary nucleus (MSO) and bilaterally to the lateral superior olivary nucleus (LSO).

Bushy cells are responsible for the precise temporal processing of acoustic information. Their firing is determined by interactions between excitatory and inhibitory inputs (Kopp-Scheinpflug et al. 2002). The physiological properties are consistent with evidence that the BC soma and dendrites receive both types of inputs from many sources (Saint Marie et al. 1986; Spirou et al. 2005; Gómez-Nieto and Rubio 2009). Spherical BCs receive the large AN endings known as the endbulbs of Held on the cell body, whereas globular BCs receive smaller AN endings (also called modified endbulbs) on the cell body (Liberman 1991; Ryugo et al. 1997). Both types of BCs also receive smaller endings of the AN on proximal and distal dendrites (Gómez-Nieto and Rubio 2009, 2011) with the inhibitory inputs terminating mainly on the cell body and proximal dendrites (Gómez-Nieto and Rubio 2009, 2011). Details of the distribution of excitatory and inhibitory inputs will be described in Sect. 3.3.

3.2.1.2 Stellate Cells (Multipolar Cells)

Two main types of stellate cells have been identified that differ in somata sizes, dendritic arborizations, axonal projections, and functions. T stellate cells correspond to type I or planar neurons, and D stellate cells correspond to type II or radiate neurons (Fig. 3.1) (Doucet and Ryugo 2006). The T stellate cells encode the amplitude envelope of sound on a slower time scale than BCs. They are glutamatergic neurons and are common throughout the VCN, have relatively high-input resistances, and are characterized by their tonic firing in response to depolarizing current (Oertel et al. 1990; Rodrigues and Oertel 2006). The T stellate cells have thin dendrites that receive most of the excitatory and inhibitory endings, whereas the soma receives very little input (Cant 1981). Their dendrites and terminal arbors align with isofrequency laminae in the tonotopically arranged VCN (Oertel et al. 1990), and they project to the inferior colliculus (Cant 1981; Alibardi 1998a). The positive correlation between rise and decay times of miniature excitatory postsynaptic currents (EPSCs) indicates that dendritic filtering occurs in T stellate cells (Gardner et al. 1999). T stellate cells, at least in cats, are tuned sharply and respond to tones near the best frequency with tonic firing at a steady rate (Rhode and Smith 1986).

D stellate cells are large and glycinergic with long, sparsely branched dendrites that spread across isofrequency lamina (Fig. 3.1). D stellate cell axons terminate widely in the ipsilateral VCN with local collaterals (Oertel et al. 1990). Their axons also project to the deeper layer of the dorsal cochlear nucleus (Ferragamo et al. 1998) and can cross the midline to innervate T stellate cells of the contralateral cochlear nucleus (Wenthold 1987; Alibardi 1998b; Doucet and Ryugo 2006). The spread of the dendrites indicates that D stellate cells are likely to be broadly tuned.

3.2.2 Neuronal Cells in the Granular Cell Domain

The granular cell domain (GCD) is an external shell of small neurons and neuropil that surrounds the magnocellular regions of the ventral and dorsal cochlear nuclei (VCN, DCN) (Fig. 3.1). There are at least four separate classes of neurons: granule cells, unipolar brush cells, Golgi cells, and chestnut cells (for anatomical and ultrastructural details of these cell types see Weedman et al. 1996). The granular cell domain is not a major target of myelinated AN fibers, although it does receive terminal branches of low spontaneous rate AN fibers (Ryugo 2008). Instead, this region is the target for a variety of nonprimary inputs (SOC; inferior colliculus, IC; auditory cortex, AC) (Brown et al. 1988b; Caicedo and Herbert 1993; Saldaña 1993). Unmyelinated Type II AN fibers, carrying information from outer hair cells of the cochlea, also terminate among granule cells (Brown et al. 1988a; Benson and Brown 2004). Additionally, the granular cell domain receives nonauditory projections from somatosensory nuclei, including the cuneate (Weinberg and Rustioni 1989) and trigeminal nuclei (Itoh et al. 1987; Shore et al. 2000) and the vestibular organ (Burian and Goesttner 1988; Kevetter and Parachio 1989). Granule cells receive mossy fiber-like endings from these inputs on their distal dendrites, so they are in a position to integrate a wide spectrum of information carrying cues about attention, head position, sound localization, or sound recognition (Shore et al. 2000). Granule cell axons project into the superficial layer of the dorsal cochlear nucleus, where they form parallel fibers that synapse on fusiform and cartwheel cells (Waterlood and Mugnaini 1984; Rubio and Wenthold 1997). Unipolar brush and chestnut cells also receive mossy fiberlike endings on distal dendrites or near the cell body, respectively, and project to the dorsal cochlear nucleus (Weedman et al. 1996), although their neuronal targets and functions are unknown. Further discussion of the synaptic connections within the dorsal cochlear nucleus can be found in Chap. 4 by Trussell and Oertel.

3.3 Intrinsic and Extrinsic Connections

The neurons of the cochlear nuclei are the first central processors of auditory information, and they provide inputs to all the major brainstem and midbrain auditory nuclei. The synaptic connectivity pattern of neural networks in each part of the cochlear nuclear complex is an essential determinant of their role in information processing. The following sections describe the most relevant intrinsic and extrinsic connections within the VCN (Figs. 3.1 and 3.2).

3.3.1 Intrinsic Connections Between Cells

The VCN has a simple intrinsic circuitry compared to the synaptic circuitry of the dorsal cochlear nucleus (see Trussell and Oertel, Chap. 4). Principal cells of the VCN (BCs and stellate cells) do not have many intrinsic connections. In BCs this probably makes sense as they encode the fine structure of sound. Convergence of AN fibers sharpens the timing, but intrinsic connections could add noise to the signal. In general, within the core region of the VCN, stellate cells (T and D) innervate each other, whereas only D stellate cells innervate BCs (Fig. 3.1). There is no evidence that BCs innervate each other or any of the stellate cell types. T stellate cells innervate other T stellate cells and have axonal collaterals that leave the VCN to innervate tuberculoventral (vertical) cells of the ipsilateral dorsal cochlear nucleus (Oertel and Wickesberg 1993). The D stellate cells innervate BCs and T stellate cells on the same (ipsilateral) side and T stellate cells on the opposite (contralateral) side (Ferragamo et al. 1998; Needham and Paolini 2006; Oertel et al. 2011). So far, there is no evidence that D stellate cells innervate other D stellate cells. Intrinsic connectivity among neuronal types in the granular cell domain is unknown.

To gain information about the spatial patterns of connectivity and the connection strengths in local circuits of the VCN, Campagnola and Manis (2014) mapped circuits in the mouse VCN. They showed that the two major types of excitatory projecting neurons (BCs and T stellate cells) received a spatially broad inhibition from D stellate cells and a spatially confined inhibition from tuberculoventral cells of the dorsal cochlear nucleus. Additionally, T stellate cells integrated D stellate inhibition from an area that spanned twice the frequency range of that integrated by BCs. The study also showed that a subset of both BCs and T stellate cells received inhibition from an unidentified cell population at the dorsomedial boundary of the VCN; a smaller subset of BCs appeared to receive an unidentified local excitation from within the VCN. The results from that study were largely consistent with previously published findings and provided new evidence that inhibitory circuits can have target-specific patterns of spatial convergence on principal cells of the AVCN. This, together with specific synaptic strength and receptor kinetics, can give rise to different spectral and temporal processing capabilities within the VCN.

3.3.2 Extrinsic Excitatory Input from the Cochlea

The main extrinsic excitatory input into the VCN is the AN fiber innervation. The AN endings are glutamatergic (Wenthold and Gulley 1977; Godfrey et al. 2000). Type I myelinated AN fibers innervate the main cell types of the VCN (Figs. 3.1 and 3.2). Type II unmyelinated AN fibers have sparse terminals in the magnocellular part of the VCN and also terminate in the granular cell domain (Benson and Brown 2004). The targets and roles of the Type I fibers are known, while those for the Type II fibers are still undetermined.

3.3.2.1 Type I Auditory Nerve Fibers

The central processes of the Type I spiral ganglion neurons in the cochlea form 95% of the AN fibers (Spoendlin 1985). Once in the cochlear nucleus, each fiber bifurcates into ascending and descending branches. The ascending branch distributes within the AVCN and innervates the main cell types in the core region. The descending branch distributes caudally within the PVCN and then turns dorsally to innervate the main cell types in the dorsal cochlear nucleus (fusiform cells, giant cells, and vertical cells).

Auditory Nerve Input on Bushy Cells (Spherical and Globular)

In slice preparations, BCs fire transiently when they are depolarized and show strong rectification before a hyperpolarizing sag develops in response to hyperpolarizing current pulses (Cao et al. 2007). The BCs have evoked EPSCs whose amplitude grows in discrete steps, indicating that they receive excitatory input from a relatively small number of AN fibers. The short and consistent synaptic delays indicate that synaptic responses are monosynaptic. In rodents, spherical BCs receive one to three AN inputs and fire more than two action potentials. Globular BCs receive more than four AN inputs and fire only one or two action potentials (Cao et al. 2007).

Auditory Nerve Input on Stellate Cells

T stellate cells receive monosynaptic inputs from about five or six AN fibers (Cao and Oertel 2010). They receive very little input at the soma; most of the Type I AN endings synapse on the dendrites (Cant 1981) (Fig. 3.2). D stellate cells also receive monosynaptic inputs from the AN fibers, although the number of fibers has not been determined. In cats, the cell body of D stellate cells is densely covered by AN terminals (Smith and Rhode 1987). In rats, the difference in innervation between T and D stellate cells is less clear (Alibardi 1998a).

3.3.3 *Sources of Extrinsic Inhibitory Inputs*

Both glycine and gamma-aminobutyric acid (GABA) mediate inhibition in the VCN circuitry (Osen et al. 1990; Xie and Manis 2014). Extrinsic inhibitory inputs on BCs and T stellate cells arise from glycinergic tuberculoventral cells in the ipsilateral dorsal cochlear nucleus (Fig. 3.1) (Wickesberg and Oertel 1988). Additional inhibitory sources to the VCN include the superior olivary complex and the contralateral cochlear nucleus (Ferragamo et al. 1998; Doucet and Ryugo 2006; Oertel et al. 2011). Proportions and detailed maps of the distribution of extrinsic inhibitory inputs on BCs and stellate cells (T and D) are still undetermined. Addressing this issue calls for complex quantitative anatomical studies combining: (1) anterograde tracer injections in the superior olivary complex and/or contralateral cochlear nucleus; (2) immunocytochemistry for localizing glycine or GABA; and (3) tracer labeling of the cell bodies and dendrites of BCs and stellate cells.

3.3.4 Modulatory Extrinsic Excitatory Innervation: Somatosensory and Cholinergic

Somatosensory and auditory signals are integrated by auditory neurons in the inferior colliculus, superior colliculus, auditory cortex, and dorsal cochlear nucleus (Aitkin et al. 1978, 1981; Itoh et al. 1987; Eliades and Wang 2005). On the other hand, the VCN has been traditionally considered a pure auditory nucleus as it receives direct excitatory input from the AN, and its synaptic circuitry specializes in precise and rapid representation of the incoming signals from the cochlear afferents (Young and Oertel 2003). Nevertheless, a number of tracer studies have shown that the VCN is also the target of a large number of somatosensory inputs that subserve tactile and kinesthetic senses, including the trigeminal ganglion (ophthalmic and mandibular subdivisions), spinal trigeminal nucleus, and nucleus cuneatus (Shore et al. 2000; Haenggeli et al. 2005). Although most of these inputs end within the granular cell domain, small axon collaterals have been described within the core of the VCN. Further evidence that the core region is a target of somatosensory innervation comes from immunocytochemical studies using the excitatory synaptic markers for vesicular glutamate transporters (VGLUT1, VGLUT2) (Fig. 3.3) together with tracer injections (Gómez-Nieto and Rubio 2009). The VGLUT1 and VGLUT2 isoforms are associated with specific pathways in the cochlear nucleus. The former is present only on AN endings, whereas the latter is present on somatosensory terminals (Zhou et al. 2007). Thus, if principal cells (or a subset of them) receive somatosensory inputs in addition to the auditory cochlear inputs, it appears that somatosensory signals can influence auditory processing even in the core of the VCN.

Cholinergic innervation to the VCN is excitatory and arises from the ventral nucleus of the trapezoid body (Fujino and Oertel 2001; Oertel and Fujino 2001). Whole-cell recordings in brain slices of the VCN have shown that most of the T stellate cells are excited by a cholinergic agonist. In contrast, the agonists have only subtle effects on the firing of BCs and have no detectable influence on D stellate cells. Oertel and Fujino (2001) suggested that cholinergic excitation of T stellate cells contributes to a shift of their acoustic dynamic ranges and increases the encoding of spectral peaks in noisy conditions. The distribution of cholinergic inputs on cell bodies and/or dendrites of principal cells of the VCN has not been determined.

3.3.5 Ionotropic ATP Purinoreceptors

ATP ionotropic purine receptors (P2XRs) are expressed in the VCN early during development, in particular in BCs of Mongolian gerbils (Milenkovic et al. 2009). Both in vitro and in vivo electrophysiological studies have shown that P2XR mediate a modulatory effect of ATP on action potential firing before and shortly after the onset of hearing (Dietz et al. 2012). ATP is endogenously released during

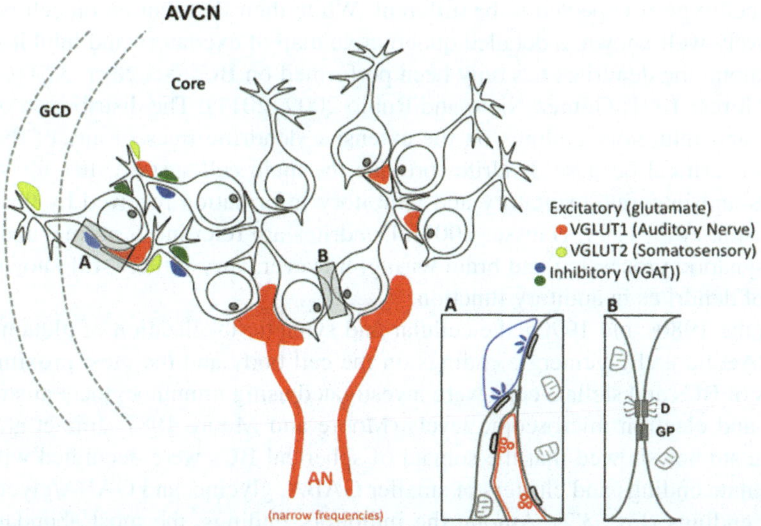

Fig. 3.3 Schematic of the organization of clusters of bushy cells (BCs) in the ventral cochlear nucleus (*VCN*) *core* and the granular cell domain (*GCD*) shell. Inputs of the auditory nerve (AN) immunolabled for VGLUT1 (*red*) were observed on the cell body and distal dendrites. Other excitatory endings that were VGLUT2 positive (*yellow*) and likely to be from somatosensory nuclei were observed on distal dendrites of BCs. Inhibitory endings immunopositive for vesicular GABA transporters (*VGAT*) were mainly observed on primary dendrites. *Gray-shaded* **A** and **B** rectangles are enlarged as insets in lower right: **(A)** *Inset* represents synaptic dyads of the auditory nerve and inhibitory endings on a cell body and dendrites of two distinct BCs. It also represents a dendrosomatic plasma membrane specialization of a BC on another BC. **(B)** *Inset* represents somasomatic plasma membrane specializations between two BCs. (*GP*, gap junction; *D*, desmosome)

development, and purinergic signaling is postulated to play a specific role in development of neuronal circuits within the VCN and upper auditory nuclei (Jovanovic et al. 2017). The role of P2XR in the mature BCs synaptic circuitry is unknown.

3.4 Excitatory and Inhibitory Endings on Bushy Cells and Multipolar/Stellate Cells

In addition to the general mapping of connectivity between neurons, it is also relevant to examine the distribution of excitatory and inhibitory inputs on each neuronal type (Fig. 3.2). Glutamate is the main excitatory neurotransmitter in the VCN; GABA and glycine mediate inhibitory neurotransmission. Modulatory neurotransmitters include acetylcholine. Synaptic endings containing glutamate, GABA, or glycine, as well as synaptic endings co-localizing GABA and glycine, have been found on the cell bodies and proximal dendrites of BCs and both T and D stellate cells. The general distribution of the excitatory and inhibitory synaptic endings on each cell type is expected to be different. While their distribution on cell bodies is relatively well known, a detailed quantitative map of excitatory and inhibitory endings along the dendrites has only been performed on BCs (see Sect. 3.5) (Ostapoff and Morest 1991; Gómez-Nieto and Rubio 2009, 2011). The distribution of excitatory and inhibitory endings on the extensive dendritic trees of any of these cell types is critical because dendrites provide the main cell-surface sites for synaptic inputs and integrate excitatory and inhibitory information received by the neuron (Johnston et al. 1996; Häusser 2001). Dendrites are relevant in normal and abnormal synaptic processing and brain wiring; however, very little is still known of the role of dendrites in auditory function.

In the 1980s and 1990s, the cellular and synaptic localization of glutamatergic, GABAergic and glycinergic endings on the cell body and the most proximal dendrites of BCs and stellate cells were investigated using immunocytochemistry at the light and electron microscopic levels (Moore and Moore 1987; Juiz et al. 1996). Those studies showed that the somata of spherical BCs were decorated with large glutamate endings and clusters of smaller GABA, glycine, and GABA/glycine synaptic endings (Fig. 3.2). Among the inhibitory endings, the most abundant were those labeled with glycine, whereas the endings that contained only GABA were less common. In contrast to spherical BCs, the somata of T stellate cells received sparse glycine or GABA inputs (Fig. 3.2), whereas the primary dendrite received mainly glycinergic synaptic endings (Juiz et al. 1996).

The distribution of synaptic inputs on D stellate cells differs from those on T stellate cells. In cats, the somata of D stellate cells are densely decorated with excitatory and inhibitory inputs with many of the excitatory inputs from AN collaterals. Ultrastructurally, glutamatergic endings are characterized as Gray type I synaptic contacts; the glycinergic and GABAergic endings are characterized as Gray type II synapses (Valdivia 1971). These anatomical studies served to emphasize that the sound-evoked spike activity of BCs is the result of the integration of acoustically driven excitatory and inhibitory inputs (Kopp-Scheinpflug et al. 2002). Anatomically this integration correlates with excitatory and inhibitory inputs from a variety of sources on the BC soma (Saint Marie et al. 1986; Wu and Oertel 1986; Spirou et al. 2005).

More recently, Gómez-Nieto and Rubio (2009) combined a number of techniques and conducted a study of the distribution of excitatory and inhibitory inputs on identified dendrites of BCs. In retrogradely labeled BC dendrites, the study showed both excitatory (VGLUT1 and VGLUT2 isoforms) and inhibitory (vesicular GABA transporters: VGAT) presynaptic markers distributed along the dendritic profile from the proximal to the most distal tufted segments. The primary dendrites receive more inhibitory endings than the distal dendrites, which are contacted by more excitatory endings. This distribution of excitatory and inhibitory endings along the dendritic tree of BCs was further confirmed through analyses of the labeled dendrites at the electron microscopic level. Proximal and distal dendrites received synaptic endings with the typical ultrastructural characteristics of excitatory or inhibitory endings.

Four main subtypes of excitatory endings and four main subtypes of inhibitory endings were identified quantitatively based on differences in the size of the ending and of the synaptic vesicles (Gómez-Nieto and Rubio 2009). Only one of the types identified as excitatory shared ultrastructural characteristics with the endbulb of Held. The other three types of excitatory terminals were identified as noncochlear. It is possible that some of these noncochlear terminals are of somatosensory origin. In addition, there were other types of synaptic endings that contained small synaptic vesicles, had clear axoplasm, and made single synaptic contacts. Those ultrastructural characteristics are typical of cholinergic inputs (Gómez-Nieto et al. 2008). The study of the distribution and proportion of these endings in different regions of the dendritic tree was consistent with the analysis of the vesicular glutamate transporters. Quantitative data showed that the percentage of excitatory endings was larger on the distal dendrites (59%) than on secondary (35%) or primary dendrites (43%). All four types of inhibitory endings were found on primary and secondary dendritic segments, whereas terminals containing large, flat synaptic vesicles tended to be concentrated on distal dendrites. The distribution of these endings on BC dendrites was similar to that found with VGAT. Putative sources of these endings are intrinsic interneurons in the VCN (D stellate cells), glycinergic interneurons in the dorsal cochlear nucleus (tuberculoventral cells), and/or descending inputs from other auditory nuclei (Wu and Oertel 1986; Schofield 1994; Ostapoff et al. 1997).

The study by Gómez-Nieto and Rubio (2009) demonstrated that the distal tufted dendrites of BCs were a major target of the incoming afferents, including the AN inputs (Figs. 3.2 and 3.3). This evidence indicates that BC dendrites, and in particular distal dendrites, provide a site for specific synaptic configurations that can influence neuronal output. Moreover, the AN innervated BC dendrites could explain the small excitatory peaks observed in paired BC–auditory nerve electrophysiological recordings (Young and Sachs 2008) in which the presence of excitatory peaks was much more frequent than expected based on the limited number of endbulbs or modified endbulbs on BCs (Ryugo and Sento 1991).

Further studies are needed to elucidate the role of dendritic integration of excitatory and inhibitory endings on BC dendrites. Additionally, studies similar to those described in this section are needed to determine the synaptic distribution and the proportion of excitatory and inhibitory endings along the entire dendritic profile of T and D stellate cells.

3.5 Divergent Multiple-Contact Synapses Among Bushy Cells

In response to low-frequency tones at high intensities, a subpopulation of BCs shows enhanced phase locking (i.e., synchronization to the stimulus frequency) relative to the AN (Joris et al. 1994a, b). One explanation for this enhancement is that the BCs serve as coincidence detectors, with an increase in firing rate occurring as the result of an increase in the correlation among the inputs to the neuron. New anatomical evidence that supports such a mechanism was reported by Gómez-Nieto and Rubio (2009, 2011) based on experiments in which they made tracer injections into the superior olivary complex to retrogradely label BCs. In the core of the VCN, the distal dendrites of BCs branch into a complex tuft, which surrounds the somata of four to five adjacent BCs forming a cluster (Fig. 3.3). Combining immunofluorescence for VGLUT1, anterograde tracer injections, and electron microscopy, they further demonstrated that an endbulb was usually found in apposition to the cell body of one BC and the distal dendrites of another BC. Ultrastructural analysis of the synaptic inputs on the labeled dendrites showed that a large number of AN endings (including endbulbs) innervated BC dendrites and made dyads and triads in the form of axodendrosomatic synaptic contacts with two or three BCs. Previous ultrastructural studies also described these synaptic connections, although the source of the dendrites was unknown (Ostapoff and Morest 1991; Ryugo and Sento 1991).

The anatomical results provide strong evidence for a mechanism by which incoming inputs might disperse activity among BCs. The same study showed that inhibitory inputs also made divergent dyadic contacts with at least two BCs (Gómez-Nieto and Rubio 2009, 2011). The existence of excitatory and inhibitory synaptic dyads and triads suggests a general mechanism to disperse activity through the nucleus. Although physiological confirmation is required, it is plausible that the divergent, multiple-contact synapses of cochlear and noncochlear inputs on BC clusters underlie a morphological substrate for the enhanced synchronization of BCs firing in the VCN. Alternatively, electrical coupling between BCs also could explain this enhanced synchronization within the VCN (see Sect. 3.7).

3.6 Chemical Synaptic Communication Within the Ventral Cochlear Nucleus

As in any nucleus in the central nervous system, neurons in the VCN communicate through chemical synapses. Neurotransmitters are released and they activate excitatory (glutamate) and inhibitory (glycine and GABA) neurotransmitter receptors. This section will describe the main neurotransmitter receptor types and subunits on principal neurons within the core of the VCN. The expressions of neurotransmitter receptors in the different neuronal types of the granular cell domain are yet to be determined.

3.6.1 Excitatory Ionotropic and Metabotropic Glutamate Receptors

The amino acid glutamate is the major chemical mediator of excitatory neurotransmission in the cochlear nucleus (Wenthold and Gulley 1977). Glutamate acts via ionotropic receptors (α-amino-3-hydroxy-5-methyl-4-isoxazolepropionic acid

[AMPA] and N-methyl-D-aspartate [NMDA] receptor types) and metabotropic receptors, which provide both rapid and slow excitatory synaptic transmission of the acoustic signal carried by AN fibers (Raman et al. 1994; Gardner et al. 1999, 2001).

3.6.1.1 Ionotropic Glutamate Receptors

As described previously, the AN drives bushy (Cant and Morest 1979) and T and D stellate cells (Cant 1981; Ferragamo et al. 1998) in the VCN. At the AN synapse, BCs contain AMPA glutamate receptor subunits with rapid kinetics (GluA3 and GluA4) (Fig. 3.4) (Wang et al. 1998; Gardner et al. 1999, 2001). The slower GluA1 subunit is not present at adult AN synapses on BCs. The GluA2 subunit, whose absence is associated with fast synapses, is expressed at very low levels at the synapses of the AN on BCs (Fig. 3.4) (Wang et al. 1998; Gardner et al. 1999, 2001).

A recent study using freeze-fracture immunogold labeling reported that the distribution of GluA3 and GluA4 subunits at AN synapses depends on the target cells (Rubio et al. 2017). The study showed that GluA3 was the main fast-gating AMPA receptor subunit present in the ultrafast AN synapse on BCs, whereas GluA4 was the main subunit present in the AN synapse on fusiform cells of the dorsal cochlear nucleus, which is a slower synapse. The study also showed that immunogold

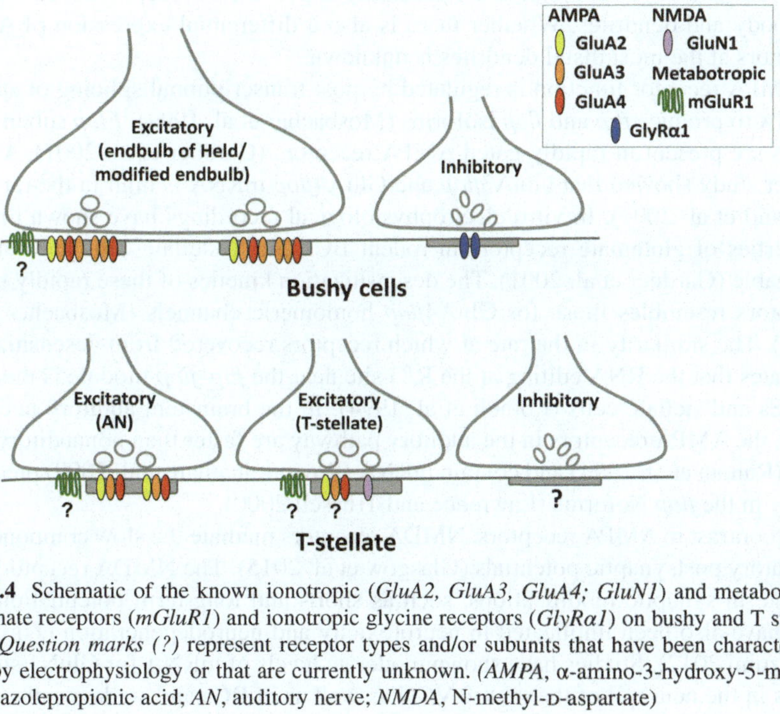

Fig. 3.4 Schematic of the known ionotropic (*GluA2, GluA3, GluA4; GluN1*) and metabotropic glutamate receptors (*mGluR1*) and ionotropic glycine receptors (*GlyRα1*) on bushy and T stellate cells. *Question marks (?)* represent receptor types and/or subunits that have been characterized only by electrophysiology or that are currently unknown. (*AMPA*, α-amino-3-hydroxy-5-methyl-4-isoxazolepropionic acid; *AN*, auditory nerve; *NMDA*, N-methyl-D-aspartate)

labeling for AMPA receptor subunits, particularly GluA3, was concentrated at the center of the synapse of the AN on BCs. On the other hand, GluA4 gold labeling was homogenously distributed along both synapse types. This differential expression and distribution of these fast kinetic AMPA receptor subunits may be important to enable the extraction of specific aspects of information transmitted by the presynaptic AN, for example, the precise temporal processing of acoustic information by BCs.

The subcellular distribution of AMPA receptor subunits in stellate cells has not been determined, but electrophysiological studies in T stellate cells have shown that AMPA receptors postsynaptic to the AN have fast decay kinetics (Gardner et al. 1999). AMPA receptor subunit composition (GluA2, GluA3, and GluA4) among BCs and T stellate cells are indistinguishable with respect to kinetics, blockage by polyamines, or sensitivity to cyclothiazide. AMPA receptors on BCs and T stellate cells contain little GluA2, although T stellate cells commonly have more GluA2 subunits than BCs (Fig. 3.4) (Gardner et al. 1999, 2001).

The AN makes synaptic contacts on the cell body but also on proximal and distal dendrites of BCs. The presence of differential AMPA receptor subunit compositions between AN synapses on the cell body and dendrites has not been studied in detail. Using postembedding immunogold labeling at the electron microscopic level, one study showed that AN synapses on proximal dendrites had lower levels of gold labeling for GluA2/3 and GluA4 than those on the cell body (Wang et al. 1998). This result suggests that, as in fusiform cells of the dorsal cochlear nucleus (Rubio and Wenthold 1997, 1999), BCs differentially target AMPA receptor subunits to the cell body and dendrites. Whether there is also a differential expression of AMPA receptors at the most distal dendrites is unknown.

AMPA receptor function is regulated by post-transcriptional splicing of subunit mRNA to produce *flip* and *flop* isoforms (Mosbacher et al. 1994). *Flop* subunit isoforms are present in rapidly gated AMPA receptors (Gardner et al. 2001). A messenger study showed that GluA3*flop* and GluA4*flop* mRNA is high in the rat VCN (Schmid et al. 2001). In vitro electrophysiological recordings have shown that the properties of glutamate receptors in rodent BCs and T stellate cells are indistinguishable (Gardner et al. 2001). The desensitization kinetics of these rapidly gating receptors resembles those for GluA4*flop* homomeric channels (Mosbacher et al. 1994). The similarity in the rate at which receptors recovered from desensitization indicates that the RNA editing at the R/G site near the *flip–flop* module is the same in BCs and stellate cells (Lomeli et al. 1994). In the brainstem auditory nuclei of birds, the AMPA receptors in the auditory pathway are faster than nonauditory neurons (Raman et al. 1994) and contain mRNA (Ravindranathan et al. 2000) predominately in the *flop* isoforms (Lawrence and Trussell 2000).

In contrast to AMPA receptors, NMDA receptors mediate the slow component of excitatory postsynaptic potentials (Glasgow et al. 2015). The NMDA receptors participate in synaptic modifications, such as short- and long-term potentiation, and they have also been implicated in neurotoxicity and neurodegeneration (Zorumski and Izumi 2012). Studies have shown moderate levels of mRNA for GluN1 subunit genes in the neuropil of the mouse VCN as well as in BCs and stellate cells (Bilak et al. 1996). At the protein level, immunohistochemical studies showed that the expression of NMDA receptors in the VCN was developmentally regulated (Caicedo and Eybalin 1999).

Electrophysiological studies of VCN brain slices of three-week-old rodents have shown a large AMPA receptor component at the AN synapse on BCs and a clear but small NMDA receptor component, which decreases by postnatal day 25 (Isaacson and Walmsley 1995; Pliss et al. 2009; Cao and Oertel 2010). This small NMDA receptor component in young adult mice may promote firing probability and improve temporal precision (Pliss et al. 2009). A recent study used freeze-fracture immunogold labeling of AN synapses on BCs and on fusiform cells of the dorsal cochlear nucleus from postnatal day 30 rats and showed that only the synapse of the AN on fusiform cells presented positive NMDA receptor label, whereas the AN synapse on BCs lacked immunolabeling for NMDA receptors (Fig. 3.4) (Rubio et al. 2014). Thus the results suggested that the NMDA receptor component of the EPSC is likely to decrease with age and that NMDA receptors in the synapse of the AN on BCs might be important for wiring inputs with approximately coincident synaptic activity during development.

The presence of NMDA receptors in T stellate cells has been addressed using electrophysiological recordings on brain slices (Ferragamo et al. 1998). This study showed that application of a specific blocker of NMDA receptors abolished the long, slow depolarization in the majority of T stellate cells. Intriguingly, the NMDA receptor-mediated slow depolarizations were generated with shock strengths greater than those required to produce maximal monosynaptic excitatory postsynaptic potentials. This finding raises the possibility that Type I AN fibers act primarily on AMPA receptors, whereas other sources of excitation, such as innervation from other T stellate cells, may contribute to the long, slow depolarization. Additional studies are needed to confirm such a possibility.

3.6.1.2 Metabotropic Glutamate Receptors

Metabotropic glutamate receptors (mGluRs) are members of the group C family of G-protein coupled receptors, and they are involved in learning and memory in the central nervous system (Balschun et al. 1999; Gladding et al. 2009). The expression of mGluRs in the VCN has only been addressed with mRNA and immunohistochemical studies at the light microscopic level (Bilak and Morest 1998). That study showed moderate expression levels for mGluR1 in BCs and stellate cells within the core of the VCN, whereas granule cells in the granular cell domain were strongly labeled. There is no electron microscopic data confirming the subcellular distribution of mGluR1 on these neuronal types.

A recent electrophysiological study, using brainstem slices in vitro, addressed the functional consequences of mGluR activation in BCs within the VCN (Chanda and Xu-Friedman 2011). Bushy cells relay the temporal information in AN spike trains to higher centers for sound localization (Grothe et al. 2010). Endbulbs show short-term depression during high-frequency activity (Oleskevich and Walmsley

2002; Wang and Manis 2008; Yang and Xu-Friedman 2008) and modulation in response to $GABA_B$ receptor activation (Chanda and Xu-Friedman 2010). Both of those processes reduce the likelihood of BC response to AN activity, which raises the question of whether there are modulatory mechanisms that maintain or enhance the response properties of BCs. Application of a specific agonist of mGluRs depolarized BCs but had no measurable effect on neurotransmitter release from endbulbs. The depolarization enhanced the response of BCs in response to AN activity, offsetting the effects of depression. Furthermore, mGluR activation largely restored spiking after $GABA_B$ receptor activation, suggesting that these two modulatory pathways could interact to tune the response properties of BCs.

3.6.2 Inhibitory Glycine and GABA Receptors

The amino acids glycine and GABA are the major chemical mediators of inhibitory neurotransmission in the cochlear nucleus (Altschuler et al. 1993; Evans and Zhao 1993; Oertel and Wickesberg 1993). Both glycine and GABA act via ionotropic receptors and provide rapid inhibitory modulation of the excitatory acoustic signal carried by AN fibers; GABA also acts via the metabotropic $GABA_B$ receptor (Brenowitz and Trussell 2001).

3.6.2.1 Glycine Receptors

Glycine acts by binding to the α1 subunit of glycine receptors (GlyRs) (Mildbrandt and Caspary 1995; Golding and Oertel 1996), although in young animals glycine binds to the α2 subunit (Malosio et al. 1991; Sato et al. 2000). This receptor has high strychnine binding affinity (Kuhse et al. 1995). In the adult cochlear nucleus the presence of the GlyRα1 subunit has been well documented by a variety of procedures, including post-embedding immunogold labeling (Altschuler et al. 1986; Piechotta et al. 2001; Rubio and Juiz 2004). As in the dorsal cochlear nucleus, GlyRα1 is expressed in the VCN at the postsynaptic densities opposite to synaptic endings that contain flattened synaptic vesicles and make symmetric synaptic contacts on BC somata (Whiting et al. 2009). The expression of the β subunits of GlyRs has not been determined. Tuberculoventral cells in the dorsal cochlear nucleus and D stellate cells in the VCN are the two major sources of glycinergic inhibition on BCs (Wickesberg and Oertel 1988, 1990; Doucet and Ryugo 2006). In the VCN the synaptically mediated conductances and kinetics of GlyRs have been extensively studied (Wu and Oertel 1986; Harty and Manis 1998). Recently, Xie and Manis (2014) showed that the time course of glycinergic inhibition is slow in BCs and fast in stellate cells, and this may contribute to the processing of temporal cues in both cell types.

3.6.2.2 GABA Ionotropic and Metabotropic Receptors

The distribution of $GABA_A$ (ionotropic) and $GABA_B$ (metabotropic) binding sites in the cochlear nucleus was determined using quantitative receptor autoradiography with [3H]GABA. In the VCN, $GABA_A$ and $GABA_B$ binding sites were concentrated in the peripheral granule cell cap with low binding levels in the central region (Juiz et al. 1989). Messenger studies showed that spherical BCs expressed $\alpha1$, $\alpha3$, $\alpha5$, $\beta3$, and $\gamma2L$ genes, whereas stellate cells expressed $\alpha1$, $\alpha5$, $\beta3$, $\gamma2L$ and δ genes (Campos et al. 2001). At the protein level, light microscopy immunocytochemistry showed that $GABA_A$ $\beta2/3$ subunits labeled the VCN weakly; they do not appear to be expressed on cell bodies of BCs or stellate cells (Wang et al. 2011). The protein expression levels for the other $GABA_A$ receptor subunits have not been determined.

Synaptically mediated conductances associated exclusively with $GABA_A$ receptors have been difficult to detect in the VCN. The most likely explanation is the small size and number of such synapses compared to glycinergic synapses (Juiz et al. 1996). GABAergic inhibition in the VCN arises from outside the nucleus, mostly from descending projections from the superior olivary complex. Taking advantage of VGAT channelrhodopsin-2 mice, a recent study demonstrated the presence of functional GABAergic inhibitory synaptic potentials and conductances in BCs and T stellate cells (Xie and Manis 2014). The results also showed that GABAergic inhibition is more prominent in T stellate cells than in BCs, and the time course of inhibition for both cells was slow relative to the fast glycinergic inhibition in T stellate cells. The time course of the synaptic conductances appears to be similar for both cell types.

GABA also acts via the metabotropic $GABA_B$ receptor. Messenger studies or immunohistochemistry at the light or electron microscopic level for $GABA_B$ receptors have not been performed. Nevertheless, in vitro electrophysiological studies suggest that they are expressed presynaptically at the AN ending where they are involved in short-term depression during high-frequency activity (Brenowitz and Trussell 2001; Chanda and Xu-Friedman 2010).

3.7 The Synchronicity Issue: Electrical Synapses Within the Ventral Cochlear Nucleus?

As described in Sect. 3.5, BCs appear to form clusters of four or five neurons in rats (Gómez-Nieto and Rubio 2009), cats (Lorente de Nó 1981; Ostapoff and Morest 1991; Ryugo and Sento 1991), and rhesus monkeys (Gómez-Nieto and Rubio 2011). When BCs are filled with neuronal tracers, their dendrites appear to form an elaborate network within the VCN (Gómez-Nieto and Rubio 2009). Confocal microscopic analyses of retrogradely labeled neurons suggest a close apposition between distal dendrites and/or somata of rat BCs (Fig. 3.3). Three-dimensional reconstruction of serial sections and electron microscopic analyses confirmed the existence of dendrosomatic and somasomatic junctions within a BC cluster. The somasomatic junctions were ultrastructurally characterized as puncta adherentia and gap junctions. Those specializations were also observed between BCs of the rhesus monkey (Gómez-Nieto and Rubio 2011). Early studies reported the existence of gap junctions in the VCN of the rat, although the cell types were not identified (Sotelo et al. 1976).

The presence of gap junctions between two BC somata would suggest that they are electronically coupled. If functional, such junctions could help to explain the enhanced synchronization of BCs (spherical BCs in particular) when compared to AN fibers (Joris et al. 1994a, b; Paolin et al. 2001; Joris and Smith 2008). So far, however, electrical coupling between BCs has not been reported, and intracellular injections of biocytin, a tracer to which connexin36-containing gap junctions are permeable, does not appear to spread between BCs (Cao et al. 2007). It is of note that the absence of dye or tracer coupling does not equate to absence of gap junctions or electrical coupling. Neurons that are well-established as electrophysiologically coupled by connexin36-containing gap junctions may nevertheless defy demonstrations of dye coupling (Gibson et al. 1999) or exhibit variability in such coupling (Curti et al. 2012). These results suggest a loss of junctional channel permeability to molecules, which may occur in the course of brain slice preparation, despite the persistence of electrical coupling. This is an important issue that needs further study, for example, with target-specific electrophysiological studies in vitro, as immunolabeling has shown the presence of connexin36 (Rash et al. 2005, 2007a, b) on BCs as well as between two BCs of the VCN (Rubio and Nagy 2015).

3.8 Experience-Dependent Plasticity and Hearing Loss

The AN synapse on principal neurons of the VCN is a secure and highly conservative synapse. For these reasons, and in contrast to excitatory synapses in other parts of the central nervous system, AN synapses lack the classical long-term synaptic plasticity mechanisms and spike-timing-dependent plasticity (Fujino and Oertel 2003; Zhao et al. 2011). On the other hand, the endbulb of Held is subject to synaptic depression (Brenowitz and Trussell 2001; Yang and Xu-Friedman 2008), the predominant form of short-term plasticity at synapses with high probability of transmitter release (Zucker 1989). This mechanism has been studied mainly at the synapse of the AN on BCs. In vitro functional studies have shown that when AN inputs carry different timing information, depression greatly improves BC precision by suppressing highly active inputs that carry little phaselocked information (Yang and Xu-Friedman 2008). Thus changes in short-term depression in response to fluctuations in AN activity would likely affect the enhancement of temporal coding observed in BCs from in vivo recordings (Joris et al. 1994a, b) and affect the postsynaptic channels regulating coincident detection (Rothman and Young 1996; Rothman et al. 1993). For example, changes in short-term depression, measured using electrophysiological recordings in vitro, have been reported in congenitally deaf mice (*dn/dn*), in early age-related hearing loss DBA/2 J mice (Oleskevich and Walmsley 2002; Wang and Manis 2005), after noise-induced hearing loss (Ngodup et al. 2015), and in response to conductive hearing loss (Zhuang et al. 2017).

Studies in animals with hearing loss also have shown effects on the postsynaptic principal neurons within the VCN, suggesting the existence of plasticity in BCs and T stellate cells. Ultrastructural studies of congenitally deaf cats and mice showed hypertrophy of the postsynaptic membrane of BCs in apposition to AN endings (Lee et al. 2003; Ryugo et al. 1997). Interestingly, after receiving a cochlear implant and electrical stimulation for 6 months, the hypertrophy reversed and the postsynaptic membranes were similar to those of normal hearing cats (Ryugo et al. 2005; O'Neil et al. 2010). In vitro electrophysiological recordings reported larger EPSC amplitude in BCs of congenitally deaf mice than in those with normal hearing (Oleskevich and Walmsley 2002). In noise-induced hearing loss studies, it has been reported that over time (after 8 months), noise exposure leads to degeneration of AN synapses and to a rewiring of the VCN synaptic circuitry (Kim et al. 2004), although what this rewiring means in terms of VCN function is still undetermined. In vitro electrophysiological recordings in mice following 2 weeks of 120 min noise-induced hearing loss have shown an increase in the miniature EPSC amplitude in T stellate cells, suggesting a postsynaptic remodeling (Rich et al. 2010). Cochlear trauma in guinea pigs induces hyperactivity in the VCN, suggesting that the VCN should be considered in relation to neural models of the genesis of tinnitus (Vogler et al. 2011).

Finally, milder forms of hearing loss have shown effects also at principal neurons receiving AN synapses. For example, in vitro electrophysiological studies of DBA/2 J mice showed less rectification of evoked EPSCs, suggesting synaptic scaling of GluA2 AMPA receptors at BC synapses (Wang and Manis 2005). Sound reduction by conductive hearing loss is known to alter the neuronal metabolic rates and protein synthesis in the ipsilateral and contralateral VCN (Tucci et al. 1999; Huston et al. 2007). Transient conductive hearing loss for 1 day led to a reversible increase in hearing thresholds and to a reversible synaptic upregulation of GluA3 AMPA receptors at the synapse of the AN on BCs. The same neurons showed a downregulation of GlyRα1 on inhibitory synapses that reversed once hearing levels were restored (Whiting et al. 2009), suggesting the existence of a homeostatic response to the reduced sound. On the other hand, ten days of monaural conductive hearing loss led to an increase in hearing thresholds and in central gain upstream of the cochlear nucleus at the level of the lateral lemniscus. There were long lasting, pre- and post-synaptic, structural and molecular effects at the endbulb of Held synapse, including a sustained upregulation of synaptic GluA3 (Clarkson et al. 2016). These findings showed that sensory-dependent evoked plasticity is more complex than what might be predicted from experiments in reduced systems. Future studies could investigate long-term hearing sensitivities of adults following conductive hearing loss from chronic otitis media with effusion, which is prevalent in children. Such studies could determine the central mechanisms for deficits in auditory perception, language acquisition, and/or educational disabilities that occur after inadequate or abnormal sensory experience.

It is evident that normal AN activity is critical for maintaining fundamental structural, molecular, and functional parameters of AN synapses, as well as the postsynaptic excitatory or inhibitory neurons within the VCN, and that hearing deficits alter those parameters. The studies mentioned previously show that there are multiple ways through which neurons respond to fluctuations in their inputs and that these responses probably relate to the etiology of hearing loss. For these reasons, more studies are needed to confirm the cellular mechanisms and to determine the common features that will help us to understand whether the reported synaptic modifications compensate for the hearing deficits or represent pathological responses to hearing loss.

3.9 Summary

The gross anatomy, cytoarchitecture, and function of the VCN have been extensively studied, and those studies have shown a relatively simple synaptic intrinsic circuitry that specializes in the precise and rapid representation of incoming signals from cochlear afferents. Nevertheless, the VCN should not be considered to be a simple relay nucleus; new evidence challenges that traditional view. For example, there is the possibility that somatosensory signals can influence auditory processing in the VCN. This appears to occur mainly in the granular cell domain, which is the major target of somatosensory inputs and Type II AN fibers, and may also occur within the core, where collaterals of somatosensory axons distribute on principal cells receiving Type I cochlear afferents. To understand the putative convergence of cochlear and somatosensory information within the granular cell domain and the core requires a better understanding of the role of the granular cell domain in auditory processing, understanding the role of Type II unmyelinated AN fibers within the VCN and identification of their neuronal targets, and determining whether there is a subpopulation of principal cells that receives both cochlear and somatosensory inputs. This convergence of auditory and somatosensory inputs is of relevance given new evidence in animal models after noise exposure (Vogler et al. 2011) and evidence from clinical studies in humans (Gu et al. 2012) indicating that the VCN is involved in the genesis of tinnitus.

Other findings that challenge the VCN as a simple relay nucleus are the existence of synaptic dyads of the AN between the cell body of one BC and the distal dendrites of another BC and the fact that pairs (at least) of BCs can be electrically coupled. Electrophysiological and computational studies that address the role of BCs in auditory processing have usually considered only the synaptic input on their cell bodies. Due to the large and strong innervation of the AN on the soma, in vitro electrophysiological studies have not demonstrated dendritic filtering. Nevertheless, the anatomical findings provide strong evidence of excitatory and inhibitory integration in dendrites of BCs. New genetically modified animal models that specifically target BCs and their dendritic arborization are part of a strategy that would allow addressing this issue with electrophysiology. Similarly, these animals also will be useful for determining the existence of functional coupling between BCs.

Last but not least, there is increasing interest in understanding the putative protective role of the descending auditory pathway in response to hearing loss (Reiter and Liberman 1995; Kirk and Smith 2003). In contrast to the ascending auditory pathway, less is known about the anatomy and function of the descending auditory pathways. Thus determining the proportion, distribution, and neuronal targets of descending auditory inputs on the principal cells of the VCN is critical. This information will also be important for understanding the synaptic microcircuits involving the cell bodies and dendrites of principal cells of the VCN.

Acknowledgments Support was provided by the National Institute for Deafness and Other Communication Disorders (grant DC013048).

Compliance with Ethics Requirements Maria E. Rubio declares that she has no conflict of interest.

References

Adams, J. C. (1986). Neuronal morphology in the human cochlear nucleus. *Archives of Otolaryngology—Head and Neck Surgery, 112*(12), 1253–1261.

Aitkin, L. M., Dickhaus, H., Schult, W., & Zimmermann, M. (1978). External nucleus of inferior colliculus: Auditory and spinal somatosensory afferents and their interactions. *Journal of Neurophysiology, 41*(4), 837–847.

Aitkin, L. M., Kenyon, C. E., & Philpott, P. (1981). The representation of the auditory and somatosensory systems in the external nucleus of the cat inferior colliculus. *The Journal of Comparative Neurology, 196*(1), 25–40.

Alibardi, L. (1998a). Ultrastructural and immunocytochemical characterization of neurons in the rat ventral cochlear nucleus projecting to the inferior colliculus. *Annals of Anatomy, 180*(5), 415–426.

Alibardi, L. (1998b). Ultrastructural and immunocytochemical characterization of commissural neurons in the ventral cochlear nucleus of the rat. *Annals of Anatomy, 180*(5), 427–438.

Altschuler, R. A., Betz, H., Parakkal, M. H., Reeks, K. A., & Wenthold, R. J. (1986). Identification of glycinergic synapses in the cochlear nucleus through immunocytochemical localization of the postsynaptic receptor. *Brain Research, 369*(1–2), 316–320.

Altschuler, R. A., Juiz, J. M., Shore, S. E., Bledsoe, S. C., Helfert, R. H., & Wenthold, R. J. (1993). Inhibitory amino acid synapses and pathways in the ventral cochlear nucleus. In M. A. Merchán, J. M. Juiz, D. A. Godfrey, & E. Mugnaini (Eds.), *The mammalian cochlear nuclei: Organization and function* (pp. 211–224). New York: Plenum Publishing Corporation.

Balschun, D., Manahan-Vaughan, D., Wagner, T., Behnisch, T., Reymann, K. G., & Wetzel, W. (1999). A specific role for group I mGluRs in hippocampal LTP and hippocampus-dependent spatial learning. *Learning and Memory, 6*(2), 138–152.

Benson, T. E., & Brown, M. C. (2004). Postsynaptic targets of type II auditory nerve fibers in the cochlear nucleus. *Journal of the Association for Research in Otolaryngology, 5*(2), 111–125.

Bilak, M. M., Bilak, S. R., & Morest, D. K. (1996). Differential expression of N-methyl-D-aspartate receptor in the cochlear nucleus of the mouse. *Neuroscience, 75*(4), 1075–1097.

Bilak, S. R., & Morest, D. K. (1998). Differential expression of the metabotropic glutamate receptor mGluR1alpha by neurons and axons in the cochlear nucleus: In situ hybridization and immunohistochemistry. *Synapse, 28*(4), 251–270.

Brenowitz, S., & Trussell, L. O. (2001). Minimizing synaptic depression by control of release probability. *The Journal of Neuroscience, 21*(6), 1857–1867.

Brown, M. C., Berglund, A. M., Kiang, N. Y., & Ryugo, D. K. (1988a). Central trajectories of type II spiral ganglion neurons. *The Journal of Comparative Neurology, 278*(4), 581–590.

Brown, M. C., Liberman, M. C., Benson, T. E., & Ryugo, D. K. (1988b). Brainstem branches from olivocochlear axons in cats and rodents. *The Journal of Comparative Neurology, 278*(4), 591–603.

Burian, M., & Goesttner, W. (1988). Projection of primary vestibular afferent fibres to the cochlear nucleus in the guinea pig. *Neuroscience Letters, 84*(1), 13–17.

Caicedo, A., & Eybalin, M. (1999). Glutamate receptor phenotypes in the auditory brainstem and midbrain of the developing rat. *European Journal of Neuroscience, 11*(1), 51–74.

Caicedo, A., & Herbert, H. (1993). Topography of descending projections from the inferior colliculus to auditory brainstem nuclei in the rat. *The Journal of Comparative Neurology, 328*(3), 377–392.

Campagnola, L., & Manis, P. B. (2014). A map of functional synaptic connectivity in the mouse anteroventral cochlear nucleus. *The Journal of Neuroscience, 34*(6), 2214–2230.

Campos, M. L., de Cabo, C., Wisden, W., Juiz, J. M., & Merlo, D. (2001). Expression of GABA(A) receptor subunits in rat brainstem auditory pathways: Cochlear nuclei, superior olivary complex and nucleus of the lateral lemniscus. *Neuroscience, 102*(3), 625–638.

Cant, N. B. (1981). The fine structure of two types of stellate cells in the anterior division of the anteroventral cochlear nucleus of the cat. *Neuroscience, 6*(12), 2643–2655.

Cant, N. B. (1992). The cochlear nucleus: Neural types and their synaptic organization. In D. B. Webster, A. N. Popper, & R. R. Fay (Eds.), *The mammalian auditory pathway: Neuroanatomy* (pp. 66–116). New York: Springer.

Cant, N. B., & Benson, C. G. (2003). Parallel auditory pathways: Projection patterns of the different neuronal populations in the dorsal and ventral cochlear nuclei. *Brain Research Bulletin, 60*(5–6), 457–474.

Cant, N. B., & Morest, D. K. (1979). Organization of the neurons in the anterior division of the anteroventral cochlear nucleus of the cat. Light-microscopic observations. *Neuroscience, 4*(12), 1909–1923.

Cao, X. J., & Oertel, D. (2010). Auditory nerve fibers excite targets through synapses that vary in convergence, strength, and short-term plasticity. *Journal of Neurophysiology, 104*, 2308–2230.

Cao, X. J., Shatadal, S., & Oertel, D. (2007). Voltage-sensitive conductances of bushy cells of the mammalian ventral cochlear nucleus. *Journal of Neurophysiology, 97*(6), 3961–3975.

Chanda, S., & Xu-Friedman, M. A. (2010). Neuromodulation by GABA converts a relay into a coincidence detector. *Journal of Neurophysiology, 104*(4), 2063–2074.

Chanda, S., & Xu-Friedman, M. A. (2011). Excitatory modulation in the cochlear nucleus through group I metabotropic glutamate receptor activation. *The Journal of Neuroscience, 31*(20), 7450–7455.

Clarkson, C., Antunes, F., & Rubio, M. E. (2016). Conductive hearing loss has long-lasting structural and molecular effects on pre- and post-synaptic structures of the auditory nerve in the cochlear nucleus. *The Journal of Neuroscience, 36*(39), 10214–10227.

Curti, S., Hoge, G., Nagy, J. I., & Pereda, A. E. (2012). Synergy between electrical coupling and membrane properties promotes strong synchronization of neurons of the mesencephalic trigeminal nucleus. *The Journal of Neuroscience, 32*(13), 4341–4359.

Dietz, B., Jovanovic, S., Wielsch, B., Nerlich, J., Rübsamen, R., & Milenkovic, I. (2012). Purinergic modulation of neuronal activity in developing auditory brainstem. *The Journal of Neuroscience, 32*(31), 10699–10712.

Doucet, J. R., & Ryugo, D. K. (2006). Structural and functional classes of multipolar cells in the ventral cochlear nucleus. *The Anatomical Records Part A: Discoveries in Molecular, Cellular, and Evolutionary Biology, 288*(4), 452–465.

Eliades, S. J., & Wang, X. (2005). Dynamics of auditory-vocal interaction in monkey auditory cortex. *Cerebral Cortex, 15*(10), 1510–1523.

Evans, E. F., & Zhao, W. (1993). Varieties of inhibition in the processing and control of processing in the mammalian cochlear nucleus. *Progress in Brain Research, 97*, 117–126.

Ferragamo, M., Golding, N. L., & Oertel, D. (1998). Synaptic inputs to stellate cells in the ventral cochlear nucleus. *Journal of Neurophysiology, 79*(1), 51–63.

Fujino, K., & Oertel, D. (2001). Cholinergic modulation of stellate cells in the mammalian ventral cochlear nucleus. *The Journal of Neuroscience, 21*(18), 7372–7383.

Fujino, K., & Oertel, D. (2003). Bidirectional synaptic plasticity in the cerebellum-like mammalian dorsal cochlear nucleus. *Proceedings of the National Academy of Sciences of the United States of America, 100*(1), 265–270.

Gardner, S. M., Trussell, L. O., & Oertel, D. (1999). Time course and permeation of synaptic AMPA receptors in cochlear nuclear neurons correlate with input. *The Journal of Neuroscience, 19*(20), 8721–8729.

Gardner, S. M., Trussell, L. O., & Oertel, D. (2001). Correlation of AMPA receptor subunit composition with synaptic input in the mammalian cochlear nuclei. *The Journal of Neuroscience, 21*(18), 7428–7437.

Gibson, J. R., Beier, M., & Connors, B. W. (1999). Two networks of electrically coupled inhibitory neurons in neocortex. *Nature, 402*(6757), 75–79.

Gladding, C. M., Fitzjohn, S. M., & Molnár, E. (2009). Metabotropic glutamate receptor-mediated long-term depression: Molecular mechanisms. *Pharmacological Reviews, 61*(4), 395–412.

Glasgow, N. G., Siegler Retchless, B., & Johnson, J. W. (2015). Molecular bases of NMDA receptor subtype-dependent properties. *Journal of Physiology, 593*(1), 83–95.

Godfrey, D. A., Farms, W. B., Godfrey, T. G., Mikesell, N. L., & Liu, J. (2000). Amino acid concentrations in rat cochlear nucleus and superior olive. *Hearing Research, 150*(1–2), 189–205.

Golding, N. L., & Oertel, D. (1996). Context-dependent synaptic action of glycinergic and GABAergic inputs in the dorsal cochlear nucleus. *The Journal of Neuroscience, 16*(7), 2208–2219.

Gómez-Nieto, R., & Rubio, M. E. (2009). A bushy cell network in the rat ventral cochlear nucleus. *The Journal of Comparative Neurology, 516*(4), 241–263.

Gómez-Nieto, R., & Rubio, M. E. (2011). Ultrastructure, synaptic organization, and molecular components of bushy cell networks in the anteroventral cochlear nucleus of the rhesus monkey. *Neuroscience, 179*, 188–207.

Gómez-Nieto, R., Rubio, M. E., & López, D. E. (2008). Cholinergic input from the ventral nucleus of the trapezoid body to cochlear root neurons in rats. *The Journal of Comparative Neurology, 506*(3), 452–468.

Grothe, B., Peck, M., & McAlpine, D. (2010). Mechanisms of sound localization in mammals. *Physiological Reviews, 90*(3), 983–1012.

Gu, J. W., Hermann, B. S., Levine, R. A., & Melcher, J. R. (2012). Brainstem auditory evoked potentials suggest a role for the ventral cochlear nucleus in tinnitus. *Journal of the Association for Research in Otolaryngology, 13*(6), 819–833.

Haenggeli, C. A., Pongstaporn, T., Doucet, J. R., & Ryugo, D. K. (2005). Projections from the spinal trigeminal nucleus to the cochlear nucleus in the rat. *The Journal of Comparative Neurology, 484*(2), 191–205.

Harty, H. T., & Manis, P. B. (1998). Kinetic analysis of glycine receptor currents in ventral cochlear nucleus. *Journal of Neurophysiology, 79*(4), 1891–1901.

Häusser, M. (2001). Synaptic function: Dendritic democracy. *Current Biology, 11*(1), R10–R12.

Heffner, H. E., & Hefner, R. S. (1985). Hearing in two cricetid rodents: Wood rat (*Neotoma floridana*) and grasshopper mouse (*Onychomys leucogaster*). *Journal of Comparative Psychology, 99*(3), 275–288.

Huston, K. A., Durham, D., & Tucci, D. L. (2007). Consequences of unilateral hearing loss: Time dependent regulation of protein synthesis in auditory brainstem nuclei. *Hearing Research, 233*(1–2), 124–134.
Isaacson, J. S., & Walmsley, B. (1995). Receptors underlying excitatory synaptic transmission in slices of the rat anteroventral cochlear nucleus. *Journal of Neurophysiology, 73*(3), 964–973.
Itoh, K., Kamiya, H., Mitani, A., Yasui, Y., Takada, M., & Mizuno, N. (1987). Direct projections from the dorsal column nuclei and the spinal trigeminal nuclei to the cochlear nuclei in the cat. *Brain Research, 400*(1), 145–150.
Johnston, D., Magee, J. C., Colbert, C. M., & Cristie, B. R. (1996). Active properties of neuronal dendrites. *Annual Review of Neuroscience, 19*, 165–186.
Joris, P. X., Carney, L. H., Smith, P. H., & Yin, T. C. (1994a). Enhancement of neural synchronization in the anteroventral cochlear nucleus. I. Responses to tones at the characteristic frequency. *Journal of Neurophysiology, 71*(3), 1022–1036.
Joris, P. X., & Smith, P. H. (2008). The volley theory and the spherical cell puzzle. *Neuroscience, 154*(1), 65–76.
Joris, P. X., Smith, P. H., & Yin, T. C. (1994b). Enhancement of neural synchronization in the anteroventral cochlear nucleus. II. Responses in the tuning curve tail. *Journal of Neurophysiology, 71*(3), 1037–1051.
Jovanovic, S., Radulovic, T., Coddou, C., Dietz, B., Nerlich, J., Stojilkovic, S. S., et al. (2017). Tonotopic action potential tuning of maturing auditory neurons through endogenous ATP. *Journal of Physiology, 595*(4), 1315–1337.
Juiz, J. M., Helfert, R. H., Bonneau, J. M., Campos, M. L., & Altschuler, R. A. (1996). Distribution of glycine and GABA immunoreactivities in the cochlear nucleus: Quantitative patterns of putative inhibitory inputs on three cell types. *Journal Für Hirnforschung, 37*(4), 561–574.
Juiz, J. M., Helfert, R. H., Wenthold, R. J., De Blas, A. L., & Altschuler, R. A. (1989). Immunocytochemical localization of the $GABA_A$/benzodiazepine receptor in the guinea pig cochlear nucleus: Evidence for receptor localization heterogeneity. *Brain Research, 504*(1), 173–179.
Kevetter, G. A., & Parachio, A. A. (1989). Projections from the sacculus to the cochlear nuclei in the Mongolian gerbil. *Brain, Behavior and Evolution, 34*(4), 193–200.
Kim, J. J., Gross, S. J., Potashner, S. J., & Morest, D. K. (2004). Fine structure of long-term changes in the cochlear nucleus after acoustic overstimulation: Chronic degeneration and new growth of synaptic endings. *The Journal of Neuroscience Research, 77*(6), 817–828.
Kirk, E. C., & Smith, D. W. (2003). Protection from acoustic trauma is not a primary function of the medial olivocochlear efferent system. *Journal of the Association for Research in Otolaryngology, 4*, 445–465.
Kopp-Scheinpflug, C., Dehmel, S., Dörrscheidt, G. J., & Rübsamen, R. (2002). Interaction of excitation and inhibition in anteroventral cochlear nucleus neurons that receive large endbulb synaptic endings. *The Journal of Neuroscience, 22*(24), 11004–11018.
Kuhse, J., Betz, H., & Kirsch, J. (1995). The inhibitory glycine receptor: Architecture, synaptic localization and molecular pathology of a postsynaptic ion-channel complex. *Current Opinion in Neurobiology, 5*(3), 318–323.
Kuwabara, N., DiCaprio, R. A., & Zook, J. M. (1991). Afferents to the medial nucleus of the trapezoid body and their collateral projections. *The Journal of Comparative Neurology, 314*(4), 684–706.
Lauer, A. M., Connelly, C. J., Graham, H., & Ryugo, D. K. (2013). Morphological characterization of bushy cells and their inputs in the laboratory mouse (*Mus musculus*) anteroventral cochlear nucleus. *PLoS One, 8*(8), e73308.
Lawrence, J. J., & Trussell, L. O. (2000). Long-term specification of AMPA receptor properties after synapse formation. *The Journal of Neuroscience, 20*(13), 4864–4870.
Lee, D. J., Cahill, H. B., & Ryugo, D. K. (2003). Effects of congenital deafness in the cochlear nuclei of shaker-2 mice: An ultrastructural analysis of synapse morphology in the endbulbs of Held. *Journal of Neurocytology, 32*(3), 229–243.

Liberman, M. C. (1991). Central projections of auditory-nerve fibers of differing spontaneous rate. I. Anteroventral cochlear nucleus. *The Journal of Comparative Neurology, 313*(2), 240–258.

Lomeli, H., Mosbacher, J., Melcher, T., Höger, T., Geiger, J. R., Kuner, T., et al. (1994). Control of kinetic properties of AMPA receptor channels by nuclear RNA editing. *Science, 266*(5191), 1709–1713.

Lorente de Nó, R. (1981). *The primary acoustic nuclei*. New York: Raven Press.

Malosio, M. L., Marqueze-Pouey, B., Kuhse, J., & Betz, H. (1991). Widespread expression of glycine receptor subunit mRNAs in the adult and developing rat brain. *EMBO Journal, 10*(9), 2401–2409.

Manis, P. B., Xie, R., Wang, Y., Narrs, G. S., & Spirou, G. A. (2012). The endbulbs of Held. In L. O. Trussell, A. N. Popper, & R. R. Fay (Eds.), *Synaptic mechanisms in the auditory system* (pp. 61–93). New York: Springer.

Mildbrandt, J. C., & Caspary, D. M. (1995). Age-related reduction of [^{3}H]strychnine binding sites in the cochlear nucleus of the Fischer 344 rat. *Neuroscience, 67*(3), 713–719.

Milenkovic, I., Rinke, I., Witte, M., Dietz, B., & Rübsamen, R. (2009). P2 receptor-mediated signaling in spherical bushy cells of the mammalian cochlear nucleus. *Journal of Neurophysiology, 102*(3), 1821–1833.

Moore, J. K., & Moore, R. Y. (1987). Glutamic acid decarboxylase-like immunoreactivity in brainstem auditory nuclei of the rat. *The Journal of Comparative Neurology, 260*, 157–174.

Moore, J. K., & Osen, K. K. (1979). The cochlear nuclei in man. *American Journal of Anatomy, 154*(3), 393–418.

Mosbacher, J., Schoepfer, R., Monyer, H., Burnashev, N., Seeburg, P. H., & Ruppersberg, J. P. (1994). A molecular determinant for submillisecond desensitization in glutamate receptors. *Science, 266*(5187), 1059–1062.

Needham, K., & Paolini, A. G. (2006). Neural timing, inhibition and the nature of stellate cell interaction in the ventral cochlear nucleus. *Hearing Research, 216*, 31–42.

Ngodup, T., Goetz, J. A., McGuire, B. C., Sun, W., Lauer, A. M., & Xu-Friedman, M. A. (2015). Activity-dependent, homeostatic regulation of neurotransmitter release from auditory nerve fibers. *Proceedings of the National Academy of Sciences of the United States of America, 112*, 6479–6484.

O'Neil, J. N., Limb, C. J., Baker, C. A., & Ryugo, D. K. (2010). Bilateral effects of unilateral cochlear implantation in congenitally deaf cats. *The Journal of Comparative Neurology, 518*(12), 2382–23404.

Oertel, D., Bal, R., Gardner, S. M., Smith, P. H., & Joris, P. X. (2000). Detection of synchrony in the activity of auditory nerve fibers by octopus cells of the mammalian cochlear nucleus. *Proceedings of the National Academy of Sciences of the United States of America, 97*(22), 11773–11779.

Oertel, D., & Fujino, K. (2001). Role of biophysical specialization in cholinergic modulation in neurons of the ventral cochlear nuclei. *Audiology and Neurootology, 6*(4), 161–166.

Oertel, D., & Wickesberg, R. E. (1993). Glycinergic inhibition in the cochlear nuclei: Evidence for tuberculoventral neurons being glycinergic. In M. A. Merchan (Ed.), *The mammalian cochlear nuclei: Organization and function* (pp. 225–237). New York: Plenum Press.

Oertel, D., Wright, S., Ferragamo, M., & Bal, R. (2011). The multiple functions of T stellate/multipolar/chopper cells in the ventral cochlear nucleus. *Hearing Research, 276*(1–2), 61–69.

Oertel, D., Wum, S. H., Garb, M. W., & Dizack, C. (1990). Morphology and physiology of cells in slice preparations of the posteroventral cochlear nucleus of mice. *The Journal of Comparative Neurology, 295*(1), 136–154.

Oleskevich, S., & Walmsley, B. (2002). Synaptic transmission in the auditory brainstem of normal and congenitally deaf mice. *Journal of Physiology, 540*(2), 447–455.

Osen, K. K. (1969). Cytoarchitecture of the cochlear nuclei in the cat. *The Journal of Comparative Neurology, 136*(4), 453–484.

Osen, K. K., Ottersen, O. P., & Storm-Mathisen, J. (1990). Colocalization of glycine-like and GABA-like immunoreactivities: A semiquantitative study of individual neurons in the dorsal

cochlear nucleus of the cat. In O. P. Ottersen & J. Storm-Mathisen (Eds.), *Glycine neurotransmission* (pp. 417–451). New York: Wiley.
Ostapoff, E. M., Benson, C. G., & Saint Marie, R. L. (1997). GABA- and glycine-immunoreactive projections from the superior olivary complex to the cochlear nucleus in guinea pig. *The Journal of Comparative Neurology, 381*(4), 500–512.
Ostapoff, E. M., & Morest, D. K. (1991). Synaptic organization of globular bushy cells in the ventral cochlear nucleus of the cat: A quantitative study. *The Journal of Comparative Neurology, 314*(3), 598–613.
Paolin, A. G., FitzGerald, J. V., Burkitt, A. N., & Clark, G. M. (2001). Temporal processing from the auditory nerve to the medial nucleus of the trapezoid body in the rat. *Hearing Research, 159*(1–2), 101–116.
Piechotta, K., Weth, F., Harvery, R. J., & Friauf, E. (2001). Localization of rat glycine receptor α1 and α 2 subunit transcripts in the developing auditory brainstem. *The Journal of Comparative Neurology, 438*(3), 336–352.
Pliss, L., Yang, H., & Xu-Friedman, M. A. (2009). Context-dependent effects of NMDA receptors on precise timing information at the endbulb of Held in the cochlear nucleus. *Journal of Neurophysiology, 102*(5), 2627–2637.
Raman, I. M., Zhang, S., & Trussell, L. O. (1994). Pathway-specific variants of AMPA receptors and their contribution to neuronal signaling. *The Journal of Neuroscience, 14*(8), 4998–5010.
Rash, J. E., Davidson, K. G. V., Kamasawa, N., Yasumura, T., Kamasawa, M., Zhang, C., et al. (2005). Ultrastructural localization of connexins (Cx36, Cx43, Cx45), glutamate receptors and aquaporin-4 in rodent olfactory mucosa, olfactory nerve, and olfactory bulb. *Journal of Neurocytology, 34*(3–5), 309–342.
Rash, J. E., Olson, C. O., Davidson, K. G. V., Yasumura, T., Kamasawa, N., & Nagy, J. I. (2007a). Identification of connexin36 in gap junctions between neurons in rodent locus coeruleus. *Neuroscience, 147*(4), 938–956.
Rash, J. E., Olson, C. O., Pouliot, W. A., Davidson, K. G. V., Yasumura, T., Furman, C. S., et al. (2007b). Connexin36, miniature neuronal gap junctions, and limited electrotonic coupling in rodent suprachiasmatic nucleus (SCN). *Neuroscience, 149*(2), 350–371.
Ravindranathan, A., Donevan, S. D., Sugden, S. G., Greig, A., Rao, M. S., & Parks, T. N. (2000). Contrasting molecular composition and channel properties of AMPA receptors on chick auditory and brainstem motor neurons. *Journal of Physiology, 523*(3), 667–684.
Reiter, E. R., & Liberman, M. C. (1995). Efferent-mediated protection from acoustic overexposure: Relation to slow effects of olivocochlear stimulation. *Journal of Neurophysiology, 73*(2), 506–514.
Rhode, W. S., & Smith, P. H. (1986). Encoding timing and intensity in the ventral cochlear nucleus of the cat. *Journal of Neurophysiology, 56*(2), 261–286.
Rich, A. W., Xie, R., & Manis, P. B. (2010). Hearing loss alters quantal release at cochlear nucleus stellate cells. *Laryngoscope, 120*(10), 2047–2053.
Richter, E. A., Norris, B. E., Fullerton, B. C., Levine, R. A., & Kiang, N. Y. (1983). Is there a medial nucleus of the trapezoid body in humans? *American Journal of Anatomy, 168*(2), 157–166.
Rodrigues, A. R., & Oertel, D. (2006). Hyperpolarization-activated currents regulate excitability in stellate cells of the mammalian ventral cochlear nucleus. *Journal of Neurophysiology, 95*(1), 76–87.
Rothman, J. S., & Young, E. D. (1996). Enhancement of neural synchronization in computational models of ventral cochlear nucleus bushy cells. *Auditory Neuroscience, 2*, 47–62.
Rothman, J. S., Young, E. D., & Manis, P. B. (1993). Convergence of auditory fibers onto bushy cells in the ventral cochlear nucleus: Implications of a computational model. *Journal of Neurophysiology, 70*(6), 2562–2583.
Rubio, M. E., Fukazawa, Y., Kamasawa, N., Clarkson, C., Molnár, E., & Shigemoto, R. (2014). Target- and input-dependent organization of AMPA and NMDA receptors in synaptic connections of the cochlear nucleus. *The Journal of Comparative Neurology, 522*(18), 4023–4042.

Rubio, M. E., & Juiz, J. M. (2004). Differential distribution of synaptic endings containing glutamate, glycine, and GABA in the rat dorsal cochlear nucleus. *The Journal of Comparative Neurology, 477*(3), 253–272.

Rubio, M. E., Matsui, K., Fukazawa, Y., Kamasawa, N., Harada, H., Itakura, M., et al. (2017). The number and distribution of AMPA receptor channels containing fast kinetic GluA3 and GluA4 subunits at auditory nerve synapses depend on the target cells. *Brain Structure and Function, 222*, 3375–3393. https://doi.org/10.1007/s00429-017-1408-0.

Rubio, M. E., & Nagy, J. (2015). Connexin36 expression in major centers of the auditory system in the CNS of mouse and rat: Evidence for neurons forming purely electrical synapses and morphologically mixed synapses. *Neuroscience, 303*, 604–629.

Rubio, M. E., & Wenthold, R. J. (1997). Glutamate receptors are selectively targeted to postsynaptic sites in neurons. *Neuron, 18*(6), 939–950.

Rubio, M. E., & Wenthold, R. J. (1999). Differential subcellular distribution of glutamate receptors in neurons. *The Journal of Neuroscience, 19*(13), 5549–5562.

Ryugo, D. K. (2008). Projections of low spontaneous rate, high threshold auditory nerve fibers to the small cell cap of the cochlear nucleus in cats. *Neuroscience, 154*(1), 114–126.

Ryugo, D. K., Kretzmer, E. A., & Niparko, J. K. (2005). Restoration of auditory nerve synapses in cats by cochlear implants. *Science, 310*(5753), 1490–1492.

Ryugo, D. K., Pongstaporn, T., Huchton, D. M., & Niparko, J. K. (1997). Ultrastructural analysis of primary endings in deaf white cats: Morphologic alterations in endbulbs of Held. *The Journal of Comparative Neurology, 385*(2), 230–244.

Ryugo, D. K., & Sento, S. (1991). Synaptic connections of the auditory nerve in cats: Relationship between endbulbs of Held and spherical bushy cells. *The Journal of Comparative Neurology, 305*(1), 35–48.

Saint Marie, R. L., Morest, D. K., & Brandon, C. J. (1986). The form and distribution of GABAergic synapses on the principal cell types of the ventral cochlear nucleus of the cat. *Hearing Research, 42*(1), 97–112.

Saldaña, E. (1993). Descending projections from the inferior colliculus to the cochlear nuclei in mammals. In M. A. Merchán, J. M. Juiz, D. A. Godfrey, & E. Mugnaini (Eds.), *The mammalian cochlear nuclei: Organization and function* (pp. 153–165). New York: Plenum Publishing Corporation.

Sato, K., Kuriyama, H., & Altschuler, R. A. (2000). Expression of glycine receptor subunit mRNAs in the rat cochlear nucleus. *Hearing Research, 144*(1–2), 47–52.

Schmid, G., Guthmann, A., Ruppersberg, J. P., & Herbert, H. (2001). Expression of AMPA receptor subunit *flip/flop* splice variants in the rat auditory brainstem and inferior colliculus. *The Journal of Comparative Neurology, 430*(2), 160–171.

Schofield, B. R. (1994). Projections to the cochlear nuclei from principal cells in the medial nucleus of the trapezoid body in guinea pigs. *The Journal of Comparative Neurology, 344*(1), 83–100.

Shore, S. E., Vass, Z., Wys, N. L., & Altschuler, R. A. (2000). Trigeminal ganglion innervates the auditory brainstem. *The Journal of Comparative Neurology, 419*(3), 271–285.

Smith, P. H., & Rhode, W. S. (1987). Characterization of HRP-labeled globular bushy cells in the cat anteroventral cochlear nucleus. *The Journal of Comparative Neurology, 266*(3), 360–375.

Sotelo, C., Gentschev, T., & Zamora, A. J. (1976). Gap junctions in ventral cochlear nucleus of the rat. A possible new example of electrotonic junctions in the mammalian central nervous system. *Neuroscience, 1*(1), 5–7.

Spirou, G. A., Rager, J., & Manis, P. B. (2005). Convergence of auditory-nerve fiber projections onto globular bushy cells. *Neuroscience, 136*(3), 843–863.

Spoendlin, H. (1985). Anatomy of cochlear innervation. *American Journal of Otolaryngology, 8*, 453–487.

Suneja, S. K., Benson, C. G., Gross, J., & Potashner, S. J. (1995). Evidence for glutamatergic projections from the cochlear nucleus to the superior olive and the ventral nucleus of the lateral lemniscus. *Journal of Neurochemistry, 64*(1), 161–171.

Tolbert, L. P., & Morest, D. K. (1982). The neuronal architecture of the anteroventral cochlear nucleus in the region of the cochlear nerve root: Golgi and Nissl methods. *Neuroscience, 7*(12), 3013–3030.

Tolbert, L. P., Morest, D. K., & Yurgelun-Todd, D. A. (1982). The neuronal architecture of the anteroventral cochlear nucleus of the cat in the region of the cochlear nerve root: Horseradish peroxidase labeling of identified cell types. *Neuroscience, 7*(12), 3031–3052.

Tucci, D. L., Cant, N. B., & Durham, D. (1999). Conductive hearing loss results in a decrease in central auditory system activity in the young gerbil. *Laryngoscope, 109*(9), 1359–1371.

Valdivia, O. (1971). Methods of fixation and the morphology of synaptic vesicles. *The Journal of Comparative Neurology, 142*(3), 257–273.

Vogler, D. P., Roberston, D., & Mulders, W. H. A. M. (2011). Hyperactivity in the ventral cochlear nucleus after cochlear trauma. *The Journal of Neuroscience, 31*(18), 6639–6645.

Wang, H., Yin, G., Rogers, K., Miralles, C., De Blas, A. L., & Rubio, M. E. (2011). Monaural conductive hearing loss alters the general expression of the GluA3 AMPA and glycine receptor α1 subunits in bushy and fusiform cells of the cochlear nucleus. *Neuroscience, 199*, 438–451.

Wang, Y., & Manis, P. B. (2005). Synaptic transmission at the cochlear nucleus endbulb synapse during age-related hearing loss in mice. *Journal of Neurophysiology, 94*(3), 1814–1824.

Wang, Y., & Manis, P. B. (2008). Short-term synaptic depression and recovery at the mature mammalian endbulb of Held synapse in mice. *Journal of Neurophysiology, 100*(3), 1255–1264.

Wang, Y. X., Wenthold, R. J., Ottersen, O. P., & Petralia, R. S. (1998). Endbulb synapses in the anteroventral cochlear nucleus express specific subset of AMPA-type glutamate receptor subunits. *The Journal of Neuroscience, 18*(3), 1148–1160.

Waterlood, F. G., & Mugnaini, E. (1984). Cartwheel neurons of the dorsal cochlear nucleus: A Golgi-electron microscopic study in rat. *The Journal of Comparative Neurology, 227*(1), 136–157.

Weedman, D. L., Pongstaporn, T., & Ryugo, D. K. (1996). Ultrastructural study of the granule cell domain of the cochlear nucleus in rats: Mossy fiber endings and their targets. *The Journal of Comparative Neurology, 369*(3), 345–360.

Weinberg, R. J., & Rustioni, A. (1989). Brainstem projections to the rat cuneate nucleus. *The Journal of Comparative Neurology, 282*(1), 142–156.

Wenthold, R. J. (1987). Evidence for a glycinergic pathway connecting the two cochlear nuclei: An immunocytochemical and retrograde transport study. *Brain Research, 415*(1), 183–187.

Wenthold, R. J., & Gulley, R. L. (1977). Aspartic acid and glutamic acid levels in the cochlear nucleus after auditory nerve lesion. *Brain Research, 138*(1), 279–284.

Whiting, B., Moiseff, A., & Rubio, M. E. (2009). Cochlear nucleus neurons redistribute synaptic AMPA and glycine receptors in response to monaural conductive hearing loss. *Neuroscience, 163*(4), 1264–1276.

Wickesberg, R. E., & Oertel, D. (1988). Tonotopic projection from the dorsal to the anteroventral cochlear nucleus of mice. *The Journal of Comparative Neurology, 268*(3), 389–399.

Wickesberg, R. E., & Oertel, D. (1990). Delayed, frequency-specific inhibition in the cochlear nuclei of mice: A mechanism for monaural echo suppression. *The Journal of Neuroscience, 10*(6), 1762–1768.

Wu, S. H., & Oertel, D. (1986). Inhibitory circuitry in the ventral cochlear nucleus is probably mediated by glycine. *The Journal of Neuroscience, 6*(9), 2691–2706.

Xie, R., & Manis, P. B. (2014). GABAergic and glycinergic inhibitory synaptic transmission in the ventral cochlear nucleus studied in VGAT channel rhodopsin-2 mice. *Frontiers in Neural Circuits, 8*, 84. https://doi.org/10.3389/fncir.2014.00084.

Yang, H., & Xu-Friedman, M. (2008). Relative roles of different mechanisms of depression at the mouse endbulb of Held. *Journal of Neurophysiology, 99*(5), 2510–2521.

Young, E. D., & Oertel, D. (2003). Cochlear nucleus. In G. M. Shepherd (Ed.), *The synaptic organization of the brain* (pp. 125–163). New York: Oxford University Press.

Young, E. D., & Oertel, D. (2010). Cochlear nucleus. In G. M. Shepherd & S. Grillner (Eds.), *Handbook of brain microcircuits* (pp. 215–223). New York: Oxford University Press.

Young, E. D., & Sachs, M. B. (2008). Auditory nerve inputs to cochlear nucleus neurons studied with cross-correlation. *Neuroscience, 154*(1), 127–138.

Zhan, X., & Ryugo, D. K. (2007). Projections of the lateral reticular nucleus to the cochlear nucleus in rats. *The Journal of Comparative Neurology, 504*(5), 583–598.

Zhao, Y., Rubio, M. E., & Tzounopoulos, T. (2011). Mechanisms underlying input-specific synaptic plasticity in the dorsal cochlear nucleus. *Hearing Research, 279*(1–2), 67–73.

Zheng, C., Shroff, H., & Shore, S. (2011). Cuneate and spinal trigeminal nucleus projections to the cochlear nucleus are differentially associated with vesicular glutamate transporter-2. *Neuroscience, 176*, 142–151.

Zhou, J., Nannapaneni, N., & Shore, S. (2007). Vesicular glutamate transporters 1 and 2 are differentially associated with auditory nerve and spinal trigeminal inputs to the cochlear nucleus. *The Journal of Comparative Neurology, 500*(4), 777–787.

Zhuang, X., Sun, W., & Xu-Friedman, M. A. (2017). Changes in properties of auditory nerve synapses following conductive hearing loss. *The Journal of Neuroscience, 37*(2), 323–332.

Zorumski, C. F., & Izumi, Y. (2012). NMDA receptors and metaplasticity: Mechanisms and possible roles in neuropsychiatric disorders. *Neuroscience and Biobehavioral Reviews, 36*(3), 989–1000.

Zucker, R. S. (1989). Short-term synaptic plasticity. *Annual Review of Neuroscience, 12*, 13–31.

Chapter 4
Microcircuits of the Dorsal Cochlear Nucleus

Laurence O. Trussell and Donata Oertel

Abstract The dorsal cochlear nucleus (DCN) integrates excitatory input from auditory and nonauditory sources. Auditory signals are conveyed to the deep layer by the auditory nerve and by excitatory interneurons in the ventral cochlear nucleus (VCN). Signals from diverse auditory, somatosensory, proprioceptive, and vestibular sources arrive through mossy fibers in the molecular layer. Thus the DCN is a multisensory integrator. Auditory and mossy inputs are processed through separate microcircuits and are then integrated and conveyed to the inferior colliculus by fusiform cells. Signals arriving from the auditory nerve and VCN in the DCN deep layer are refined by inhibitory neurons that give the acoustic responses of the principal cells a striking nonlinearity as a function of sound intensity and inhibitory sidebands in the spectral domain. Mossy inputs are preprocessed by local circuits in a granule cell region and further refined in the molecular layer. Unlike the auditory signals in the deep layer, signals in the molecular layer exhibit diverse forms of long-term synaptic plasticity. The function of the DCN is not fully understood. The sensitivity of the DCN to spectral notches suggests a role in sound localization using monoaural cues. Input associated with pinna muscles and the trigeminal nerve suggests that the DCN relates head orientation to incoming sounds. The anatomical and physiological similarity of the DCN to structures in electric fish that sensitize the fish to novel signals in the environment has led to the idea that the DCN cancels self-generated and expected features of sounds.

Keywords Cerebellum · Glutamate · Glycine · Multisensory · Sound localization · Synaptic transmission · Synaptic plasticity · Tinnitus

---

L. O. Trussell (✉)
Oregon Hearing Research Center and Vollum Institute, Oregon Health and Science University, Portland, OR, USA
e-mail: trussell@ohsu.edu

D. Oertel
Department of Neuroscience, University of Wisconsin, Madison, WI, USA
e-mail: doertel@wisc.edu

© Springer International Publishing AG 2018

D. L. Oliver et al. (eds.), *The Mammalian Auditory Pathways*,
Springer Handbook of Auditory Research 65,
https://doi.org/10.1007/978-3-319-71798-2_4

4.1 Introduction

The dorsal cochlear nucleus (DCN), like the ventral cochlear nucleus (VCN; see all abbreviations in Table 4.1), receives projections directly from the auditory nerve, but these two nuclei are strikingly different in organization and function (also see Chap. 3 by Rubio). The unlayered VCN conveys acoustic information along the lemniscal auditory pathway in all vertebrates; however, the layered DCN is unique to mammals. Two features are special to the DCN: its role in multisensory integration and the presence of elaborate systems for neuromodulation and synaptic plasticity. These features serve needs that arose with the evolution of high-frequency hearing in mammals (Musicant et al. 1990). The principal cells of the DCN are exquisitely sensitive to the spectral notches that allow mammals to localize sounds in elevation, to distinguish front from back, and to localize sounds monaurally (Spirou and Young 1991; Nelken and Young 1994). While the function of multisensory integration remains only partly understood, one can readily rationalize the need for such integration. Auditory perception must be placed in the context of movements of the body or of the sound source, may be tuned according to different behavioral states, and may be sensitized to differentiate between novel and expected sounds.

Table 4.1 Abbreviations

AChR	Acetylcholine receptor
AMPA	α-Amino-3-hydroxy-5-methyl-4-isoxazolepropionic acid
AVCN	Anteroventral cochlear nucleus
DCN	Dorsal cochlear nucleus
D stellate	Stellate cells whose axons exit dorsalward through the intermediate acoustic stria
ELL	Electrosensory lobe
EPSC	Excitatory postsynaptic current
EPSP	Excitatory postsynaptic potential
GABA	Gamma-aminobutyric acid
GluA	AMPA class of glutamate receptors
GPCR	G-protein coupled receptors
IPSP	Inhibitory postsynaptic potential
LTD	Long-term depression
LTP	Long-term potentiation
mGluR	Metabotropic glutamate receptors
NMDA	N-methyl-D-aspartate
PVCN	Posteroventral cochlear nucleus
T stellate	Stellate cells whose axons exit through the trapezoid body
UBC	Unipolar brush cells
VCN	Ventral cochlear nucleus
VGLUT	Vesicular glutamate transporter

4.2 Proposed Functions and Organization of the Dorsal Cochlear Nucleus

The DCN forms a curved ridge of neural tissue that overlies the VCN most laterally and wraps around the inferior cerebellar peduncle medially. This shape inspired the name "acoustic tubercle" or "tuberculum acousticum" used by Cajal and Lorente de Nó (Ramón y Cajal 1909; Lorente de Nó 1933, 1981). The dorsolateral surface of the DCN is covered by an ependymal layer that forms the floor of the fourth ventricle.

A cross section through the DCN of mammals reveals a layered structure (Fig. 4.1). Three layers are distinct. The outermost layer immediately beneath the ependyma contains few cell bodies and is known as the *molecular layer*. The *pyramidal cell layer* or *fusiform cell layer* is densely populated with cell bodies not only of fusiform cells but also of cartwheel, granule, Golgi, and unipolar brush cells. Finally, the *deep layer* contains numerous cell bodies of tuberculoventral and giant cells that are less densely packed (Osen 1969; Brawer et al. 1974; Wickesberg and Oertel 1988; Mugnaini et al. 1997). All of the excitatory and some of the inhibitory inputs, as well as many of the dendrites in the deep layer of the DCN, lie in tonotopically organized isofrequency laminae (Wickesberg and Oertel 1988). Isofrequency

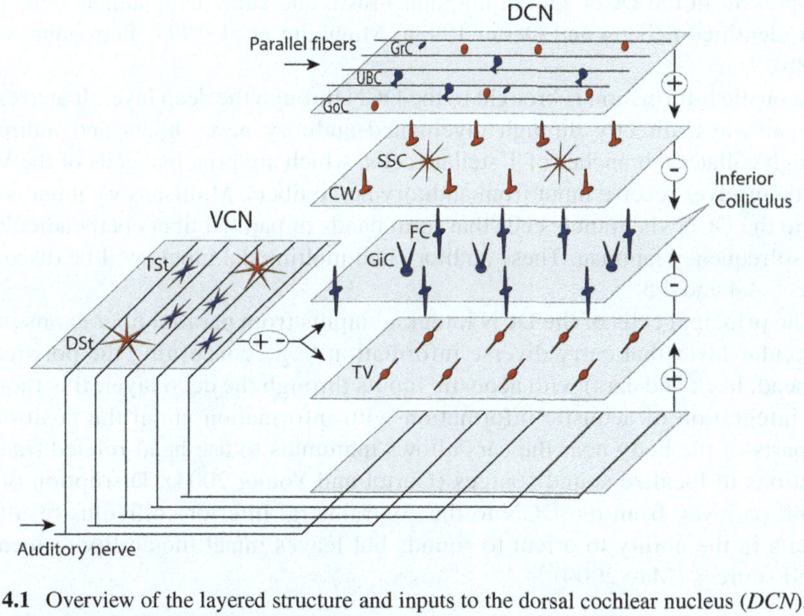

Fig. 4.1 Overview of the layered structure and inputs to the dorsal cochlear nucleus (*DCN*). *Left side* of the diagram illustrates the ventral cochlear nucleus (*VCN*) neurons that project to the DCN. (*CW*, cartwheel cell; *DSt*, D stellate cell; *FC*, fusiform cell; *GiC*, giant cell; *GoC*, Golgi cell; *GrC*, granule cell; *SSC*, superficial stellate cell; *TSt,* T stellate cell; *TV*, tuberculoventral cell; *UBC*, unipolar brush cell)

laminae run approximately rostrocaudally, perpendicular to the lateroventral–dorsomedial tonotopic axis that has low frequencies represented ventrally and high frequencies represented dorsally.

Primates have a large DCN, but in many primates the layering is less clear than in other mammals (Moore and Osen 1979; Adams 1986). Primates have relatively fewer granule cells and a correspondingly thin molecular layer. In the galago (Galagidae) and slow loris (Lorisidae) the layers are more distinct than in owl monkeys (Aotidae), squirrel monkeys (Saimirinae), gibbons (Hylobatidae), chimpanzees (Hominidae), and humans (Moore and Osen 1979; Heiman-Patterson and Strominger 1985). A recent study of primates with immunohistochemical and electron microscopic methods revealed that although the layering is less obvious than in other mammals, the corresponding neuronal types can be identified (Rubio et al. 2008). Comparisons of immunohistochemical labeling patterns between species confirm the conclusion that the human DCN is layered, but the less obvious layering changes some of the anatomical features of the neurons, making those neurons more difficult to identify compared to those of nonprimate mammals (Baizer et al. 2014).

The DCN lies immediately adjacent to the cerebellum and bears considerable resemblance to it. As in the cerebellum, granule cells send their unmyelinated axons to the molecular layer (Mugnaini et al. 1980). One of their targets, the cartwheel cell, resembles cerebellar Purkinje cells in shape, in genetic markers, and in their intrinsic electrical properties (Mugnaini et al. 1987; Zhang and Oertel 1993a; Manis et al. 1994). Most of the other types of neurons recognized in the cerebellum are also present in the DCN; Golgi, unipolar brush, and superficial stellate cells have been identified (Zhang and Oertel 1993a; Mugnaini et al. 1997; Ferragamo et al. 1998b).

Acoustic information is brought to the DCN through the deep layer. It arrives via two pathways: directly through myelinated auditory nerve fibers and indirectly through collateral branches of T stellate cells, which are principal cells of the VCN that themselves receive input from auditory nerve fibers. Multisensory input is carried to the DCN via granule cells that form bands of parallel fibers perpendicular to the isofrequency laminae. These auditory and multimodal inputs will be discussed in Sect. 4.4 and 4.5.

The principal cells of the DCN integrate inputs from parallel fiber axons in the molecular layer that carry diverse information (e.g., concerning the position of the head, neck and ears) with acoustic inputs through the deep layer. It is thought that integration of acoustic information with information about the position of the parts of the body near the ears allows mammals to use head-related transfer functions to localize sound sources (Oertel and Young 2004). Disruption of the output pathway from the DCN to the contralateral inferior colliculus results in deficits in the ability to orient to sounds but leaves intact the ability to localize sound sources (May 2000).

4.3 Principal Cells

4.3.1 Fusiform Cells

Fusiform cell bodies lie in the cell layer that is interposed between the superficial molecular layer and the deep layer (Figs. 4.1 and 4.2). Early studies that used the Golgi technique showed that fusiform cells (also called pyramidal cells by some authors) integrate two systems of inputs through separate tufts of dendrites (Lorente de Nó 1933; Osen 1969; Brawer et al. 1974) and project to the contralateral inferior colliculus (Adams and Warr 1976). Intracellular labeling of neurons with horseradish peroxidase, biocytin, or with fluorescent dyes allows visualization of whole neurons with their dendritic and axonal arbors. These methods showed that the tuft of apical dendrites in the molecular layer is densely covered with spines while the tuft of basal dendrites is smooth. Furthermore, the ability to reconstruct the morphology of neurons in their entirety showed that the smooth basal dendrites lie parallel to, and presumably within, isofrequency laminae. The morphology of fusiform cells has been examined in adult cats (Rhode et al. 1983), mice (Zhang and Oertel 1994), and gerbils (Hancock and Voigt 2002a). In primates, fusiform cells have a less stereotyped morphology with only a thin molecular layer, they are less clearly oriented, and the fusiform cells are more difficult to identify unambiguously (Moore and Osen 1979; Adams 1986; Baizer et al. 2014).

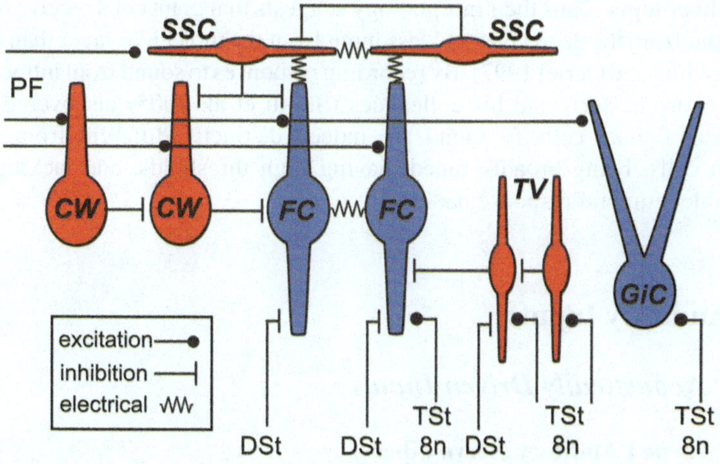

Fig. 4.2 Detailed circuit diagram of the molecular and auditory input domains of the dorsal cochlear nucleus (DCN). Cell types are labeled in ***bold italicized text***, input sources in standard text. A recurring theme is the presence of synaptic contacts between interneurons as well as between interneurons and principal cells. Little is known about inhibition to giant cells, but such contacts are likely given that the cells traverse the body of the DCN. Only fusiform and giant cells are excitatory. Anatomically, these domains include some granule cells (diagrammed in Fig. 4.3). (*8n*, auditory nerve; *CW*, cartwheel cell; *DSt*, D stellate cell; *FC*, fusiform cell; *GiC*, giant cell; *PF*, parallel fibers; *SSC*, superficial stellate cell; *TSt*, T stellate cell; *TV*, tuberculoventral cell)

Responses of identified fusiform cells to sound reflect the integration of excitation and inhibition from both the molecular and deep layers. Fusiform cells have low thresholds and are sharply tuned. They fire with *pauser* or *buildup* peristimulus-time histogram patterns. Nonmonotonicity as a function of level is one indication of the prominence of inhibition (Rhode et al. 1983; Hancock and Voigt 2002b; Smith et al. 2005). Inhibition is even more prominent in the absence of anesthesia; type IV units (presumptive fusiform cells) are excited by broadband noise, but they are inhibited by noise that has a notch near the best frequency (Spirou and Young 1991). The presence of a source of narrowband inhibition and a separate source of wideband inhibition accounts for the responses of fusiform cells to tones and noise (Nelken and Young 1994).

4.3.2 Giant Cells

Giant cells integrate inputs from the molecular and deep layers as fusiform cells do, but they do so differently. The cell bodies of giant cells lie in the deep layer, but in contrast with fusiform cells, their dendrites are not confined to an isofrequency lamina (Zhang and Oertel 1993b; Kecskes et al. 2013). The spread of their dendrites suggests that giant cells are more broadly tuned than fusiform cells because they span a larger section of the tonotopic axis. Also, only the tips of a few of their dendrites are spiny and reach into the molecular layer, presumably catching some parallel fiber input. Thus their morphology suggests that giant cells receive relatively more input from the deep layer and less input from the molecular layer than fusiform cells (Golding and Oertel 1997). By recording responses to sound from intracellularly labeled neurons, Joris and his colleagues (Smith et al. 2005) discovered that the responses of giant cells to sound are indeed distinctly different from those of fusiform cells, being broadly tuned, having high thresholds, and lacking clearly identifiable temporal response patterns.

4.4 Auditory Inputs

4.4.1 Acoustically Driven Inputs

4.4.1.1 Type I Auditory Nerve Fibers

The classical work of Osen revealed the pattern of innervation of the DCN by auditory nerve fibers (Osen 1970). Large diameter, myelinated auditory nerve fibers enter the VCN and bifurcate in the nerve root. The descending branches of auditory nerve fibers pass through the posteroventral cochlear nucleus (PVCN) and aggregate into a tight bundle as they pass through the octopus cell area (Lorente de Nó 1933; Fekete et al. 1984). Fibers then leave the bundle to project caudorostrally along an isofrequency lamina, remaining within the deep layer of the DCN (Leake and Snyder 1989; Wickesberg et al. 1991). Those fibers impose a tonotopic organization on the DCN with fibers encoding low frequencies terminating most ventrally and those that encode higher frequencies terminating more dorsally. Auditory nerve terminals contact cell bodies and dendrites in the deep layer (Liberman 1993).

Exactly what the targets of Type I auditory nerve fibers are in the DCN remains uncertain since labeled fibers have not been traced to fusiform cells. Fibers that were labeled intracellularly had very few terminals adjacent to the cell bodies of fusiform and giant cells, but that does not eliminate the possibility that their dendrites are more heavily innervated (Liberman 1993). It has been assumed, based on the similarity of vesicle morphology, that terminals containing large vesicles that contact the basal dendrites are associated with auditory nerve fibers, but this assumption has not been demonstrated directly (Smith and Rhode 1985). Those authors found a second group of terminals with smaller round vesicles that they designated as "fusiform-collateral-like", whose origin has not been identified. Responses of auditory nerve fibers to sounds have been studied extensively (Liberman 1982; Ruggero 1992). They are sharply tuned and their firing patterns are by definition "primary-like."

4.4.1.2 Type II Auditory Nerve Fibers

Small diameter, unmyelinated Type II auditory nerve fibers are less common than the larger Type I fibers, constituting only about 7% of the total. Descending branches of Type II fibers initially follow Type I fibers but then skirt the octopus cell area rostrally and laterally to terminate in the granule cell lamina (Berglund and Brown 1994). Those Type II fibers that continue into the DCN terminate in the fusiform cell layer. Species differ in how Type II fibers innervate the DCN. All Type II fibers continue into the DCN in gerbils, but in mice only about one third of the Type II fibers reach the DCN. It has proven difficult to record responses to sound in Type II auditory nerve fibers (Brown et al. 1988). Recently, the suggestion was made that Type II fibers respond only to noxious sounds and not to innocuous sounds, acting somewhat like pain fibers (Flores et al. 2015).

4.4.1.3 T Stellate Cells

The possibility that T stellate cells (also known as T multipolar, planar stellate, and chopper neurons) provide substantial acoustic input to the DCN has been appreciated recently (Young and Oertel 2010; Oertel et al. 2011). The T stellate cells reside in the multipolar cell area of the PVCN and in the anteroventral cochlear nucleus (AVCN), intermingled with bushy cells. While the main axon leaves the cochlear nuclei through the ventral acoustic stria and ultimately reaches the contralateral inferior colliculus and thalamus (Schofield et al. 2014), axon collaterals innervate the VCN, DCN, and numerous brain stem nuclei (Oertel et al. 2011). Intracellular labeling of T stellate cells revealed that they make substantial projections to the DCN (Smith and Rhode 1989; Oertel et al. 1990; Doucet and Ryugo 2003). The axon terminals lie within a band about 100 μm wide that spans the deep layer of the DCN and defines an isofrequency lamina. The profusion of T stellate cell terminals in the deep layer of the DCN, relative to the sparse terminals of auditory nerve fibers, supports the possibility that T stellate cells carry a substantial portion of acoustic input to the DCN.

Individual T stellate cells respond to sound with regular firing for the duration of the sound, producing *chopping patterns*. They encode the envelopes of sounds over narrow frequency bands (Blackburn and Sachs 1989; Smith and Rhode 1989; Frisina et al. 1990; May et al. 1998), and these neurons have strong inhibitory sidebands (Rhode and Greenberg 1994). As a population, choppers provide the brain with an ongoing representation of spectrum.

4.4.2 Interneurons That Modify Acoustic Responses of Principal Cells

4.4.2.1 D Stellate Cells

The identification of D stellate cells (also called type II, stellate multipolar, onset choppers, wide band inhibitors) followed the finding by multiple groups that stellate cells in the VCN are of two distinct types (Cant 1981; Smith and Rhode 1989; Oertel et al. 1990). The dendrites of D stellate cells spread widely across isofrequency laminae and receive input from numerous auditory nerve fibers, which is consistent with the observations that they are broadly tuned (Smith and Rhode 1989; Oertel et al. 1990; Doucet and Ryugo 1997; Xie and Manis, 2017). The D stellate cells were named for having a main axon that projects dorsalward, in contrast with T stellate cells whose axons exit through the trapezoid body (Oertel et al. 1990). The main axon projects around the restiform body and then crosses to the contralateral cochlear nucleus through the intermediate acoustic stria (Wenthold 1987; Alibardi 1998). Through collateral branches, D stellate cells project locally over broad regions of the VCN and over wide regions of the deep layer of the ipsilateral DCN (Smith and Rhode 1989; Oertel et al. 1990; Doucet and Ryugo 1997).

D stellate cells are inhibitory and glycinergic. The first clue that they are glycinergic came from their strong labeling for glycine, an inhibitory neurotransmitter (Wenthold 1987). Also, at the ultrastructural level their terminals have the flattened vesicles typical of inhibitory neurons (Smith and Rhode 1989; Alibardi 1998). Inhibitory input to the cochlear nuclei from the contralateral side was first noted in responses to sound (Mast 1970) and was later confirmed with intracellular recordings (Needham and Paolini 2003). Ipsilateral inhibition in the VCN and DCN could also be attributed to D stellate cells in vivo (Nelken and Young 1994) and in vitro (Ferragamo et al. 1998a).

As expected from their morphology, D stellate cells are broadly tuned. Tones evoke *onset chopper* response patterns that involve a sharply timed initial action potential followed by somewhat regular firing, and the firing rate adapts (Smith and Rhode 1989). It is likely that these neurons correspond to the "wideband inhibitor" proposed by Nelken and Young (1994) to help generate the Type IV response of fusiform cells.

4.4.2.2 Tuberculoventral (Vertical) Cells

In sections not cut parallel to isofrequency laminae in the DCN, Golgi labeling revealed segments of neurons that have dendrites and axons extending throughout the vertical spread of the deep layer of the DCN, inspiring the term "vertical cells" (Osen 1969; Osen et al. 1990). Some of these neurons were shown to have axons that project to the VCN (Lorente de Nó 1981). Tuberculoventral neurons were named for their projection from the tuberculum acousticum to the VCN and for having axons in the ventrotubercular tract (Lorente de Nó 1981). They were identified in mice by their topographic projection from the DCN to the VCN (Wickesberg and Oertel 1988, 1990; Campagnola and Manis 2014). While single cell labeling in mice consistently reveals an axon that projects to the VCN, some vertical cells in cats seem to lack an axonal projection to the VCN (Rhode 1999). Thus the question about whether all vertical cells correspond to tuberculoventral cells or whether some vertical cells in some species have only a local projection in the DCN remains unresolved. We will consider them to be a single group of neurons.

Single cell labeling in mice showed that dendrites of tuberculoventral cells spread across the deep layer of the DCN within a band of approximately 100 μm, which is the width of an isofrequency lamina defined by the terminals of T stellate cells. Their axons projected to an isofrequency lamina in the VCN and DCN (Zhang and Oertel 1993c). The auditory nerve fibers that innervate tuberculoventral cells also innervated their targets, indicating that the inhibition is specific to a narrow frequency range (Wickesberg and Oertel 1988; Muniak and Ryugo 2014).

Units designated as type II (Evans and Nelson 1973) can be driven antidromically from the VCN but not from the dorsal acoustic stria, which is consistent with what is known about the projections of tuberculoventral neurons (Young 1980). They have little spontaneous firing in quiet, respond vigorously to narrowband sounds, but respond little or not at all to broadband noise (Young and Voigt 1982). The presence of inhibition is evident from nonmonotonicity in the rate-level functions and from weak or absent responses to broadband noise (Spirou et al. 1999). Recordings from labeled tuberculoventral cells in anesthetized cats indicated that tuberculoventral cells respond to tones with a response rate that falls over about 50 ms, which is termed an "onset-graded" response (Rhode 1999). The finding that these neurons respond to narrowband and not broadband sound stimuli indicates that these onset-graded cells are the same cells that are type II in unanesthetized cats.

Tuberculoventral cells are inhibitory and glycinergic (Wickesberg and Oertel 1990; Wickesberg et al. 1994; Campagnola and Manis 2014). In the DCN, fusiform cells (type IV units) and other tuberculoventral cells are likely to be the major synaptic targets. In the VCN, bushy and T stellate cells are the targets (Voigt and Young 1980; Wickesberg and Oertel 1990; Kuo et al. 2012). Tuberculoventral cells are thought to contribute to the ability of fusiform cells to encode the sharp spectral notches that enable mammals to use spectral cues in head-related transfer functions for localizing sounds (Spirou and Young 1991; Rice et al. 1992).

4.5 Multisensory Input to DCN

4.5.1 Multimodal Inputs

The DCN also receives a system of multisensory, excitatory inputs through its superficial layers, thereby making the DCN a multimodal processor. Fibers of diverse origin terminate on granule, Golgi, and unipolar brush cells. In the cerebellum, *mossy fibers* refer to input fibers of diverse origins that terminate in the granule cell layer as large terminal clusters or swellings. In the cochlear nuclei, the fibers ending in the auditory granule cell domain are often termed mossy fibers by analogy to the cerebellum. However the cerebellum analogy may be misleading, as inputs to the auditory granule cell domain can have a mossy terminal appearance but also may be bouton-like (Schofield and Coomes 2005; Zhan et al. 2006; Zhou et al. 2007). Moreover, the larger mossy inputs vary in ultrastructure (Dunn et al. 1996), perhaps reflecting different functional properties of the mossy input. Therefore, for purposes of this review, we will use the term *multimodal input* to refer to this broader class of glutamatergic fibers.

The multimodal input is remarkable in its variety. Retrograde or anterograde tracer studies have described inputs to the cochlear nucleus from the vestibular nuclei and ganglion (Burian and Gstoettner 1988; Bukowska 2002; Barker et al. 2012), dorsal column nuclei (Itoh et al. 1987; Weinberg and Rustioni 1987; Young et al. 1995), dorsal root ganglia (Zhan et al. 2006), trigeminal ganglia and trigeminal nucleus (Shore et al. 2000; Zhou et al. 2007; Zeng et al. 2011), inferior colliculus (Conlee and Kane 1982; Shore et al. 1991; Alibardi 2002), pontine nuclei (Ohlrogge et al. 2001), auditory nerve (Benson and Brown 2004), and auditory cortex (Schofield and Coomes 2005; Meltzer and Ryugo 2006; Schofield et al. 2006). Moreover, octopus cells of the VCN also provide input to the granule cell domain in mice (Golding et al. 1995). Medial olivocochlear efferent neurons, which are likely cholinergic, also terminate in the vicinity of granule cells, but whether granule cells are their targets has not been confirmed (Osen et al. 1984; Benson and Brown 1990). Thus, multimodal input includes motor and multisensory signals, and the sensory signals range from primary afferents to cortical pyramidal cells.

Putative targets of multimodal inputs are granule cells, unipolar brush cells, and Golgi cells; which brain regions specifically target which kind of neuron is not clear. Moreover, it remains to be seen whether different sources of input are specific to a given subregion of the auditory granule cell domain, which, as described in the following section, is spread out. Understanding this level of anatomical specificity will lead to a better understanding of the function of multisensory processing.

The nonauditory input to the cochlear nucleus can also be distinguished by transmitter and terminal morphology, as well as by fiber source. Multiple neuromodulatory systems impact the DCN. Glutamatergic inputs can terminate as mossy fibers or en passant boutons. These also differ in the subtype of vesicular glutamate transporters (VGLUT) that they use to package glutamate. For example, mossy fibers in the cuneate and spinal trigeminal nuclei predominantly express VGLUT2 and terminate in the granule cell domain. The trigeminal nucleus also sends fibers that terminate as boutons in broad regions of the DCN and express less VGLUT2 (Zhou et al. 2007). These results highlight that multimodal input probably varies significantly in its physiological consequences in the DCN.

4.5.2 Targets of Multimodal Input

4.5.2.1 Granule Cells

Auditory granule cells are the smallest neurons of the auditory system, and they resemble the granule cells of the cerebellar cortex (Osen 1969; Mugnaini et al. 1980; Balakrishnan and Trussell 2008). Morphological similarities include their small size, the presence of several short dendrites (often with claw-like endings), and the emission of a long and fine axon that makes en passant boutons within the molecular layer. Auditory granule cells are somewhat larger than their cerebellar counterparts and have rather smaller dendritic claws. Also, the auditory granule cells are not distributed in a central layer but are in multiple subregions: cell clusters overlying the VCN (shell), between the DCN and VCN (lamina), several regions medial to the cochlear nucleus, and, lastly, a sparse distribution of cells within the DCN itself (Osen 1969; Mugnaini et al. 1980). These multiple homes for granule cells are collectively called the *granule cell domain* and encompass a wide area that includes diverse cell types in addition to the granule cells. The axons from all these granule cell clusters make their way to the DCN molecular layer (Fig. 4.3), which is effectively surrounded by granule cells; for this reason Mugnaini et al. (1980) described the parallel fibers as "sunk in a granule cell pool".

There are few studies of the physiology of synapses between the multimodal fibers and auditory target cells. Electrical stimulation of excitatory inputs to granule cells produced modestly depressing EPSCs mediated by AMPA

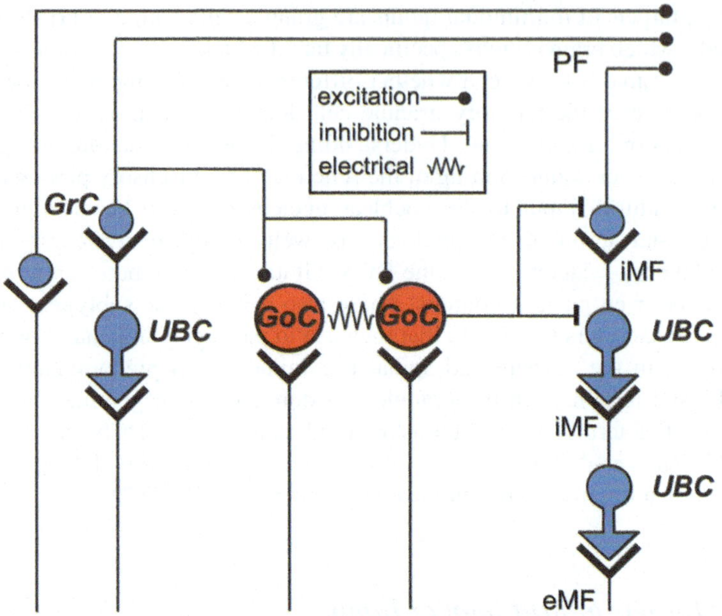

Fig. 4.3 Detailed circuit diagram of the multisensory, granule cell domains of the dorsal cochlear nucleus (DCN). Cell types are labeled in ***bold italicized text***; input sources in standard text. Only Golgi cells are inhibitory (indicated by red fill). (*eMF*, extrinsic mossy fiber; *GoC*, Golgi cell; *GrC*, granule cell; *iMF*, intrinsic mossy fiber; *PF*, parallel fibers; *UBC*, unipolar brush cell)

(α-amino-3-hydroxy-5-methyl-4-isoxazolepropionic acid) and NMDA (N-methyl-D-aspartate) receptors, similar to that seen in the cerebellum (Balakrishnan and Trussell 2008). However such stimulation may select for stronger, more obvious inputs whose origin is not certain, possibly including intrinsic mossy fibers that originate from unipolar brush cells. Of great interest is the extent to which individual granule cells collect signals from single or multiple modalities. There is now compelling evidence that single cerebellar granule cells integrate diverse modalities, even sensory and motor, each terminating on its own dendrite (Huang et al. 2013; Chabrol et al. 2015). Understanding whether or not single auditory granule cells integrate different modalities will help define the function of multimodal circuits in DCN.

4.5.2.2 Golgi Cells

Golgi cells are found in the granule cell domains that extend their dendrites and axons broadly within those domains (Wouterlood et al. 1984; Ferragamo et al. 1998b). The axons of Golgi cells branch profusely, contacting a contiguous group of granule cells that may form a functional unit (Fig. 4.3). They receive glutamatergic excitatory input from auditory nerve fibers and from granule cells (Ferragamo et al. 1998b; Irie et al. 2006; Yaeger and Trussell 2015), and they make inhibitory synapses (GABAergic and glycinergic) onto granule cells (Yaeger and Trussell 2015) and probably also onto brush cells. Accordingly they mediate both feedforward and feedback inhibition. While the number of granule cells far outweighs the number of Golgi cells, there is a surprisingly high degree of connectivity between them, such that single granule cells may be potent enough to drive spikes in a Golgi cell, which in turn effectively inhibits many granule cells (Yaeger and Trussell 2015). Stimulation of auditory nerve fibers generates excitatory postsynaptic potentials (EPSPs) and excitatory postsynaptic currents (EPSCs) of moderate-to-long latency, suggesting that they are mediated by unmyelinated fibers and/or polysynaptic pathways (Ferragamo et al. 1998b; Irie et al. 2006). As with the Golgi cells of the cerebellum, auditory Golgi cells are electrically coupled to one another, and these connections mainly mediate inhibitory control of neighboring Golgi cells by transmitting the slow afterhyperpolarization of the action potential (Yaeger and Trussell 2016). Thus, Golgi cells form a network that serves to temper and even block multimodal signals entering the auditory system.

4.5.2.3 Unipolar Brush Cells

Unipolar brush cells (UBCs) are glutamatergic interneurons found primarily in the fusiform and deep layers of DCN and in the vestibulocerebellum (Mugnaini et al. 1997). They receive mossy fiber input onto the large brush-like dendrite that gives these cells their name (Fig. 4.3). Moreover, the axons of UBCs form intrinsic mossy terminals on granule cells and other UBCs. Two subtypes of UBCs have been identified based on the expression of cell markers and on the response to mossy fiber stimulation (Diño et al. 1999; Borges-Merjane and Trussell 2015). The *ON* UBCs express mGluR1α and respond to mossy fiber activity with a slow AMPA receptor-mediated and mGluR1α-mediated excitatory current (Rossi et al. 1995; Borges-Merjane and Trussell 2015). The *OFF* UBCs express the calcium-binding protein calretinin and respond to glutamate with a slow, inhibitory mGluR2-mediated hyperpolarization generated by G-protein dependent, inward rectifier potassium channels (Borges-Merjane and Trussell 2015). The synaptic response, which often evokes many action potentials and can last for hundreds of milliseconds after a single mossy fiber action potential, amplifies the multisensory input and draws it out in time. Also, the opposing actions of glutamate in the two types of UBCs suggest that UBCs may form a push–pull mechanism for transforming multimodal inputs to the auditory system. These complex transformations are presumably necessary for meaningful convergence of different modalities, as has been suggested for UBCs in the electrosensory system of mormyrid electric fish (Kennedy et al. 2014).

4.6 Molecular Layer Circuitry

In addition to Golgi cells, the synaptic targets of granule cells include cartwheel (Wouterlood and Mugnaini 1984; Zhang and Oertel 1993a; Manis et al. 1994), fusiform (Blackstad et al. 1984; Manis 1989; Zhang and Oertel 1994), superficial stellate (Wouterlood et al. 1984; Zhang and Oertel 1993a), and probably giant cells (Zhang and Oertel 1993b; Golding and Oertel 1997) (Fig. 4.2). These glutamatergic synapses may activate both AMPA and NMDA receptors as well as metabotropic glutamate receptors. As discussed in Sect. 4.7, activation of NMDA receptors may play a role in long-term plasticity. Short-term plasticity in the form of synaptic facilitation is characteristic of all parallel fiber synapses examined (Roberts and Trussell 2010; Apostolides and Trussell 2014a; Yaeger and Trussell 2015), and the presence of facilitation implies that parallel fibers transmit most effectively as a burst or train of spikes unless a large number of fibers are synchronously active. Most of the depolarization is probably mediated by the AMPA receptors (Manis and Molitor 1996; Fujino and Oertel 2003).

Interestingly, the time course of the EPSCs generated at these synapses is determined by the target cells. The EPSCs in cartwheel and fusiform cells decay much more slowly than at synapses onto Golgi or superficial stellate cells (Gardner et al. 1999; Apostolides and Trussell 2014a; Yaeger and Trussell 2015). Moreover, the latter two cell types use a different subtype of AMPA receptor that lacks the GluA2 subunit and are permeable to Ca^{2+}. Thus, parallel fiber targets may respond differently to the same input signal based on their profile of receptor expression. This principle is mirrored by EPSC duration and receptor subunit expression in auditory nerve targets, which also vary in a target cell-dependent manner (Gardner et al. 1999, 2001). A recent ultrastructural analysis shows that these synapses differ not only in receptor subunit composition but also in the number and placement of receptors in the subsynaptic membrane (Rubio et al. 2014).

Even within a single cell type, the types of glutamate receptors may differ at different synapses. For example, the subtypes of receptors are not identical in the apical and basal dendrites, with the GluA4 subunit being predominant in the basal dendrite (Rubio and Wenthold 1997; Gardner et al. 2001). This molecular distinction suggests that multimodal (apical) and auditory (basal) input signals might generate different types of responses or be differentially regulated.

The arrangement and morphology of dendrites of parallel fiber targets in the molecular layer are distinct. Fusiform cells receive parallel fiber boutons on their spiny apical dendrites. In mice, activity from about sixteen parallel fibers is required to drive spikes in a resting fusiform cell, although fewer are needed to modulate spontaneous activity (Roberts and Trussell 2010). However, there are probably hundreds of parallel fiber synapses on a fusiform neuron; therefore, fusiform cells may be driven by different subsets of granule cells and, by extension, different combinations of sensory signals.

4.6.1 Cartwheel Cells

Cartwheel cells are glycinergic/GABAergic interneurons that reside in the fusiform and molecular layer. They extend their densely spiny dendrites across the molecular layer to the ependymal layer that covers the DCN. Their name derives from the "looping back around" arrangement of their dendrites in the plane of the molecular layer that was observed in cat (Brawer et al. 1974). Cartwheel cells are also excited by subsets of parallel fibers and send potent inhibitory signals to fusiform cells and to one another (Golding and Oertel 1997; Roberts et al. 2008; Roberts and Trussell 2010). This circuit relationship implies that cartwheel cells mediate feedforward inhibition. However, in mice it is likely that the parallel fibers that excite a given cartwheel cell are not the same as those that excite the fusiform cell to which that interneuron is connected (Roberts and Trussell 2010). This raises the possibility that cartwheel cells play a role in lateral inhibition as well, perhaps shutting down some fusiform cells in favor of others for a given stream of sensory input.

Cartwheel cells are characterized by the generation of complex spikes that are made up of a brief cluster of Na^+ action potentials riding on a slow Ca^{2+}-dependent depolarization (Zhang and Oertel 1993a; Manis et al. 1994; Golding and Oertel 1997; Bender and Trussell 2009). This waveform, whose Ca^{2+} and Na^+ components initiate firing within the axon initial segment, is unique in the cochlear nucleus and, as such, is diagnostic for cartwheel cells in extracellular recordings.

4.6.2 Superficial Stellate Cells

Superficial stellate cells are tiny glycinergic/GABAergic interneurons in the molecular layer (Wouterlood et al. 1984; Zhang and Oertel 1993a). They are excited by parallel fibers and their axons terminate on fusiform cells, cartwheel cells, and on one another (Apostolides and Trussell 2014a, 2014c). Indeed, even autaptic contacts are not uncommon in superficial stellate cells as in stellate cells of the cerebellar molecular layer. As with their cerebellar counterparts, superficial stellate cells communicate with each other via electrical, gap-junction-mediated synapses. Remarkably, they also make electrical synapses with fusiform cells (Apostolides and Trussell 2013, 2014b). As a result, excitatory synaptic activity in fusiform cells can trigger action potentials in superficial stellate cells and lead to inhibition of stellate cell targets. Since fusiform cells are driven by auditory stimuli, this result implies that auditory activity can modify processing of multimodal signals in the molecular layer.

4.6.3 Molecular Layer Responses to Sensory Input

Given these synaptic relationships, what is the effect of multimodal activity on fusiform cells and on the response of fusiform cells to acoustic stimuli? This question has been particularly difficult to address with classical recording techniques because the natural stimuli that activate the multimodal systems produce instability in the recordings. Stimulation of a variety of brain regions elicits responses in the DCN, suggestive of activation of multimodal inputs. Prominent among these stimuli are mechanical stimulation of pinna muscles that support proprioception or electrical stimulation of dorsal column nuclei that relay signals from these fibers (Young et al. 1995; Kanold and Young 2001; Kanold et al. 2011). Such activity elicits facilitating field potentials in parallel fibers and a complex set of postsynaptic responses in type IV cells, which are putative fusiform neurons (Davis et al. 1996; Kanold et al. 2011). These responses in the fusiform cell consist of a shortlatency inhibition of spontaneous activity, a brief period of excitation, and then a longerlasting inhibition. The early inhibition slightly precedes the parallel fiber response. Increases in activity of cartwheel cells driven by dorsal column stimulation excite neighboring cartwheel cells and may be associated with the longer-lasting inhibition in fusiform cells. The source of the early inhibition is currently unresolved. It may result from inhibition from Golgi cells that are activated before excitation of parallel fibers reaches the molecular layer; however, action potentials in different groups of parallel fibers may reach inhibitory neurons and fusiform cells at slightly different times. In any case, the results show that suppression of fusiform cell spiking is a major consequence of proprioceptive input to DCN.

Coactivation of dorsal column nuclei with acoustic stimuli leads to inhibition of early phases of acoustic responses, indicating that somatosensory signals can effectively modify the earliest levels of auditory processing in the brain (Kanold et al. 2011). Other sources of input to the granule cell domain are the trigeminal ganglion and spinal trigeminal nuclei. Activation of either region can inhibit or enhance the response of fusiform cells to acoustic stimuli (Shore et al. 2008; Koehler and Shore 2013b). Interestingly, these studies show that multimodal inputs that do not themselves affect baseline firing in fusiform cells can also regulate acoustic responses. A particularly compelling study has demonstrated that principal cells of mouse DCN can actively suppress neural responses to the sound the mouse makes while licking, using a mechanism that requires trigeminal inputs (Singla et al. 2017). Finally, such multimodal inputs can also provide a trigger for synaptic plasticity, as described in the next section.

4.7 Plasticity

Different forms of long-term plasticity occur at parallel fiber synapses in the DCN molecular layer. Fujino and Oertel (2003) showed in brain slices from mice that patterns of parallel fiber and postsynaptic coactivation that lead to long-term potentiation and depression (LTP and LTD, respectively) in higher brain regions can also trigger plasticity in fusiform and cartwheel cells. Moreover, this plasticity was not seen when auditory inputs were activated, indicating that it is modality specific. Interestingly, a similar profile of modality-specific plasticity was observed in the DCN-like electrosensory lobe (ELL) of mormyrid electric fish (Han et al. 2000). In that system, it was proposed that plasticity served to create "negative images" of corollary discharges needed to mask sensory responses to self-generated sound. Thus, the presence of such plasticity in the DCN suggests that a similar masking function may occur in the mammalian auditory system.

Further parallels in the ELL and DCN were revealed in studies of spike timing dependent plasticity, in which LTP or LTD is triggered by a precise timing relationship between relatively lowfrequency pre- and postsynaptic activity (Tzounopoulos et al. 2004). Timing plots for potentiation and depression as a function of pre/post spike timing differences were opposite for cartwheel and fusiform cells, which indicated that plasticity might result in differential regulation of EPSP and IPSP (inhibitory postsynaptic potential) sequences in fusiform cells. Remarkably, the timing plot for cartwheel cells in particular was identical to that at the Purkinje-like cells in the fish ELL.

The biochemical mechanisms also differed between fusiform and cartwheel cells. Whereas plasticity in fusiform cells required NMDA receptors and postsynaptic CAMKII (calcium-calmodulin-dependent kinase II), plasticity in cartwheel cells depended on Ca^{2+}-dependent release of an endocannabinoid, probably 2-arachidonoyl-glycerol, that acted presynaptically on parallel fiber boutons (Tzounopoulos et al. 2007; Zhao et al. 2009). Thus, a refined, cell type-specific set of mechanisms has evolved in the lowest levels of auditory processing that refines the impact of multimodal input.

Since fusiform and cartwheel cells can be activated by auditory signals, it is possible that plasticity can be induced by critical timing relationships between acoustic and multimodal activity. This prediction has been confirmed in vivo in guinea pigs, in which spike-timing relationships between the different modalities were discovered that are quite similar to those described in the mouse brain slice studies (Koehler and Shore 2013a).

4.8 Neuromodulation

Neuromodulatory systems involve transmitter–receptor interactions that mediate slow changes in excitability or synaptic function. Most such systems use G-protein coupled receptors (GPCRs). The DCN is remarkably rich in this regard. Group I mGluR agonists depressed postsynaptic field responses to parallel fiber stimulation and had a direct depolarizing action on fusiform and cartwheel cells (Molitor and Manis 1997; Fujino and Oertel 2003). Trains of stimuli to parallel fibers induced a slow current that followed the AMPA and NMDA receptor components of the response, and this was blocked by an mGluR antagonist. Thus is it likely that glutamate released by parallel fibers acts on its targets over different time courses. Glutamate release was also modulated by endocannabinoids released by cartwheel cells during depolarization, which may generate a short-term depression of transmission (Zhao et al. 2011). Release of endocannabinoid after parallel fiber stimulation required not only the depolarizing action of glutamate but also a secondary modulation by Zn^{2+}, which is coreleased by parallel fibers along with glutamate and binds to its own GPCR (Perez-Rosello et al. 2013). Moreover, the release of endocannabinoids can be regulated by yet another neuromodulator, acetylcholine, acting on postsynaptic M1/3 mAChR (acetylcholine receptor) in fusiform cells (Zhao and Tzounopoulos 2011). When these receptors are activated, endocannabinoid release is enhanced, resulting in conversion from long-term potentiation to long-term depression.

The DCN is also innervated by dopaminergic, noradrenergic, and serotonergic fibers, each with distinct actions on spontaneous spike activity. Dopamine binds to D3 dopamine receptors at the axon initial segment of cartwheel cells and reduces complex spike generation by inhibiting T-type Ca^{2+} channels (Bender et al. 2010, 2012). As a result, the pattern of spontaneous spike activity in these cells is reorganized from period clusters of spikes to more regular, simple spikes. Noradrenaline acts in a completely different manner by inhibiting spontaneous spiking in cartwheel cells completely, probably by hyperpolarizing the neurons (Kuo and Trussell 2011). Since spontaneous spiking leads to depression of cartwheel glycinergic synapses, the result of inhibiting such spiking is to both enhance glycine release and reduce background synaptic noise. Serotonin, by contrast, does not act on cartwheel cells but rather enhances excitability of fusiform and tuberculoventral cells by activating hyperpolarization and cyclic-nucleotide-gated cation channels (HCN) (Tang and Trussell 2015, 2017). Two receptor classes appear to mediate this same effect, $5HT_{2C}R$ and $5HT_7R$, and the result is selective enhancement of multimodal signaling relative to auditory signaling in fusiform cells. Taken together with the cholinergic effects described above, major circuit elements of the DCN are affected by a wide variety of neuromodulatory transmitters in distinct ways, suggesting the possibility that complex changes in the behavioral state of an animal may affect the earliest levels of auditory processing.

4.9 Summary

The functional organization of the DCN highlights three physiological features: monaural sound source localization, integration of auditory signals with those of other sensory modalities, and the short- and long-term regulation of DCN circuitry with synaptic plasticity and neuromodulators. The sensitivity to sound source location arises from a tonotopic array of principal cells whose activity is sculpted by various subsets of inhibitory neurons. This allows mammals to use high-frequency sounds to localize sound sources in the vertical plane and to distinguish whether they arise from the front or from behind. The ongoing sculpting of activity in the population of principal cells by multimodal sensory signals, through a cerebellum-like granule cell circuit involving six different cell types, indicates a surprising level of complexity for such an early level of sensory processing. The role of such processing may be to differentiate changes in sound, for example, an animal *expects* as it moves its ears or its body from those *unexpected changes* that are of greater biological significance. Lastly, the rich capacity of the DCN for modulation and plasticity implies sensitivity to behavioral state and changes in the environment.

Future studies of the DCN will have to address several anatomical and physiological gaps in our understanding. How is multimodal input distributed in the DCN, onto what cell types, and with what efficacy? How do different inputs vary in their capacity to alter the response to behaviorally relevant acoustic stimuli? What is the behavioral consequence of eliminating relevant input to the DCN? How does the growing wealth of information about DCN circuitry and its plasticity shed light on the role of the DCN in tinnitus and its potential alleviation? Renewed efforts to examine the DCN through modern anatomical and in vivo approaches may provide a way forward.

Acknowledgments Our work was supported by NIH grants DC00176 (DO) and DC004450 (LT).

Compliance with Ethics Requirements Laurence Trussell and Donata Oertel declare that they have no conflict of interest.

References

Adams, J. C. (1986). Neuronal morphology in the human cochlear nucleus. *Archives of Otolaryngology–Head and Neck Surgery, 112*(12), 1253–1261.

Adams, J. C., & Warr, W. B. (1976). Origins of axons in the cat's acoustic striae determined by injection of horseradish peroxidase into severed tracts. *Journal of Comparative Neurology, 170*(1), 107–121. https://doi.org/10.1002/cne.901700108.

Alibardi, L. (1998). Ultrastructural and immunocytochemical characterization of commissural neurons in the ventral cochlear nucleus of the rat. *Annals of Anatomy—Anatomischer Anzeiger, 180*(5), 427–438. https://doi.org/10.1016/S0940-9602(98)80103-9.

Alibardi, L. (2002). Putative inhibitory collicular boutons contact large neurons and their dendrites in the dorsal cochlear nucleus of the rat. *Journal of Submicroscopic Cytology and Pathology, 34*(4), 433–446.

Apostolides, P. F., & Trussell, L. O. (2013). Regulation of interneuron excitability by gap junction coupling with principal cells. *Nature Neuroscience, 16*(12), 1764–1772. https://doi.org/10.1038/nn.3569.

Apostolides, P. F., & Trussell, L. O. (2014a). Chemical synaptic transmission onto superficial stellate cells of the mouse dorsal cochlear nucleus. *Journal of Neurophysiology, 111*(9), 1812–1822. https://doi.org/10.1152/jn.00821.2013.

Apostolides, P. F., & Trussell, L. O. (2014b). Control of interneuron firing by subthreshold synaptic potentials in principal cells of the dorsal cochlear nucleus. *Neuron, 83*(2), 324–330. https://doi.org/10.1016/j.neuron.2014.06.008.

Apostolides, P. F., & Trussell, L. O. (2014c). Superficial stellate cells of the dorsal cochlear nucleus. *Frontiers in Neural Circuits, 8*, 63. https://doi.org/10.3389/fncir.2014.00063.

Baizer, J. S., Wong, K. M., Paolone, N. A., Weinstock, N., Salvi, R. J., Manohar, S., et al. (2014). Laminar and neurochemical organization of the dorsal cochlear nucleus of the human, monkey, cat, and rodents. *Anatomical Record, 297*(10), 1865–1884. https://doi.org/10.1002/ar.23000.

Balakrishnan, V., & Trussell, L. O. (2008). Synaptic inputs to granule cells of the dorsal cochlear nucleus. *Journal of Neurophysiology, 99*(1), 208–219. https://doi.org/10.1152/jn.00971.2007.

Barker, M., Solinski, H. J., Hashimoto, H., Tagoe, T., Pilati, N., & Hamann, M. (2012). Acoustic overexposure increases the expression of VGLUT-2 mediated projections from the lateral vestibular nucleus to the dorsal cochlear nucleus. *PLoS One, 7*(5), e35955. https://doi.org/10.1371/journal.pone.0035955.

Bender, K. J., Ford, C. P., & Trussell, L. O. (2010). Dopaminergic modulation of axon initial segment calcium channels regulates action potential initiation. *Neuron, 68*(3), 500–511. https://doi.org/10.1016/j.neuron.2010.09.026.

Bender, K. J., & Trussell, L. O. (2009). Axon initial segment Ca^{2+} channels influence action potential generation and timing. *Neuron, 61*(2), 259–271. https://doi.org/10.1016/j.neuron.2008.12.004.

Bender, K. J., Uebele, V. N., Renger, J. J., & Trussell, L. O. (2012). Control of firing patterns through modulation of axon initial segment T-type calcium channels. *Journal of Physiology, 590*(1), 109–118. https://doi.org/10.1113/jphysiol.2011.218768.

Benson, T. E., & Brown, M. C. (1990). Synapses formed by olivocochlear axon branches in the mouse cochlear nucleus. *The Journal of Comparative Neurology, 295*(1), 52–70. https://doi.org/10.1002/cne.902950106.

Benson, T. E., & Brown, M. C. (2004). Postsynaptic targets of type II auditory nerve fibers in the cochlear nucleus. *Journal of the Association for Research in Otolaryngology, 5*(2), 111–125. https://doi.org/10.1007/s10162-003-4012-3.

Berglund, A. M., & Brown, M. C. (1994). Central trajectories of type II spiral ganglion cells from various cochlear regions in mice. *Hearing Research, 75*(1–2), 121–130.

Blackburn, C. C., & Sachs, M. B. (1989). Classification of unit types in the anteroventral cochlear nucleus: PST histograms and regularity analysis. *Journal of Neurophysiology, 62*(6), 1303–1329.

Blackstad, T. W., Osen, K. K., & Mugnaini, E. (1984). Pyramidal neurones of the dorsal cochlear nucleus: A Golgi and computer reconstruction study in cat. *Neuroscience, 13*(3), 827–854.

Borges-Merjane, C., & Trussell, L. O. (2015). ON and OFF unipolar brush cells transform multisensory inputs to the auditory system. *Neuron, 85*(5), 1029–1042. https://doi.org/10.1016/j.neuron.2015.02.009.

Brawer, J. R., Morest, D. K., & Kane, E. C. (1974). The neuronal architecture of the cochlear nucleus of the cat. *The Journal of Comparative Neurology, 155*(3), 251–300. https://doi.org/10.1002/cne.901550302.

Brown, M. C., Berglund, A. M., Kiang, N. Y., & Ryugo, D. K. (1988). Central trajectories of type II spiral ganglion neurons. *The Journal of Comparative Neurology, 278*(4), 581–590. https://doi.org/10.1002/cne.902780409.

Bukowska, D. (2002). Morphological evidence for secondary vestibular afferent connections to the dorsal cochlear nucleus in the rabbit. *Cells, Tissues, Organs, 170*(1), 61–68. http://doi.org/47921.

Burian, M., & Gstoettner, W. (1988). Projection of primary vestibular afferent fibres to the cochlear nucleus in the guinea pig. *Neuroscience Letters, 84*(1), 13–17.

Campagnola, L., & Manis, P. B. (2014). A map of functional synaptic connectivity in the mouse anteroventral cochlear nucleus. *Journal of Neuroscience, 34*(6), 2214–2230. https://doi.org/10.1523/JNEUROSCI.4669-13.2014.

Cant, N. B. (1981). The fine structure of two types of stellate cells in the anterior division of the anteroventral cochlear nucleus of the cat. *Neuroscience, 6*(12), 2643–2655.

Chabrol, F. P., Arenz, A., Wiechert, M. T., Margrie, T. W., & DiGregorio, D. A. (2015). Synaptic diversity enables temporal coding of coincident multisensory inputs in single neurons. *Nature Neuroscience, 18*(5), 718–727.

Conlee, J. W., & Kane, E. S. (1982). Descending projections from the inferior colliculus to the dorsal cochlear nucleus in the cat: An autoradiographic study. *Neuroscience, 7*(1), 161–178.

Davis, K. A., Miller, R. L., & Young, E. D. (1996). Effects of somatosensory and parallel-fiber stimulation on neurons in dorsal cochlear nucleus. *Journal of Neurophysiology, 76*(5), 3012–3024.

Diño, M. R., Willard, F. H., & Mugnaini, E. (1999). Distribution of unipolar brush cells and other calretinin immunoreactive components in the mammalian cerebellar cortex. *Journal of Neurocytology, 28*(2), 99–123.

Doucet, J. R., & Ryugo, D. K. (1997). Projections from the ventral cochlear nucleus to the dorsal cochlear nucleus in rats. *The Journal of Comparative Neurology, 385*(2), 245–264.

Doucet, J. R., & Ryugo, D. K. (2003). Axonal pathways to the lateral superior olive labeled with biotinylated dextran amine injections in the dorsal cochlear nucleus of rats. *The Journal of Comparative Neurology, 461*(4), 452–465. https://doi.org/10.1002/cne.10722.

Dunn, M. E., Vetter, D. L., Berrebi, A. S., Krider, H. M., & Mugnaini, E. (1996). The mossy fiber-granule cell-cartwheel cell system in the mammalian cochlear nuclear complex. *Advances in Speech, Hearing and Language Processing, 3*(A), 63–87.

Evans, E. F., & Nelson, P. G. (1973). The responses of single neurones in the cochlear nucleus of the cat as a function of their location and the anaesthetic state. *Experimental Brain Research, 17*(4), 402–427.

Fekete, D. M., Rouiller, E. M., Liberman, M. C., & Ryugo, D. K. (1984). The central projections of intracellularly labeled auditory nerve fibers in cats. *The Journal of Comparative Neurology, 229*(3), 432–450. https://doi.org/10.1002/cne.902290311.

Ferragamo, M. J., Golding, N. L., Gardner, S. M., & Oertel, D. (1998). Golgi cells in the superficial granule cell domain overlying the ventral cochlear nucleus: Morphology and electrophysiology in slices. *The Journal of Comparative Neurology, 400*(4), 519–528.

Ferragamo, M. J., Golding, N. L., & Oertel, D. (1998). Synaptic inputs to stellate cells in the ventral cochlear nucleus. *Journal of Neurophysiology, 79*(1), 51–63.

Flores, E. N., Duggan, A., Madathany, T., Hogan, A. K., Márquez, F. G., Kumar, G., et al. (2015). A non-canonical pathway from cochlea to brain signals tissue-damaging noise. *Current Biology, 25*(5), 606–612. https://doi.org/10.1016/j.cub.2015.01.009.

Frisina, R. D., Smith, R. L., & Chamberlain, S. C. (1990). Encoding of amplitude modulation in the gerbil cochlear nucleus: II. Possible neural mechanisms. *Hearing Research, 44*(2–3), 123–141.

Fujino, K., & Oertel, D. (2003). Bidirectional synaptic plasticity in the cerebellum-like mammalian dorsal cochlear nucleus. *Proceedings of the National Academy of Sciences of the United States of America, 100*(1), 265–270. https://doi.org/10.1073/pnas.0135345100.

Gardner, S. M., Trussell, L. O., & Oertel, D. (1999). Time course and permeation of synaptic AMPA receptors in cochlear nuclear neurons correlate with input. *The Journal of Neuroscience, 19*(20), 8721–8729.

Gardner, S. M., Trussell, L. O., & Oertel, D. (2001). Correlation of AMPA receptor subunit composition with synaptic input in the mammalian cochlear nuclei. *The Journal of Neuroscience, 21*(18), 7428–7437.

Golding, N. L., & Oertel, D. (1997). Physiological identification of the targets of cartwheel cells in the dorsal cochlear nucleus. *Journal of Neurophysiology, 78*(1), 248–260.

Golding, N. L., Robertson, D., & Oertel, D. (1995). Recordings from slices indicate that octopus cells of the cochlear nucleus detect coincident firing of auditory nerve fibers with temporal precision. *The Journal of Neuroscience, 15*(4), 3138–3153.

Han, V. Z., Grant, K., & Bell, C. C. (2000). Reversible associative depression and nonassociative potentiation at a parallel fiber synapse. *Neuron, 27*(3), 611–622.

Hancock, K. E., & Voigt, H. F. (2002a). Intracellularly labeled fusiform cells in dorsal cochlear nucleus of the gerbil. II. Comparison of physiology and anatomy. *Journal of Neurophysiology, 87*(5), 2520–2530.

Hancock, K. E., & Voigt, H. F. (2002b). Intracellularly labeled fusiform cells in dorsal cochlear nucleus of the gerbil. I. Physiological response properties. *Journal of Neurophysiology, 87*(5), 2505–2519.

Heiman-Patterson, T. D., & Strominger, N. L. (1985). Morphological changes in the cochlear nuclear complex in primate phylogeny and development. *Journal of Morphology, 186*(3), 289–306. https://doi.org/10.1002/jmor.1051860306.

Huang, C.-C., Sugino, K., Shima, Y., Guo, C., Bai, S., Mensh, B. D., et al. (2013). Convergence of pontine and proprioceptive streams onto multimodal cerebellar granule cells. *eLife, 2*, e00400. https://doi.org/10.7554/eLife.00400.

Irie, T., Fukui, I., & Ohmori, H. (2006). Activation of GIRK channels by muscarinic receptors and group II metabotropic glutamate receptors suppresses Golgi cell activity in the cochlear nucleus of mice. *Journal of Neurophysiology, 96*(5), 2633–2644. https://doi.org/10.1152/jn.00396.2006.

Itoh, K., Kamiya, H., Mitani, A., Yasui, Y., Takada, M., & Mizuno, N. (1987). Direct projections from the dorsal column nuclei and the spinal trigeminal nuclei to the cochlear nuclei in the cat. *Brain Research, 400*(1), 145–150.

Kanold, P. O., Davis, K. A., & Young, E. D. (2011). Somatosensory context alters auditory responses in the cochlear nucleus. *Journal of Neurophysiology, 105*(3), 1063–1070. https://doi.org/10.1152/jn.00807.2010.

Kanold, P. O., & Young, E. D. (2001). Proprioceptive information from the pinna provides somatosensory input to cat dorsal cochlear nucleus. *Journal of Neuroscience, 21*(19), 7848–7858.

Kecskes, S., Kőszeghy, Á., Szücs, G., Rusznák, Z., Matesz, C., & Birinyi, A. (2013). Three-dimensional reconstruction and quantitative morphometric analysis of pyramidal and giant neurons of the rat dorsal cochlear nucleus. *Brain Structure and Function, 218*(5), 1279–1292. https://doi.org/10.1007/s00429-012-0457-7.

Kennedy, A., Wayne, G., Kaifosh, P., Alviña, K., Abbott, L. F., & Sawtell, N. B. (2014). A temporal basis for predicting the sensory consequences of motor commands in an electric fish. *Nature Neuroscience, 17*(3), 416–422. https://doi.org/10.1038/nn.3650.

Koehler, S. D., & Shore, S. E. (2013a). Stimulus-timing dependent multisensory plasticity in the guinea pig dorsal cochlear nucleus. *PLoS One, 8*(3), e59828. https://doi.org/10.1371/journal.pone.0059828.

Koehler, S. D., & Shore, S. E. (2013b). Stimulus timing-dependent plasticity in dorsal cochlear nucleus is altered in tinnitus. *Journal of Neuroscience, 33*(50), 19647–19656. https://doi.org/10.1523/JNEUROSCI.2788-13.2013.

Kuo, S. P., Lu, H.-W., & Trussell, L. O. (2012). Intrinsic and synaptic properties of vertical cells of the mouse dorsal cochlear nucleus. *Journal of Neurophysiology, 108*(4), 1186–1198.

Kuo, S. P., & Trussell, L. O. (2011). Spontaneous spiking and synaptic depression underlie noradrenergic control of feed-forward inhibition. *Neuron, 71*(2), 306–318. https://doi.org/10.1016/j.neuron.2011.05.039.

Leake, P. A., & Snyder, R. L. (1989). Topographic organization of the central projections of the spiral ganglion in cats. *Journal of Comparative Neurology, 281*(4), 612–629. https://doi.org/10.1002/cne.902810410.

Liberman, M. C. (1982). The cochlear frequency map for the cat: Labeling auditory-nerve fibers of known characteristic frequency. *The Journal of the Acoustical Society of America, 72*(5), 1441–1449.

Liberman, M. C. (1993). Central projections of auditory nerve fibers of differing spontaneous rate. II. Posteroventral and dorsal cochlear nuclei. *The Journal of Comparative Neurology, 327*(1), 17–36. https://doi.org/10.1002/cne.903270103.

Lorente de Nó, R. (1933). Anatomy of the eighth nerve. III. General plan of structure of the primary cochlear nuclei. *Laryngoscope, 43*, 327–350.

Lorente de Nó, R. (1981). *The primary acoustic nuclei*. New York: Raven Press.

Manis, P. B. (1989). Responses to parallel fiber stimulation in the guinea pig dorsal cochlear nucleus in vitro. *Journal of Neurophysiology, 61*(1), 149–161.

Manis, P. B., & Molitor, S. C. (1996). N-methyl-D-aspartate receptors at parallel fiber synapses in the dorsal cochlear nucleus. *Journal of Neurophysiology, 76*(3), 1639–1656.

Manis, P. B., Spirou, G. A., Wright, D. D., Paydar, S., & Ryugo, D. K. (1994). Physiology and morphology of complex spiking neurons in the guinea pig dorsal cochlear nucleus. *The Journal of Comparative Neurology, 348*(2), 261–276. https://doi.org/10.1002/cne.903480208.

Mast, T. E. (1970). Binaural interaction and contralateral inhibition in dorsal cochlear nucleus of the chinchilla. *Journal of Neurophysiology, 33*(1), 108–115.

May, B. J. (2000). Role of the dorsal cochlear nucleus in the sound localization behavior of cats. *Hearing Research, 148*(1–2), 74–87.

May, B. J., Presll, G. S., & Sachs, M. B. (1998). Vowel representations in the ventral cochlear nucleus of the cat: Effects of level, background noise, and behavioral state. *Journal of Neurophysiology, 79*(4), 1755–1767.

Meltzer, N. E., & Ryugo, D. K. (2006). Projections from auditory cortex to cochlear nucleus: A comparative analysis of rat and mouse. *The Anatomical Record Part A: Discoveries in Molecular, Cellular, and Evolutionary Biology, 288(*(4), 397–408. https://doi.org/10.1002/ar.a.20300.

Molitor, S. C., & Manis, P. B. (1997). Evidence for functional metabotropic glutamate receptors in the dorsal cochlear nucleus. *Journal of Neurophysiology, 77*(4), 1889–1905.

Moore, J. K., & Osen, K. K. (1979). The cochlear nuclei in man. *The American Journal of Anatomy, 154*(3), 393–418. https://doi.org/10.1002/aja.1001540306.

Mugnaini, E., Berrebi, A. S., Dahl, A. L., & Morgan, J. I. (1987). The polypeptide PEP-19 is a marker for Purkinje neurons in cerebellar cortex and cartwheel neurons in the dorsal cochlear nucleus. *Archives Italiennes de Biologie, 126*(1), 41–67.

Mugnaini, E., Diño, M. R., & Jaarsma, D. (1997). The unipolar brush cells of the mammalian cerebellum and cochlear nucleus: Cytology and microcircuitry. *Progress in Brain Research, 114*, 131–150.

Mugnaini, E., Warr, W. B., & Osen, K. K. (1980). Distribution and light microscopic features of granule cells in the cochlear nuclei of cat, rat, and mouse. *The Journal of Comparative Neurology, 191*(4), 581–606. https://doi.org/10.1002/cne.901910406.

Muniak, M. A., & Ryugo, D. K. (2014). Tonotopic organization of vertical cells in the dorsal cochlear nucleus of the CBA/J mouse. *The Journal of Comparative Neurology, 522*(4), 937–949. https://doi.org/10.1002/cne.23454.

Musicant, A. D., Chan, J. C., & Hind, J. E. (1990). Direction-dependent spectral properties of cat external ear: New data and cross-species comparisons. *The Journal of the Acoustical Society of America, 87*(2), 757–781.

Needham, K., & Paolini, A. G. (2003). Fast inhibition underlies the transmission of auditory information between cochlear nuclei. *The Journal of Neuroscience, 23*(15), 6357–6361.

Nelken, I., & Young, E. D. (1994). Two separate inhibitory mechanisms shape the responses of dorsal cochlear nucleus type IV units to narrowband and wideband stimuli. *Journal of Neurophysiology, 71*(6), 2446–2462.

Oertel, D., Wright, S., Cao, X.-J., Ferragamo, M., & Bal, R. (2011). The multiple functions of T stellate/multipolar/chopper cells in the ventral cochlear nucleus. *Hearing Research, 276*(1–2), 61–69. https://doi.org/10.1016/j.heares.2010.10.018.

Oertel, D., Wu, S. H., Garb, M. W., & Dizack, C. (1990). Morphology and physiology of cells in slice preparations of the posteroventral cochlear nucleus of mice. *The Journal of Comparative Neurology, 295*(1), 136–154. https://doi.org/10.1002/cne.902950112.

Oertel, D., & Young, E. D. (2004). What's a cerebellar circuit doing in the auditory system? *Trends in Neurosciences, 27*(2), 104–110.

Ohlrogge, M., Doucet, J. R., & Ryugo, D. K. (2001). Projections of the pontine nuclei to the cochlear nucleus in rats. *The Journal of Comparative Neurology, 436*(3), 290–303.

Osen, K. K. (1969). Cytoarchitecture of the cochlear nuclei in the cat. *Journal of Comparative Neurology, 136*(4), 453–484. https://doi.org/10.1002/cne.901360407.

Osen, K. K. (1970). Course and termination of the primary afferents in the cochlear nuclei of the cat. An experimental anatomical study. *Archives Italiennes de Biologie, 108*(1), 21–51.

Osen, K. K., Mugnaini, E., Dahl, A. L., & Christiansen, A. H. (1984). Histochemical localization of acetylcholinesterase in the cochlear and superior olivary nuclei. A reappraisal with emphasis on the cochlear granule cell system. *Archives Italiennes de Biologie, 122*(3), 169–212.

Osen, K. K., Ottersen, O. P., & Storm-Mathisen, J. (1990). Colocalization of glycine-like and GABA-like immunoreactivities: A semiquantitative study of individual neurons in the dorsal cochlear nucleus of cat. In O. P. Ottersen & J. Storm-Mathiesen (Eds.), *Glycine neurotransmission* (pp. 415–451). New York: Wiley.

Perez-Rosello, T., Anderson, C. T., Schopfer, F. J., Zhao, Y., Gilad, D., Salvatore, S. R., et al. (2013). Synaptic Zn^{2+} inhibits neurotransmitter release by promoting endocannabinoid synthesis. *The Journal of Neuroscience, 33*(22), 9259–9272. https://doi.org/10.1523/JNEUROSCI.0237-13.2013.

Ramón y Cajal, S. (1909). *Histologie du Systeme Nerveux de l'Homme et des Vertebres* (Vol. 1A). Paris: A. Maloine.

Rhode, W. S. (1999). Vertical cell responses to sound in cat dorsal cochlear nucleus. *Journal of Neurophysiology, 82*(2), 1019–1032.

Rhode, W. S., & Greenberg, S. (1994). Lateral suppression and inhibition in the cochlear nucleus of the cat. *Journal of Neurophysiology, 71*(2), 493–514.

Rhode, W. S., Smith, P. H., & Oertel, D. (1983). Physiological response properties of cells labeled intracellularly with horseradish peroxidase in cat dorsal cochlear nucleus. *The Journal of Comparative Neurology, 213*(4), 426–447. https://doi.org/10.1002/cne.902130407.

Rice, J. J., May, B. J., Spirou, G. A., & Young, E. D. (1992). Pinna-based spectral cues for sound localization in cat. *Hearing Research, 58*(2), 132–152.

Roberts, M. T., Bender, K. J., & Trussell, L. O. (2008). Fidelity of complex spike-mediated synaptic transmission between inhibitory interneurons. *Journal of Neuroscience, 28*(38), 9440–9450. https://doi.org/10.1523/JNEUROSCI.2226-08.2008.

Roberts, M. T., & Trussell, L. O. (2010). Molecular layer inhibitory interneurons provide feedforward and lateral inhibition in the dorsal cochlear nucleus. *Journal of Neurophysiology, 104*(5), 2462–2473. https://doi.org/10.1152/jn.00312.2010.

Rossi, D. J., Alford, S., Mugnaini, E., & Slater, N. T. (1995). Properties of transmission at a giant glutamatergic synapse in cerebellum: The mossy fiber-unipolar brush cell synapse. *Journal of Neurophysiology, 74*(1), 24–42.

Rubio, M. E., Fukazawa, Y., Kamasawa, N., Clarkson, C., Molnár, E., & Shigemoto, R. (2014). Target- and input-dependent organization of AMPA and NMDA receptors in synaptic connections of the cochlear nucleus. *The Journal of Comparative Neurology, 522*(18), 4023–4042. https://doi.org/10.1002/cne.23654.

Rubio, M. E., Gudsnuk, K. A., Smith, Y., & Ryugo, D. K. (2008). Revealing the molecular layer of the primate dorsal cochlear nucleus. *Neuroscience, 154*(1), 99–113. https://doi.org/10.1016/j.neuroscience.2007.12.016.

Rubio, M. E., & Wenthold, R. J. (1997). Glutamate receptors are selectively targeted to postsynaptic sites in neurons. *Neuron, 18*(6), 939–950.

Ruggero, M. A. (1992). Physiology and coding of sound in the auditory nerve. In A. N. Popper & R. R. Fay (Eds.), *The mammalian auditory pathway: Neurophysiology* (pp. 34–93). New York: Springer.

Schofield, B. R., & Coomes, D. L. (2005). Auditory cortical projections to the cochlear nucleus in guinea pigs. *Hearing Research, 199*(1–2), 89–102. https://doi.org/10.1016/j.heares.2004.08.003.

Schofield, B. R., Coomes, D. L., & Schofield, R. M. (2006). Cells in auditory cortex that project to the cochlear nucleus in guinea pigs. *Journal of the Association for Research in Otolaryngology, 7*(2), 95–109. https://doi.org/10.1007/s10162-005-0025-4.

Schofield, B. R., Motts, S. D., Mellott, J. G., & Foster, N. L. (2014). Projections from the dorsal and ventral cochlear nuclei to the medial geniculate body. *Frontiers in Neuroanatomy, 8*, 10. https://doi.org/10.3389/fnana.2014.00010.

Shore, S. E., Helfert, R. H., Bledsoe, S. C., Altschuler, R. A., & Godfrey, D. A. (1991). Descending projections to the dorsal and ventral divisions of the cochlear nucleus in guinea pig. *Hearing Research, 52*(1), 255–268.

Shore, S. E., Koehler, S., Oldakowski, M., Hughes, L. F., & Syed, S. (2008). Dorsal cochlear nucleus responses to somatosensory stimulation are enhanced after noise-induced hearing loss. *The European Journal of Neuroscience, 27*(1), 155–168. https://doi.org/10.1111/j.1460-9568.2007.05983.x.

Shore, S. E., Vass, Z., Wys, N. L., & Altschuler, R. A. (2000). Trigeminal ganglion innervates the auditory brainstem. *The Journal of Comparative Neurology, 419*(3), 271–285.

Singla, S., Dempsey, C., Warren, R., Enikolopov, A. G., & Sawtell, N. B. (2017). A cerebellum-like circuit in the auditory system cancels responses to self-generated sounds. *Nature Neuroscience, 20*, 943. https://doi.org/10.1038/nn.4567.

Smith, P. H., Massie, A., & Joris, P. X. (2005). Acoustic stria: Anatomy of physiologically characterized cells and their axonal projection patterns. *The Journal of Comparative Neurology, 482*(4), 349–371. https://doi.org/10.1002/cne.20407.

Smith, P. H., & Rhode, W. S. (1985). Electron microscopic features of physiologically characterized, HRP-labeled fusiform cells in the cat dorsal cochlear nucleus. *The Journal of Comparative Neurology, 237*(1), 127–143.

Smith, P. H., & Rhode, W. S. (1989). Structural and functional properties distinguish two types of multipolar cells in the ventral cochlear nucleus. *The Journal of Comparative Neurology, 282*(4), 595–616. https://doi.org/10.1002/cne.902820410.

Spirou, G. A., Davis, K. A., Nelken, I., & Young, E. D. (1999). Spectral integration by type II interneurons in dorsal cochlear nucleus. *Journal of Neurophysiology, 82*(2), 648–663.

Spirou, G. A., & Young, E. D. (1991). Organization of dorsal cochlear nucleus type IV unit response maps and their relationship to activation by bandlimited noise. *Journal of Neurophysiology, 66*(5), 1750–1768.

Tang, Z.-Q., & Trussell, L. O. (2015). Serotonergic regulation of excitability of principal cells of the dorsal cochlear nucleus. *Journal of Neuroscience, 35*(11), 4540–4551. https://doi.org/10.1523/JNEUROSCI.4825-14.2015.

Tang, Z.-Q., & Trussell, L. O. (2017). Serotonergic modulation of sensory representation in a central multisensory circuit is pathway specific. *Cell Reports, 20*(8), 1844–1854. https://doi.org/10.1016/j.celrep.2017.07.079.

Tzounopoulos, T., Kim, Y., Oertel, D., & Trussell, L. O. (2004). Cell-specific, spike timing-dependent plasticities in the dorsal cochlear nucleus. *Nature Neuroscience, 7*(7), 719–725. https://doi.org/10.1038/nn1272.

Tzounopoulos, T., Rubio, M. E., Keen, J. E., & Trussell, L. O. (2007). Coactivation of pre- and postsynaptic signaling mechanisms determines cell-specific spike-timing-dependent plasticity. *Neuron, 54*(2), 291–301. https://doi.org/10.1016/j.neuron.2007.03.026.

Voigt, H. F., & Young, E. D. (1980). Evidence of inhibitory interactions between neurons in dorsal cochlear nucleus. *Journal of Neurophysiology, 44*(1), 76–96.

Weinberg, R. J., & Rustioni, A. (1987). A cuneocochlear pathway in the rat. *Neuroscience, 20*(1), 209–219.

Wenthold, R. J. (1987). Evidence for a glycinergic pathway connecting the two cochlear nuclei: An immunocytochemical and retrograde transport study. *Brain Research, 415*(1), 183–187.

Wickesberg, R. E., & Oertel, D. (1988). Tonotopic projection from the dorsal to the anteroventral cochlear nucleus of mice. *The Journal of Comparative Neurology, 268*(3), 389–399. https://doi.org/10.1002/cne.902680308.

Wickesberg, R. E., & Oertel, D. (1990). Delayed, frequency-specific inhibition in the cochlear nuclei of mice: A mechanism for monaural echo suppression. *Journal of Neuroscience, 10*(6), 1762–1768.

Wickesberg, R. E., Whitlon, D., & Oertel, D. (1991). Tuberculoventral neurons project to the multipolar cell area but not to the octopus cell area of the posteroventral cochlear nucleus. *The Journal of Comparative Neurology, 313*(3), 457–468. https://doi.org/10.1002/cne.903130306.

Wickesberg, R. E., Whitlon, D., & Oertel, D. (1994). In vitro modulation of somatic glycine-like immunoreactivity in presumed glycinergic neurons. *The Journal of Comparative Neurology, 339*(3), 311–327. https://doi.org/10.1002/cne.903390302.

Wouterlood, F. G., & Mugnaini, E. (1984). Cartwheel neurons of the dorsal cochlear nucleus: A Golgi-electron microscopic study in rat. *The Journal of Comparative Neurology, 227*(1), 136–157. https://doi.org/10.1002/cne.902270114.

Wouterlood, F. G., Mugnaini, E., Osen, K. K., & Dahl, A. L. (1984). Stellate neurons in rat dorsal cochlear nucleus studies with combined Golgi impregnation and electron microscopy: Synaptic connections and mutual coupling by gap junctions. *Journal of Neurocytology, 13*(4), 639–664.

Xie, R. & Manis, P.B. (2017). Radiate and planar multipolar neurons of the mouse anteroventral cochlear nucleus: Intrinsic excitability and characterization of their auditory nerve input. Frontiers in Neural Circuits, 1,77. https://doi:10.3389/fncir.2017.00077.

Yaeger, D. B., & Trussell, L. O. (2015). Single granule cells excite Golgi cells and evoke feedback inhibition in the cochlear nucleus. *The Journal of Neuroscience, 35*(11), 4741–4750.

Yaeger, D. B., & Trussell, L. O. (2016). Auditory Golgi cells are interconnected predominantly by electrical synapses. *Journal of Neurophysiology, 116*(2), 540–551.

Young, E. D. (1980). Identification of response properties of ascending axons from dorsal cochlear nucleus. *Brain Research, 200*(1), 23–37.

Young, E. D., Nelken, I., & Conley, R. A. (1995). Somatosensory effects on neurons in dorsal cochlear nucleus. *Journal of Neurophysiology, 73*(2), 743–765.

Young, E. D., & Oertel, D. (2010). Cochlear nucleus. In G. M. Shepherd & S. Grillner (Eds.), *Handbook of brain microcircuits* (pp. 215–223). New York: Oxford University Press.

Young, E. D., & Voigt, H. F. (1982). Response properties of type II and type III units in dorsal cochlear nucleus. *Hearing Research, 6*(2), 153–169.

Zeng, C., Shroff, H., & Shore, S. E. (2011). Cuneate and spinal trigeminal nucleus projections to the cochlear nucleus are differentially associated with vesicular glutamate transporter-2. *Neuroscience, 176*, 142–151. https://doi.org/10.1016/j.neuroscience.2010.12.010.

Zhan, X., Pongstaporn, T., & Ryugo, D. K. (2006). Projections of the second cervical dorsal root ganglion to the cochlear nucleus in rats. *The Journal of Comparative Neurology, 496*(3), 335–348. https://doi.org/10.1002/cne.20917.

Zhang, S., & Oertel, D. (1993a). Cartwheel and superficial stellate cells of the dorsal cochlear nucleus of mice: Intracellular recordings in slices. *Journal of Neurophysiology, 69*(5), 1384–1397.

Zhang, S., & Oertel, D. (1993b). Giant cells of the dorsal cochlear nucleus of mice: Intracellular recordings in slices. *Journal of Neurophysiology, 69*(5), 1398–1408.

Zhang, S., & Oertel, D. (1993c). Tuberculoventral cells of the dorsal cochlear nucleus of mice: Intracellular recordings in slices. *Journal of Neurophysiology, 69*(5), 1409–1421.

Zhang, S., & Oertel, D. (1994). Neuronal circuits associated with the output of the dorsal cochlear nucleus through fusiform cells. *Journal of Neurophysiology, 71*(3), 914–930.

Zhao, Y., Rubio, M. E., & Tzounopoulos, T. (2009). Distinct functional and anatomical architecture of the endocannabinoid system in the auditory brainstem. *Journal of Neurophysiology, 101*(5), 2434–2446. https://doi.org/10.1152/jn.00047.2009.

Zhao, Y., Rubio, M. E., & Tzounopoulos, T. (2011). Mechanisms underlying input-specific expression of endocannabinoid-mediated synaptic plasticity in the dorsal cochlear nucleus. *Hearing Research, 279*(1–2), 67–73. https://doi.org/10.1016/j.heares.2011.03.007.

Zhao, Y., & Tzounopoulos, T. (2011). Physiological activation of cholinergic inputs controls associative synaptic plasticity via modulation of endocannabinoid signaling. *The Journal of Neuroscience, 31*(9), 3158–3168. https://doi.org/10.1523/JNEUROSCI.5303-10.2011.

Zhou, J., Nannapaneni, N., & Shore, S. (2007). Vesicular glutamate transporters 1 and 2 are differentially associated with auditory nerve and spinal trigeminal inputs to the cochlear nucleus. *The Journal of Comparative Neurology, 500*(4), 777–787. https://doi.org/10.1002/cne.21208.

Chapter 5
Integration of Synaptic and Intrinsic Conductances Shapes Microcircuits in the Superior Olivary Complex

Conny Kopp-Scheinpflug and Ian D. Forsythe

Abstract The superior olivary complex is a group of interconnected brainstem nuclei that receive and integrate binaural auditory input. Each nucleus forms part of local microcircuits subserving multiple complimentary roles in auditory processing, including sound localization, detection of signals in noise, and gap detection. The three nuclei of the trapezoid body (medial, lateral, and ventral) provide indirect inhibitory local projections that are integrated with direct excitatory inputs from the cochlear nuclei at the three output nuclei (the medial and lateral superior olivary nuclei and the superior paraolivary nucleus). Each nucleus expresses a different spectrum of ionic conductances that determine the intrinsic excitability of their principal neurons and adapt how the microcircuit integrates the binaural excitatory and inhibitory synaptic inputs. Specialized synapses, such as the calyx of Held, help maintain temporal information and minimize jitter, while the location of synapses on specific dendrites or somatic regions provides further refinement of the microcircuit. This chapter also includes how the principal neurons of each nucleus express differing densities of ionic conductance by which they exhibit a unique threshold, action potential waveform, and characteristic firing properties. A broad perspective will be provided on how each of these functional elements come together to sculpt the local neuronal microcircuit into performing specific physiological roles for interaural timing discrimination, interaural level discrimination, and gap detection.

Keywords Calyx of Held · Lateral nucleus of the trapezoid body · Lateral superior olive · Medial nucleus of the trapezoid body · Medial superior olive · Superior paraolivary nucleus · Ventral nucleus of the trapezoid body

---

C. Kopp-Scheinpflug (✉)
Division of Neurobiology, Department Biology II, Ludwig-Maximilians-University Munich, Planegg-Martinsried, Germany
e-mail: cks@bio.lmu.de

I. D. Forsythe
Department of Neuroscience, Psychology and Behaviour, University of Leicester, Leicester, UK
e-mail: idf@le.ac.uk

© Springer International Publishing AG 2018

D. L. Oliver et al. (eds.), *The Mammalian Auditory Pathways*,
Springer Handbook of Auditory Research 65,
https://doi.org/10.1007/978-3-319-71798-2_5

5.1 Introduction

Many areas of the brain have local microcircuits that involve excitatory and inhibitory synaptic transmission to achieve specific computations that are then transmitted to downstream targets. Generally, the projection neurons in microcircuits are excitatory and the interneurons are inhibitory. Multiple forms of synaptic plasticity shape the balance between these interacting inputs to fine-tune the signal processing. The biophysical properties of voltage-gated ionic conductances within each respective neuron regulate the integration of these synaptic inputs and determine the overall firing pattern of the target neuron.

The superior olivary complex (SOC; see Table 5.1 for all abbreviations) in the mammalian brainstem is a cluster of interacting nuclei serving related aspects of auditory processing that require both temporal precision and binaural integration (Fig. 5.1). In this chapter, the medial (MNTB), ventral (VNTB), and the lateral (LNTB) nuclei of the trapezoid body will be introduced as the sources for local inhibition. The medial and lateral superior olive (MSO and LSO, respectively) as well as the superior paraolivary nucleus (SPN) are highly specialized processors of binaural information that integrate this inhibition with excitation originating from the cochlear nuclei. The common features of these microcircuits include fast conducting myelinated axons, giant synapses, low threshold voltage-gated potassium currents, and precise physical locations of presynaptic terminals that target somata or dendritic fields. Each of the SOC nuclei has dominant principal neurons, but other neuronal types (nonprincipal) are also present. These nonprincipal neurons are not always local interneurons, and they can serve other physiological functions in auditory signal processing, for instance, as efferent projection neurons in the olivocochlear system.

Rodent models such as mouse, rat, and gerbil are most commonly used, and although all of the SOC nuclei can be recognized in each of these species, there are differences in size and specialization. The MSO, for example, is particularly small in mammals of small head size that possess high frequency hearing, such as mice, but the MSO is considerably more significant in the gerbil (Fischl et al. 2016). Another example is the MNTB, which is very prominent in all rodents, but it appears more dispersed in humans (Kulesza and Grothe 2015), perhaps due to the vast size of the corticospinal tract in humans.

The characteristic output of the SOC is shaped by local microcircuits that are integrating excitation with inhibitory inputs arising from within the SOC and by intrinsic conductances that differ among each of the SOC nuclei. This chapter develops the concept of the SOC as a unified structure in which projection neurons in the LSO, MSO, and SPN serve specific physiological functions and where inhibitory interneurons are grouped together in the nuclei of the trapezoid body (NTBs). Table 5.2 is a summary of the general properties of the principal cells in each of the SOC nuclei.

Table 5.1 Abbreviations

AMPA AMPAR	[alpha]-amino-3-hydroxy-5-methyl-4-isoxazole propionic acid AMPA receptor
AVCN	Anteroventral cochlear nucleus
CaV DMPO E	Voltage-gated calcium channels Dorsomedial paraolivary nucleus Excitation
E_{Cl}	Chloride reversal potential
EPSC	Excitatory postsynaptic current
EPSP	Excitatory postsynaptic potentials
ERG	Ether-á-go-go-related gene channels
GABA	Gamma-aminobutyric acid
GluA	Glutamate receptor subunit types
GLYT2	Glycine transporter type 2
HCN	Hyperpolarization activated nonspecific cation channel
I IC	Inhibition Inferior colliculus
I_H	Hyperpolarization-activated nonspecific cationic current
ILD	Interaural level differences
IPSC	Inhibitory postsynaptic current
IPSP	Inhibitory postsynaptic potentials
I_{TCa}	T-type calcium current
ITD	Interaural time differences
KCC2	Potassium-chloride co-transporter type 2
Kv	Voltage-gated potassium channels
LNTB	Lateral nucleus of the trapezoid body
LOC	Lateral olivocochlear circuit
LSO	Lateral superior olive
MNTB	Medial nucleus of the trapezoid body
MOC	Medial olivocochlear circuit
MSO	Medial superior olive
Nav	Voltage-gated sodium channels
NMDA NMDAR	N-methyl-D-aspartate NMDA receptor
nNOS	Neuronal nitric oxide synthase
NO	Nitric oxide
NTBs	Nuclei of the trapezoid body
PSTH	Post-stimulus time histograms
SOC	Superior olivary complex
SPN	Superior paraolivary nucleus
VCN	Ventral cochlear nucleus
VNTB	Ventral nucleus of the trapezoid body

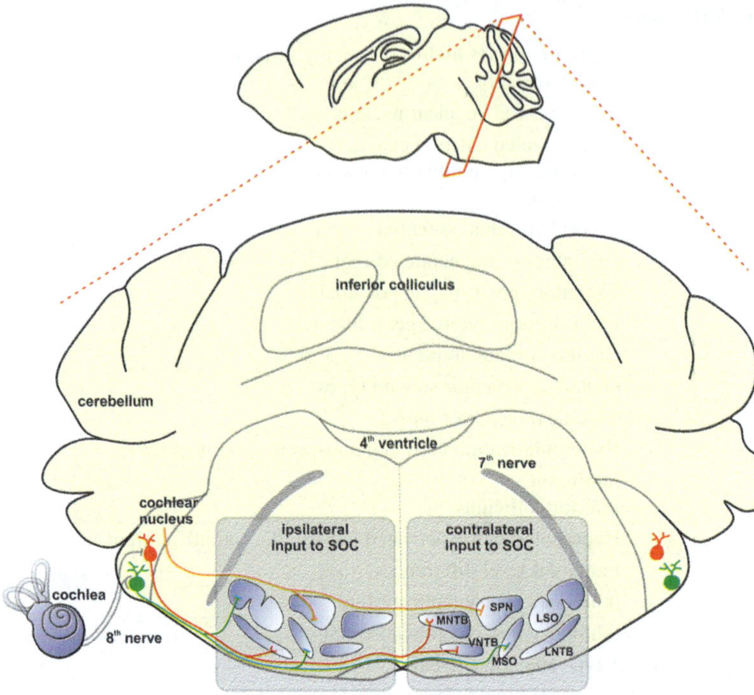

Fig. 5.1 Schematic of the superior olivary complex (*SOC*) (*Top*) Sagittal view of a rodent brain that depicts the coronal section shown below. The SOC is a mirrored structure (*two gray-shaded boxes*) on both sides of the brainstem that receives ipsilateral and contralateral input from outside the SOC. The different colored cells in the cochlear nucleus represent spherical bushy cells (*green*) and globular bushy cells (*red*) and their respective projections to the SOC. Inputs into the SOC of uncertain origin are drawn in *orange*. (*LNTB*, lateral nucleus of trapezoid body; *LSO*, lateral superior olive; *MNTB*, medial nucleus of trapezoid body; *MSO*, medial superior olive; *SPN*, superior paraolivary nucleus; *VNTB*, ventral nucleus of trapezoid body)

5.2 The Nuclei of the Trapezoid Body Provide Local Inhibition

There are three nuclei of the trapezoid body (Fig. 5.2): the medial nucleus of the trapezoid body (MNTB), the ventral nucleus of the trapezoid body (VNTB), and the lateral nucleus of the trapezoid body (LNTB). For the purposes of this chapter, the three nuclei together are called the NTBs. The majority of the neurons in these nuclei provide glycinergic projections (Albrecht et al. 2014; Roberts et al. 2014). Cholinergic (acetylcholine) and some GABAergic (gamma-aminobutyric acid) neurons have also been described, which are included in Sects. 5.2.1 and 5.2.3.

The task of the NTBs is to deliver fast inhibition to their target nuclei, which must arrive in temporal register with the excitatory inputs that originate directly

Table 5.2 General properties of principal neurons in the superior olivary complex[a] represented as a consensus of neuronal properties: resting properties, in vivo response to sound,[b] and in vitro firing

	Resting state				Response to Sound		Current step	
	RMP (mV)	Input Res. (MOhms)	Spont. firing? *in vivo*	*in vitro*	Sound stim.	PSTH to a sound	Current Inj AP Firing	AP firing rate Max. (Hz)
MNTB	-65 to -75	120-200	Yes	No	E0	Primary + notch	Single	~700
MSO	-55 to -60	5-20	Yes	No	EE EI, IE	Primary +/- notch	Single	~400
SPN	-50 to -60	50-70	Yes	No	IE, I0, II	Off-Period /chopper	Train	~200
LSO	-55 to -62	15-30	Yes	No	IE	chopper	Single & Train	~200
LNTB	-55 to -65	55-70	Yes	No	IE, 0E, EE	chopper	Train	~400
VNTB	-60 to -65	300-400	Yes	No	EE	chopper	Train	~125

Data were taken at developmental times beyond opening of the auditory canal (>P12 in rodents) and from studies on the most mature animals (P16-P30) when available

[a]LNTB, lateral nucleus of trapezoid body; LSO, lateral superior olive; MNTB, medial nucleus of trapezoid body; MSO, medial superior olive; SPN, superior paraolivary nucleus; VNTB, ventral nucleus of trapezoid body

[b]Responses to sound indicate nature of response to sound stimulus for contralateral, ipsilateral stimuli; 0, no stimulus on that side; E excitation; I, inhibition; PSTH, peri-stimulus time histogram

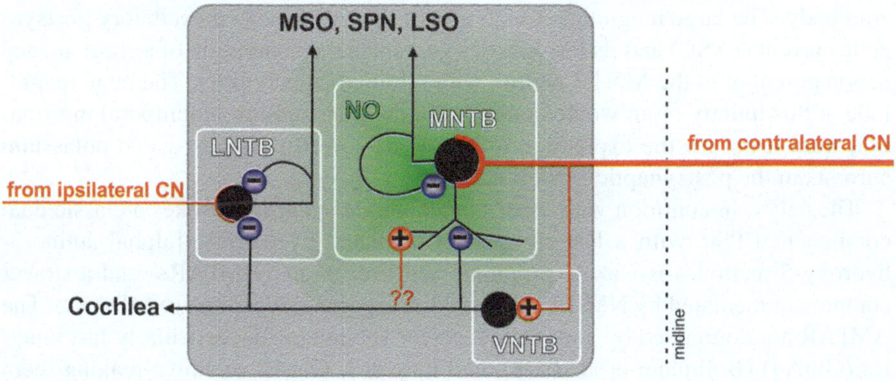

Fig. 5.2 The nuclei of the trapezoid body (*NTBs*) form an inhibitory hub within the superior olivary complex (*SOC*). The medial nucleus (*MNTB*) and the lateral nucleus (*LNTB*) both receive large somatic (calyceal) excitatory synapses as well as feedforward (via the ventral nucleus, *VNTB*) and recurrent feedback inhibition. MNTB neurons release the volume transmitter nitric oxide (*NO*) in an activity dependent manner that acts in addition to the wired connections. (+ excitatory; – inhibitory; *CN*, cochlear nucleus; *SPN*, superior paraolivary nucleus)

from the cochlear nuclei. Thus, the NTBs have an additional synaptic delay of about 0.5 ms, which must be accommodated in the overall network computation if the inhibition is to integrate with the more direct excitatory projections. The afferent pathways into the NTBs are characterized by large caliber, highly myelinated axons that provide fast conduction velocity (gerbil: Ford et al. 2015; Seidl and Rubel 2016; mouse: Sinclair et al. 2017a). The fibers innervating the MNTB also provide axon collaterals into the VNTB on the same side. Additional anatomical specializations for speed are found in the MNTB (Held 1893; Forsythe 1994) and the LNTB (Spirou and Berrebi 1996), where the excitatory input is mediated by large calyx-type somatic synapses arising from either the contralateral (for the MNTB) or the ipsilateral (for the LNTB) cochlear nucleus. Interestingly, glycinergic neurons in all three NTBs also receive glycinergic inhibition either from within the same nucleus or from other NTBs.

5.2.1 The Medial Nucleus of the Trapezoid Body

The MNTB is probably most famous as the target of the calyx of Held synapses, which are often referred to as the largest synapses in the mammalian central nervous system. Each MNTB neuron receives only one calyx and no convergence with other inputs is required to trigger a postsynaptic action potential. The calyx forms around the cell body, covering over half of the surface area (Forsythe 1994; Holcomb et al. 2013). The calyx possesses hundreds of transmitter release sites that are triggered to release glutamate by a single action potential arriving along the axon in the trapezoid body. The large magnitude (3–20 nA) of the calyx-evoked excitatory postsynaptic current (EPSC) and its fast kinetics guarantee the generation of a short latency action potential in the MNTB neuron with minimal latency jitter. The huge magnitude of this unitary input would compromise the transmission of temporal information were it not for the expression of a suite of powerful voltage-gated potassium currents in the postsynaptic MNTB neuron.

The calyx, in common with other glutamatergic synapses, evokes a classic dual component EPSC with a fast component mediated by AMPA ([alpha]-amino-3-hydroxy-5-methyl-4-isoxazole propionic acid) receptors (AMPARs) and a slower component mediated by NMDA (N-methyl-D-aspartate) receptors (NMDARs). The AMPAR are dominated by glutamate receptor subunits with particularly fast kinetics (GluA4) (Bollmann et al. 1998), and they lack GluA2 subunits, making them calcium permeable (Koike-Tani et al. 2005). The NMDAR component shows a strong developmental profile with an amplitude that peaks around the onset of hearing and then declines. As the synapse matures, the decay kinetics accelerate from about 2 ms to 0.5 ms for the AMPAR component (Koike-Tani et al. 2005) and from about 50 ms to less than 15 ms for the NMDAR component (Steinert et al. 2010). Interestingly, the somatic location of the calyx means that the synaptic calcium influx is only microns from the nucleus, suggesting the intriguing possibility that synaptic calcium may have very direct influences on gene expression. This contrasts with many other excitatory synapses that form on distal dendrites. In addition to the calyx input, there are slow small-amplitude excitatory inputs on MNTB principal neurons (Hamann et al. 2003).

Principal neurons of the MNTB have an almost spherical soma with one primary dendrite reaching far ventral toward the VNTB, which provides glycinergic inhibition to the MNTB (see Sect. 5.2.3). The MNTB neurons send glycinergic axon collaterals to neighboring MNTB neurons as well as to VNTB neurons (Kuwabara and Zook 1991; Dondzillo et al. 2016). Inhibition in the MNTB develops with a slight delay compared to the excitatory input around the 4th postnatal week (Awatramani et al. 2005) and, once matured, helps maintain sharp spectral tuning (Kopp-Scheinpflug et al. 2003; Koka and Tollin 2014). However, the main function of inhibition in the MNTB is to further increase temporal precision, especially during later time points in an ongoing stimulation (Kopp-Scheinpflug et al. 2008; Tolnai et al. 2008).

Voltage-gated ion channels have been characterized at both presynaptic (calyx of Held) and postsynaptic somata of the MNTB principle neurons but have been less explored at the other nuclei of the SOC. The calyx of Held arises from the axons of globular bushy cells in the anteroventral cochlear nucleus (AVCN); further details can be obtained from several reviews (von Gersdorff and Borst 2002; Schneggenburger and Forsythe 2006). In mature animals, action potential invasion of the calyx heminode subsequently activates P/Q-type voltage-gated calcium channels (CaV2.1 encoded by the Cacna1a gene), which are located at transmitter release sites and trigger exocytosis of synaptic vesicles that contain the excitatory neurotransmitter glutamate. A developmental switch in CaV channel expression occurs at the calyx around onset of hearing (P12) so that the P/Q-type dominates and N-type channels (CaV2.2 encoded by the Cacna1b gene) are no longer significantly coupled to exocytosis.

At the downstream inhibitory synapses of the MNTB in the LSO, L-type (CaV1.2), N-type, and P/Q-type channels mediate exocytosis in young mice (Giugovaz-Tropper et al. 2011; Alamilla and Gillespie 2013), but this is refined to predominantly P/Q channels after the onset of hearing with minor contributions from L- and N-type channels (Alamilla and Gillespie 2013). A similar P/Q > N calcium channel contribution to trigger exocytosis has been noted for the inhibitory postsynaptic current (IPSC) in the MSO of rat (Barnes-Davies et al. 2001). There is an interesting dependence on the expression of CaV1.3 for development of the brainstem auditory pathway. The transgenic knockout is deaf (due to lack of Cav1.3 in the inner hair cell), and this is associated with structural defects in the auditory brainstem, particularly the LSO (Hirtz et al. 2011). Postsynaptic calcium influx in the MNTB is mediated by CaV1.2, CaV2.1, CaV2.2, and CaV2.3 (L-, P/Q-, N-, and R-types) (Barnes-Davies et al. 2001; Tozer et al. 2012). There is little expression of CaV3 (T-type) calcium channels in the SOC, but SPN principal neurons are an exception (Felix et al. 2011; Kopp-Scheinpflug et al. 2011) where T-type channels contribute to driving rebound spiking.

Voltage-gated sodium channels (Nav) mediate the axonal action potential and salutatory conduction is sustained in conjunction with Kv3 and Kv1 voltage-gated potassium channels (located at nodes of Ranvier and juxtaparanodal regions, respectively). The Nav1.1 sodium channels are expressed at the postsynaptic MNTB neuron, and there is evidence for Nav1.6 subunits in young mice or following induction of deafness (Leao et al. 2006). Resurgent sodium currents have also been described in the presynaptic calyx of Held (Kim et al. 2010).

The soma and initial segment are densely packed with potassium channels from multiple families (Kv1: Wang et al. 1993; Brew and Forsythe 1995; Kv2: Johnston et al. 2008; Tong et al. 2013; Kv3: Wang et al. 1998; Li et al. 2001; Kv4: Oertel 2009; Johnston et al. 2010). Other related channels include the slow hyperpolarization-activated nonspecific cation channel (I_H: Banks et al. 1993) dominated by HCN2 (Koch et al. 2004; Kopp-Scheinpflug et al. 2015) and HCN4 subunits (Leao et al. 2005), sodium-dependent potassium channels (slick/slack: Yang et al. 2007), and ether-á-go-go-related gene channels (ERG: Hardman and Forsythe 2009). The MNTB displays a medial to lateral high- to low-frequency tonotopic organization, which is accompanied by subtle spatial gradients in potassium channel expression. The best-established gradient is for Kv3.1 (Li et al. 2001; Brew and Forsythe 2005). Each of the major voltage-gated potassium channels serves key functions in determining firing threshold and action potential repolarization: Kv1 channels determine threshold and suppress multiple action potential firing, Kv2 speeds recovery of voltage-gated sodium channels from inactivation during high frequency firing, and Kv3/Kv4 accelerate action potential repolarization. As a result, only fast and strong depolarizations will reach action potential threshold without being outrun and dampened by the potassium conductances. These features allow MNTB neurons to follow the firing pattern of their presynaptic calyceal synaptic input with high temporal precision up to very high firing frequencies. All together these conductances ensure a single, fast, temporally precise action potential in response to each presynaptic action potential.

The output of MNTB neurons can be measured as the number of and timing of action potentials from either intracellular or extracellular recordings. Intracellular or patch-clamp recordings allow the study of subthreshold synaptic inputs and the voltage-gated conductances underlying the intrinsic excitability of MNTB neurons (Banks and Smith 1992; Brew and Forsythe 1995). Extracellular or cell-attached recordings, on the other hand, allow simultaneous monitoring of presynaptic and postsynaptic firing activity (avoiding the dialysis inherent in whole-cell patching) and thus provide information about firing and the security of information transmission between the calyx synapse and the MNTB principal neuron. This extracellular assessment of synaptic transmission is only possible because of the large size of the synapse that generates a *prepotential* (or field potential) due to the presynaptic action potential (Guinan and Li 1990; Kopp-Scheinpflug et al. 2003). Thus, the extracellular potential recorded from MNTB neurons has a compound waveform: a prepotential, then a synaptic delay followed by the postsynaptic action potential.

The axon of each MNTB principal neuron gives a glycinergic inhibitory projection with multiple diverging collaterals into several nuclei, including the MSO and LSO, the SPN, and probably the ventral nucleus of the lateral lemniscus (Banks and Smith 1992; Sommer et al. 1993). As a whole, the MNTB projection exhibits a high degree of convergence, in other words, the projections of multiple MNTB neurons converge onto single target neurons with about 3–6 inputs onto LSO neurons (Kim and Kandler 2003), about 1–4 inputs for MSO neurons (Couchman et al. 2010; Roberts et al. 2014), and about 10–15 inputs for SPN neurons (Kopp-Scheinpflug et al. 2011). Each of these different target nuclei is specialized to use this inhibition in different computations.

5.2.2 *The Lateral Nucleus of the Trapezoid Body*

The LNTB is situated mediolateral to the LSO (Spirou et al. 1998; Albrecht et al. 2014) and contains glycinergic neurons. Fate mapping indicates that LNTB and MNTB neurons have different origins (Jalabi et al. 2013), but similar to the MNTB, LNTB neurons also receive excitatory input from the ventral cochlear nucleus (VCN) via giant calyceal somatic synapses (Spirou et al. 1998). These calyceal synapses provide fast synaptic transmission and minimal temporal jitter, allowing the main glycinergic inhibition to reach their MSO target neurons in temporal register with, or even before, the excitatory inputs (Myoga et al. 2014; Roberts et al. 2014). The LNTB also has intrinsic recurrent glycinergic axon collaterals (Roberts et al. 2014). This inhibition arrives with a delay that suggests a disynaptic pathway (Roberts et al. 2014) and may serve to improve temporal contrast between the stimulus onset and tonic components.

LNTB neurons have a fast membrane time constant of around 4 ms. In response to hyperpolarizing current injections, LNTB neurons exhibit a strong sag in membrane potential indicative of an I_H current. This contributes to a more depolarized resting membrane potential of about −60 mV (Roberts et al. 2014) compared to the −70 mV described for MNTB (see Table 5.2). In response to sustained depolarizing current injections, LNTB neurons fire multiple action potentials, suggesting little contribution from low voltage-activated potassium currents of the Kv1 family. Multiple action potentials in response to a sustained depolarization might constrain the LNTB neurons to fire repetitively in response to high-frequency trains of depolarizing stimuli (or fast transients in acoustic stimuli). Indeed, an accommodation in action potential firing rates has been shown for LNTB neurons for high-frequency trains (Roberts et al. 2014).

5.2.3 *The Ventral Nucleus of the Trapezoid Body*

The VNTB is located medioventral to the MNTB. Its neurons can be divided into at least two populations: the choline acetyltransferase positive neurons that are part of the medial olivocochlear (MOC) efferent feedback system, and the glycinergic neurons that serve as inhibitory interneurons within the SOC microcircuits together with neurons of the LNTB and MNTB. The medial olivocochlear circuit system sends a protective projection to the outer hair cells in the cochlea (Darrow et al. 2006). For further information on efferent feedback in the auditory system, see the chapters by Ryugo et al. (2011).

In response to hyperpolarizing current injections, VNTB neurons do not show a membrane potential sag (Sinclair et al. 2017b) and thus are comparable to the efferent neurons of the lateral olivocochlear circuit (LOC), which are intermingled in the LSO (Sterenborg et al. 2010). In response to small, prolonged depolarizing current injections, VNTB neurons respond with multiple action potentials whose frequency increases with increasing current amplitude (Fujino et al. 1997), which is consistent with relatively little contribution from low voltage-activated Kv1 channels. VNTB neurons faithfully follow trains of depolarizing stimuli only up to 200 Hz (Tong et al. 2013).

Based on fate mapping studies, glycinergic neurons of the MNTB and the VNTB share a common origin (Jalabi et al. 2013; Altieri et al. 2014). The VNTB neurons receive excitatory innervations from axon collaterals of the globular bushy cells, whose main branch projects into the MNTB and forms the calyx synapses (Kuwabara et al. 1991), but no calyces have been reported in the VNTB. Glycinergic projections have been described from the VNTB to the LSO and LNTB of the contralateral SOC (Warr and Beck 1996; Thompson and Schofield 2000), and a reciprocal inhibitory connection has also been suggested between MNTB and VNTB (Kuwabara and Zook 1991; Dondzillo et al. 2016).

A key marker for VNTB neurons is their unusually high expression of Kv2.2 potassium channels (Tong et al. 2013) with the only other SOC site expressing Kv2.2 being the initial segments of the MNTB axons (Johnston et al. 2008). The half-activation voltage for Kv2.2 channels is around −10 mV, making it intermediate between the classical low voltage-activated Kv1 channels and the high voltage-activated Kv3 channels. In contrast to Kv1 and Kv3 currents, Kv2.2 currents have rather slow activation kinetics and are mostly involved in re-establishing the resting membrane potential and excitability between action potentials produced in a train. Indeed, whole-cell patch recordings in the VNTB show that these Kv2.2 channels are important to maintain repetitive firing in response to trains of stimuli. The absence of Kv2.2 potassium channels reduces VNTB firing rates, weakens the efferent feedback system, and increases the vulnerability to noise over exposure (Tong et al. 2013). While there is some evidence that the acetylcholine-positive efferent cells also express Kv2.2, the extent that these efferent neurons overlap with the glycine-positive neuron population is unknown. It is possible that there is no overlap between these two populations and that the glycinergic VNTB neurons are an additional source of feedforward inhibition within the SOC (Albrecht et al. 2014).

Both the VNTB and the LNTB also contain GABAergic neurons (Roberts and Ribak 1987; Albrecht et al. 2014). The extent to which GABA and glycine are coexpressed and the function this might serve in the mature animal are unknown.

5.3 The Lateral Superior Olive

The LSO is a key site for detection of interaural level differences (ILD), which provide a measure of the azimuth location of a sound source by integrating information from both cochleae (i.e., binaural inputs; Fig. 5.3). Precisely timed ipsilateral excitatory postsynaptic potentials (EPSPs) are integrated with inhibitory postsynaptic

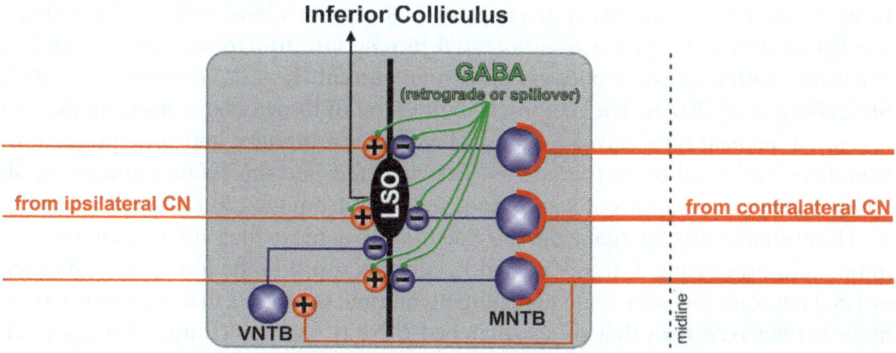

Fig. 5.3 The lateral superior olive (*LSO*) receives major external excitatory input and internal inhibition driven primarily by the contralateral ear. Different theories exist for the origin of the GABAergic modulation of LSO signaling. (+ excitatory; – inhibitory; *CN*, cochlear nucleus; *MNTB*, medial nucleus of the trapezoid body; *VNTB*, ventral nucleus of the trapezoid body)

potentials (IPSPs) arising from the contralateral ear (arriving from the MNTB). This early integration of information from opposite sides of the head provides many interesting insights into mechanisms of sound localization. The LSO also contains the cell bodies of the lateral olivocochlear system, which provides an efferent projection from the brain to the cochlea.

5.3.1 *Intrinsic Properties of Neurons in the Lateral Superior Olive*

There are at least two types of LSO neurons. Principal neurons, as the name implies, are the predominant and major cell type. Lateral olivocochlear (LOC) neurons are part of the efferent projection to the cochlea, and they form axo-axonic synapses on the spiral ganglion afferent processes that receive the output from the inner hair cells of the cochlea. LOC neuron characteristics have yet to be described in vivo. In vitro, LSO principal cells and LOC cells can be distinguished by their action potential firing in response to depolarizing and hyperpolarizing current injections, and up to five classes of neurons have been described on morphological grounds (Helfert and Schwartz 1987). Current-clamp recordings of LSO principal neurons show a single action potential or burst firing in response to depolarizing current injection (Adam et al. 1999). In the rat, this spectrum of principal cell firing patterns showed a tonotopic distribution with the single-spiking phenotype correlating with larger Kv1 currents present in the more lateral neurons (Barnes-Davies et al. 2004); however, this tonotopic distribution was not observed in the mouse LSO (Sterenborg et al. 2010; Karcz et al. 2011). Hyperpolarization of principal neurons causes a voltage sag consistent with I_H (Barnes-Davies et al. 2004; Hassfurth et al. 2009;

Kopp-Scheinpflug et al. 2015). In contrast, LOC neurons have no I_H and no voltage sag but exhibit a delayed action potential in response to depolarization, which is consistent with a transient outward potassium current (Kv4, I_A) (Fujino et al. 1997; Sterenborg et al. 2010). The reciprocal expression of these conductances in the two respective neuron types gives a robust classification in vitro, and their projections have been validated by retrograde markers injected into the cochlea (Fujino et al. 1997) and by observations of anterograde transport from the LSO (Warr et al. 1997).

These characteristics shape the LSO neurons to achieve their integrative function in binaural processing. Large Kv1 and I_H currents result in the low input resistance of LSO neurons and contribute to resting membrane potentials that are about 10 mV more depolarized than that of age-matched MNTB neurons (Barnes-Davies et al. 2004; Hassfurth et al. 2009). The membrane time constants for LSO neurons accelerate to about 2 ms with increasing postnatal age (Hassfurth et al. 2009), which further limits temporal integration times and refines integration of coincident inputs.

5.3.2 Synaptic Inputs to Neurons in the Lateral Superior Olive

Principle neurons in the LSO have a bipolar orientation with dendrites aligned orthogonal to the tonotopic axis. They receive excitatory projections from the spherical bushy cells of the ipsilateral AVCN and inhibitory inputs from the contralateral ear via the ipsilateral MNTB. The excitatory synaptic projection terminates on the dendrites of the principle neuron, with EPSPs being mediated by AMPAR (Case and Gillespie 2011) that exhibit little rectification and include calcium-impermeable GluA2 subunits (Case et al. 2011). Intriguingly, the composition of the synaptic AMPAR is changed by exposure to loud sounds, reflecting temporary expression of AMPAR subunits with slower kinetics (Pilati et al. 2016). The excitatory input also activates a NMDAR-mediated response that declines in amplitude and accelerates in time course with maturation (Case et al. 2011; Pilati et al. 2016).

The LSO receives inhibitory input via the MNTB, which is driven by contralateral sound stimulation. Inhibition into the LSO is mediated predominantly by glycine receptors in mature animals, but $GABA_A$ receptors make significant contributions early in development. In fact, a large body of work demonstrates the refinement and development of inhibitory connections in the LSO (reviewed by Kandler et al. 2009). Mature MNTB neurons continue to co-release GABA along with glycine (Weisz et al. 2016), adding another level of modulation in this microcircuit. Upon maturation, the decay time constant of the IPSC is as fast as 0.7 ms, matching that of the EPSC (Pilati et al. 2016).

The intrinsic characteristics described in Sect. 5.3.1 enable LSO neurons to integrate inhibitory and excitatory inputs within a highly restricted temporal window. Several mechanisms have been described that interrupt or modulate this integration, for example, the activation of presynaptic metabotropic GABA receptors will preferentially reduce excitatory inputs (Magnusson et al. 2008), and the lack of Kv1 channels (Karcz et al. 2011) or exposure to loud sounds will increase the temporal integration window (Pilati et al. 2016).

In vivo recordings of LSO neurons exhibit temporal firing patterns described as *onset* or *chopper* based on their post-stimulus time histograms (PSTHs). With respect to their response to binaural sound stimulation, cells in the LSO are referred to as "IE" cells. Such cells are inhibited from sound stimulation to the contralateral ear and excited by ipsilateral stimulation (see Table 5.2). LSO neurons are thus integrating information about sound received on one side of the head with the sound received on the opposite side. The inhibitory input has not only travelled a greater distance but has also passed through an additional synapse. This integration of information from both cochleae permits computation of sound location based on the ILD, since the ear which is on the far side of the head in relation to the sound's origin (in the sound shadow of the head) will perceive a sound of lower level (volume) due to reflection and refraction.

5.3.3 Output and Function of Neurons in the Lateral Superior Olive

The main function of LSO neurons is the computation of a population rate code for interaural level differences. The basic mechanism to achieve this is the synaptic integration of the excitatory (E) and inhibitory (I) synaptic projections to the LSO (from opposite cochlear nuclei), which are matched along the tonotopic axis so that there is a spectral and temporal convergence of inputs (Tollin 2003).

A simplified concept of how an LSO neuron generates an ILD function is that it starts with a maximum firing rate to purely ipsilateral, excitatory sound stimulation, which then progressively decreases as inhibition increases as the sound source moves along the azimuth toward the contralateral hemifield. The azimuth location at which the initial firing rate is reduced to 50% of the maximum firing rate is referred to as ILD_{50}. Individual LSO neurons have different $ILD_{50s,}$ and the LSO population information of the sound location is transmitted to the contralateral inferior colliculus where it is integrated with the information from the opposite LSO (Li and Kelly 1992; Tsai et al. 2010) and with other projections from the dorsal nucleus of the lateral lemniscus (Siveke et al. 2006).

5.4 The Medial Superior Olive

The MSO forms a bilateral ribbon-like vertical structure in the brainstem consisting of the aligned cell bodies of bipolar neurons. The computation of ILDs in the LSO, as described previously, is used to locate high frequency sound sources in azimuth. However, for low frequencies, ILDs are small, or even absent, simply because longer wavelengths are not shadowed by the head. For low frequencies, interaural time differences (ITDs) are more effective for determining sound location. The extent to which a particular species can utilize ITDs or ILDs (or both) depends on the frequency range of their audiogram and their head size, since ITDs become ambiguous at high frequencies with a small head. The combination of intrinsic and synaptic conductances that make MSO neurons so well suited to perform their binaural coincidence detection task are described in more detail in the next sections.

5.4.1 Intrinsic Properties of Neurons in the Medial Superior Olive

MSO neurons have a very low input resistance that is caused by multiple conductances that are open around the resting membrane potential (Scott et al. 2005; Couchman et al. 2010). This *leak* raises the apparent action potential threshold and refines coincidence detection because the respective inputs must summate for action potential generation in the MSO neuron. The two conductances contributing to low input resistance are low voltage-activated Kv1 currents (Mathews et al. 2010) and I_H currents, with the latter being dominated by the fast HCN1 subunit (Baumann et al. 2013; Kopp-Scheinpflug et al. 2015). These two conductances act in a reciprocal manner with Kv1 currents generating an outward current and I_H giving an inward current. In other areas of the brain I_H currents function to scale EPSPs coming from distal dendrites, but in the MSO the I_H current is regulating the resting conductance and membrane potential. The MSO Kv1 currents serve a similar function to that in the MNTB in restricting the generation of action potentials and refining the coincidence window during which the EPSCs can summate to trigger an action potential. Consistent with this, the default single action potential response of MSO neurons to prolonged injections of depolarizing current is similar to the respective response in MNTB neurons (see Table 5.2). Another effect of strong Kv1 currents would be the restriction of the action potential amplitude, which decreases as Kv1 expression increases during development (Scott et al. 2005). An additional element in the MSO is that action potentials are initiated along the axon at relatively remote sites from the cell body; so the amplitude measured by a somatic recording will be reduced by the local space constant as the action potential back-propagates to the recording site (Lehnert et al. 2014).

5.4.2 Synaptic Inputs to Neurons in the Medial Superior Olive

MSO neurons have a bipolar orientation, extending their two primary dendrites laterally in opposite directions toward the cochlear nuclei on either side of the head (Fig. 5.4). Each MSO neuron receives a few excitatory glutamatergic inputs with the contralateral inputs contacting the medial dendrites and the ipsilateral inputs contacting the lateral dendrites (Couchman et al. 2012). A famous model postulated by

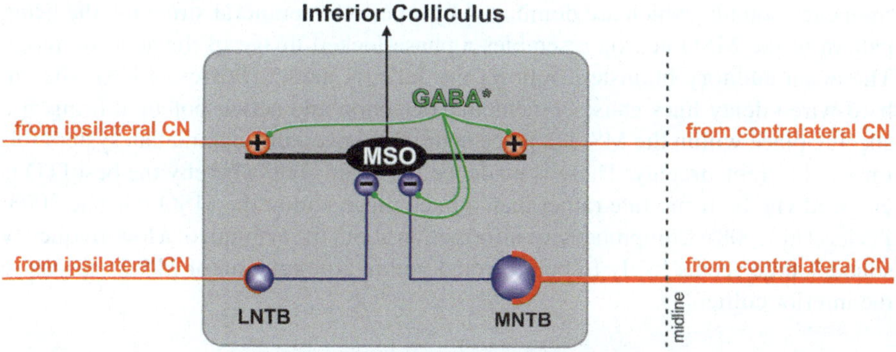

Fig. 5.4 The medial superior olive (*MSO*) receives bilateral external excitatory input. Two additional inhibitory inputs are conveyed via especially fast, calyx-type synapses to arrive in temporal register with the excitation. GABAergic signaling plays a role, but the origin of the GABA (gamma-aminobutyric acid) is not known. (+ excitatory; – inhibitory; *CN*, cochlear nuclei; *LNTB* lateral nucleus of the trapezoid body; *MNTB*, medial nucleus of the trapezoid body; *MSO* medial superior olive)

Loyd Jeffress in 1948 (Jeffress 1948) proposed axonal *delay lines* that compensate for the differing latencies of the ipsilateral versus contralateral inputs, which arise as the sound propagates from one to the other ear. Such physical delay lines have indeed been described in some avian species (Carr and Konishi 1988), but delay lines have not been observed in the mammalian SOC. It is suggested that mammals may generate *virtual* or *intrinsic delay lines* by other processes. The mechanisms are still under debate, but there is evidence for an active role of glycinergic inhibition in this process (Grothe et al. 2010). These glycinergic inputs originate from LNTB and MNTB (Roberts et al. 2014) and are predominantly found at the somata of MSO neurons (Kapfer et al. 2002). This large and fast somatic inhibition (Smith et al. 2000; Couchman et al. 2012) forms part of the overall conductance mechanism to refine coincidence detection (Brand et al. 2002; Grothe and Pecka 2014) along with strong contributions of I_H/Kv1 (Mathews et al. 2010). The hyperpolarizing inhibition both activates and increases the driving force for I_H, which can be observed as a large sag in membrane potential in response to hyperpolarizing current steps. Such enhancement of I_H conductance by inhibition further shapes the MSO ITD response. Additional, activity-dependent modulation of ITD processing acts via metabotropic $GABA_B$ receptors (Fischl et al. 2012; Stange et al. 2013), although synaptic GABA inputs have not been observed (Couchman et al. 2012).

5.4.3 *Output and Function of Neurons in the Medial Superior Olive*

The intrinsic properties of MSO neurons govern integration of the excitatory and inhibitory inputs and, hence, govern the generation of action potentials that propagate to the ipsilateral inferior colliculus (IC). Since MSO neurons are tuned to low frequency sounds, which are dominated by a periodic temporal structure, the firing pattern of the MSO neuron resembles a phase-locked image of the acoustic input. The avian auditory brainstem follows the Jeffress model (Jeffress 1948) wherein hard-wired delay lines cause coincidence detection and action potential firing at a specific place within the MSO (a place code). However, the mammalian system follows a different strategy. There is evidence for a rate code whereby the best ITD is encoded via the firing rate rather than a fixed place within the MSO (Grothe 2003; Pecka et al. 2008). Comprehensive information about the azimuth of a low-frequency sound source seems likely to be projected within a population of MSO neurons to the inferior colliculus.

5.5 The Superior Paraolivary Nucleus

The superior paraolivary nucleus (SPN or SPON) is a large and well-preserved structure across many mammalian species, including humans (Kulesza and Berrebi 2000). The SPN is regarded as homologous to the dorsomedial paraolivary nucleus (DMPO) of guinea pigs, cats, and dogs (Schofield and Cant 1991). The SPN neurons are a significant source of ascending GABAergic projections (Roberts and Ribak 1987) to the IC (Schofield and Cant 1992) and the thalamus (Schofield et al. 2014) and descending projections to the VCN and possibly the cochlea (Schofield and Cant 1991; Thompson and Thompson 1991). This section reviews the intrinsic ionic conductances, origin of inputs, neurochemical content in vesicles of SPN neurons, output firing patterns, and downstream projections, although the functional roles of the SPN in processing auditory information are not yet fully resolved.

5.5.1 *Intrinsic Properties of Neurons in the Superior Paraolivary Nucleus*

SPN neurons have large somata with diameters of 25–40 μm and a multipolar dendritic organization (Fig. 5.5) (guinea pig: Schofield 1991; mouse: Kopp-Scheinpflug et al. 2011; Felix et al. 2013). The HCN1 subunit is highly expressed and mediates an I_H current that is active around the resting membrane potential, causing a relatively depolarized resting membrane potential (Koch et al. 2004; Kopp-Scheinpflug et al. 2015). Hyperpolarization induced by incoming glycinergic IPSPs further activates HCN channels. Neurons in the SPN receive an especially powerful inhibitory drive, so that during sound stimulation the neurons are inhibited from firing, but at the end of the sound, their intrinsic excitability converts this inhibition into precisely timed action potential firing, signaling sound termination (rat: Kadner et al. 2006; mouse: Kopp-Scheinpflug et al. 2011; Felix et al. 2011). The concept of action potential firing that follows hyperpolarization is referred to as *rebound firing* and

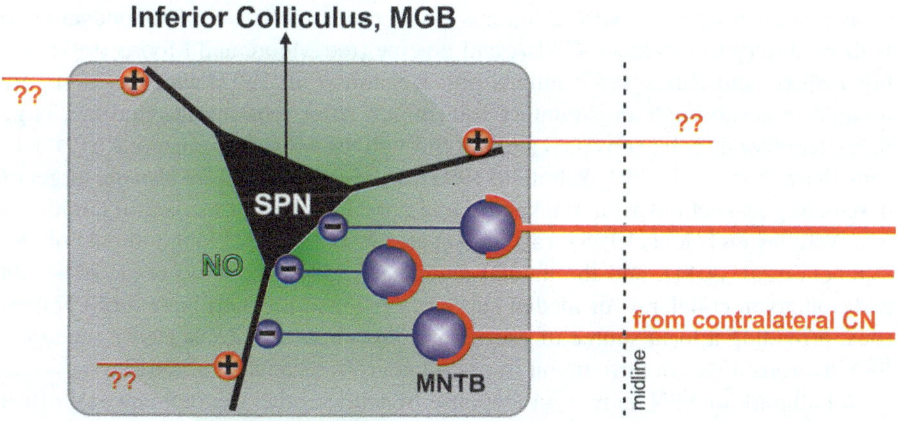

Fig. 5.5 The superior paraolivary nucleus (*SPN*) seems to be part of a monaural processing pathway. Most of its inputs are driven by the contralateral ear. Excitatory inputs driven by the ipsilateral or contralateral ear have been described by in vivo physiology but their origins are not fully understood, as indicated by the *question marks* (*??*). Descending inputs have been reported but are not shown here. SPN neurons (like neurons of the medial nucleus of the trapezoid body, *MNTB*) express nitric oxide (*NO*) synthase, which enables them to modulate their output firing activity via volume transmission. *CN*, cochlear nucleus; *GABA*, gamma-aminobutyric acid; *VNTB*, ventral nucleus of the trapezoid body)

provides a fascinating complement to the computational repertoire of action potential generation by EPSPs. Rebound firing emerges through integration of large IPSPs driven by an extremely negative chloride reversal potential (E_{Cl}) combined with a large hyperpolarization-activated nonspecific cationic current (I_H), with a contribution from a T-type calcium current (I_{TCa}). On activation by the IPSP, I_H potently accelerates the membrane time constant, so when the sound ceases, a rapid repolarization triggers multiple offset action potentials, which match the timing accuracy of the input spikes from the MNTB (Kopp-Scheinpflug et al. 2011). Therefore, the intrinsic conductances of the SPN allow the inhibitory input to function in two different ways: it can suppress spontaneous action potential firing during a sound, and it can generate a short-latency burst of action potentials after the sound ceases.

5.5.2 Synaptic Inputs to Neurons of the Superior Paraolivary Nucleus

The SPN receives bilateral (predominantly contralateral) input from octopus and multipolar cells of the VCN (rat: Friauf and Ostwald 1988; gerbil, mouse: Kuwabara et al. 1991; guinea pig: Thompson and Thompson 1991; Schofield 1995). Inhibitory projections arising from multipolar cells send their axons through the intermediate acoustic stria toward the SPN. Afferents from octopus cells are immunonegative to both inhibitory transmitters, GABA and glycine (rat: Moore and Moore 1987; gerbil: Roberts and Ribak 1987; guinea pig: Kolston et al. 1992) and, therefore, are thought to act through an excitatory transmitter. After crossing the midline, large diameter octopus cell axons pass straight through the MNTB to innervate SPN neurons (Kuwabara et al. 1991; Schofield 1995). The SPN neurons are also the target of descending projections from the ventral and dorsal nucleus of the lateral lemniscus (cat: Whitley and Henkel 1984; rat: Bajo et al. 1993), the tectal longitudinal column (Vinuela et al. 2011), and the medial geniculate body (Schofield et al. 2014). In addition, axon collaterals of medial superior olive neurons terminate on SPN neurons, providing a local source of excitation (Kuwabara and Zook 1999), while the SPN neurons might in turn inhibit the MSO neurons (Stange et al. 2013).

A hallmark of SPN activity in response to sound is the predominance of offset firing patterns (Kulesza et al. 2003; Kadner and Berrebi 2008), which are characterized by complete inhibition of action potentials for the duration of a sound stimulus followed by a rebound burst of 3–5 action potentials after sound termination (Felix et al. 2011; Kopp-Scheinpflug et al. 2011). This inhibitory input is driven by acoustic stimulation of the contralateral ear and is mediated via the ipsilateral MNTB (Kuwabara and Zook 1991; Banks and Smith 1992). Intracellular labeling of MNTB neurons shows a tonotopically organized projection from MNTB to SPN (Fig. 4 in Banks and Smith 1992). Within each isofrequency domain, single MNTB neurons project to multiple SPN neurons (Fig. 2B in Banks and Smith 1992). In vitro synaptic stimulation and computational modeling suggest a strong convergence of glycinergic inputs onto each SPN neuron (Kopp-Scheinpflug et al. 2011). Minimal stimulation of just one glycinergic input gives rise to an inhibitory postsynaptic current with a fast decay time constant of 2.5 ± 0.39 ms (Jalabi et al. 2013).

5.5.3 Outputs and Functions of Neurons in the Superior Paraolivary Nucleus

The output of SPN neurons is created by the integration of their synaptic inputs and intrinsic ionic conductances. In vivo recordings from SPN neurons in multiple species reveal broad frequency tuning arising from mixed inputs, such as bilateral excitation mixed with contralateral inhibition as well as sole excitation or inhibition (gerbil: Behrend et al. 2002; Dehmel et al. 2002; rat: Kulesza et al. 2003; Kadner et al. 2006; mouse: Felix et al. 2011; Kopp-Scheinpflug et al. 2011; bat: Grothe et al. 2001). Despite possible species-specific adaptations in the input and output firing patterns, all previous studies concur that SPN neurons achieve precise encoding of temporal transients by cross-frequency integration of unilateral or bilateral inputs.

The output firing pattern of SPN neurons is either off, on-off, or on/on-sustained and is determined by the weighted sum of excitatory and inhibitory inputs. These different temporal patterns seem to be topographically grouped within the boundaries of the SPN, with offset responders located dorsolateral and on-off responders or on-sustained responders showing a more widespread distribution (Kopp-Scheinpflug et al. 2011; Felix et al. 2013). Since onset responses are generally driven by excitatory input and offset responses in SPN are driven by inhibitory input, one would expect to find the inhibitory synaptic input exclusively dorsolateral in the SPN and excitatory synaptic input in the remaining parts of the SPN. However, intracellular labeling of MNTB neurons shows that inhibitory inputs into the SPN are tonotopically organized but are not restricted to a subdivision of the SPN (Banks and Smith 1992). Similarly, immunocytochemical labeling of the glycine transporter type 2 (GLYT2), signifying the MNTB input, consistently demonstrates that the whole of the SPN receives glycinergic input (Jalabi et al. 2013; Yassin et al. 2014). The ability to fire short latency offset responses enables SPN neurons to provide information about stimulus duration onto duration-tuned neurons in the midbrain (Aubie et al. 2012) or permits the downstream auditory pathway to detect gaps in sound (Yassin et al. 2014; Anderson and Linden 2016).

5.6 Conclusions and Thoughts for the Future

The superior olivary complex network acts in concert to perform a series of computations that require high-fidelity transmission of information from one side of the head to the other and then integrates this information to perform a range of physiological roles in sound localization and binaural auditory processing. SOC neurons have multiple specializations to promote speed of transmission, including large caliber, heavily myelinated axons, reliable synaptic transmission, fast receptor kinetics, and other adaptations to control target neuron excitability when receiving large and/or high-frequency inputs. In comparative terms, the SOC is highly conserved across mammalian species, but it is also adapted to the particular needs of each species (Grothe 2000; Grothe and Pecka 2014) in terms of the audiogram and monaural versus binaural processing with respect to head size.

This chapter highlighted the essential elements of the SOC circuit at a cellular and molecular level to provide insights into the physiological roles of the SOC nuclei and the roles of particular genes and their protein products. For the future, increased attention to central mechanisms of hearing loss is required, and a better understanding of the role of activity-dependent modulation in adapting auditory processing through changes in gene expression is needed. There is increasing interest in how over-exposure to sound or deafness cause changes in auditory brainstem function, and there is continuing interest in the extent to which different forms of synaptic and intrinsic plasticity can modify local auditory processing (and perhaps contribute to mechanisms of tinnitus). Also of increasing importance is understanding the roles of the olivocochlear system and mechanisms by which efferent pathways influence hearing through feedback from structures higher in the auditory pathway. Most networks are considered from a synaptic perspective, but there is increasing evidence for signaling beyond the synapse, for example, in terms of volume transmission when diffusible factors are released and can act locally beyond the synapse to modulate network function. One example is nitric oxide (NO). The prevalence of neuronal nitric oxide synthase (nNOS) in neurons of the SOC reinforces the possibility that NO acts here as a volume transmitter, in parallel with conventional excitatory and inhibitory signaling, to regulate the relative strength of inhibition through modulation of the potassium-chloride cotransporter type 2 (KCC2) activity (Yassin et al. 2014). Alternatively, NO may change neuronal excitability by regulation of potassium (Steinert et al. 2011) and HCN channels (Kopp-Scheinpflug et al. 2015) or through retrograde signaling at excitatory synapses.

Compliance with Ethics Requirements Conny Kopp-Scheinpflug declares that she has no conflict of interest.

Ian D. Forsythe declares that he has no conflict of interest.

References

Adam, T. J., Schwarz, D. W., & Finlayson, P. G. (1999). Firing properties of chopper and delay neurons in the lateral superior olive of the rat. *Experimental Brain Research, 124*(4), 489–502.

Alamilla, J., & Gillespie, D. C. (2013). Maturation of calcium-dependent GABA, glycine, and glutamate release in the glycinergic MNTB-LSO pathway. *PLoS One, 8*(9), e75688.

Albrecht, O., Dondzillo, A., Mayer, F., Thompson, J. A., & Klug, A. (2014). Inhibitory projections from the ventral nucleus of the trapezoid body to the medial nucleus of the trapezoid body in the mouse. *Frontiers in Neural Circuits, 8*, 83.

Altieri, S. C., Zhao, T., Jalabi, W., & Maricich, S. M. (2014). Development of glycinergic innervation to the murine LSO and SPN in the presence and absence of the MNTB. *Frontiers in Neural Circuits, 8*, 109.

Anderson, L. A., & Linden, J. F. (2016). Mind the gap: Two dissociable mechanisms of temporal processing in the auditory system. *The Journal of Neuroscience, 366*(6), 1977–1995.

Aubie, B., Sayegh, R., & Faure, P. A. (2012). Duration tuning across vertebrates. *The Journal of Neuroscience, 32*(18), 6373–6390.

Awatramani, G. B., Turecek, R., & Trussell, L. O. (2005). Staggered development of GABAergic and glycinergic transmission in the MNTB. *Journal of Neurophysiology, 93*(2), 819–828.

Bajo, V. M., Merchan, M. A., Lopez, D. E., & Rouiller, E. M. (1993). Neuronal morphology and efferent projections of the dorsal nucleus of the lateral lemniscus in the rat. *Journal of Comparative Neurology, 334*(2), 241–262.

Banks, M. I., Pearce, R. A., & Smith, P. H. (1993). Hyperpolarization-activated cation current (I_h) in neurons of the medial nucleus of the trapezoid body: Voltage-clamp analysis and enhancement by norepinephrine and cAMP suggest a modulatory mechanism in the auditory brain stem. *Journal of Neurophysiology, 70*(4), 1420–1432.

Banks, M. I., & Smith, P. H. (1992). Intracellular recordings from neurobiotin-labeled cells in brain slices of the rat medial nucleus of the trapezoid body. *The Journal of Neuroscience, 12*(7), 2819–2837.

Barnes-Davies, M., Barker, M. C., Osmani, F., & Forsythe, I. D. (2004). Kv1 currents mediate a gradient of principal neuron excitability across the tonotopic axis in the rat lateral superior olive. *European Journal of Neuroscience, 19*(2), 325–333.

Barnes-Davies, M., Owens, S., & Forsythe, I. D. (2001). Calcium channels triggering transmitter release in the rat medial superior olive. *Hearing Research, 1662*(1–2), 134–145.

Baumann, V. J., Lehnert, S., Leibold, C., & Koch, U. (2013). Tonotopic organization of the hyperpolarization-activated current (I_h) in the mammalian medial superior olive. *Frontiers in Neural Circuits, 7*, 117. https://doi.org/10.3389/fncir.2013.00117.

Behrend, O., Brand, A., Kapfer, C., & Grothe, B. (2002). Auditory response properties in the superior paraolivary nucleus of the gerbil. *Journal of Neurophysiology, 87*(6), 2915–2928.

Bollmann, J. H., Helmchen, F., Borst, J. G., & Sakmann, B. (1998). Postsynaptic Ca^{2+} influx mediated by three different pathways during synaptic transmission at a calyx-type synapse. *The Journal of Neuroscience, 18*(24), 10409–10419.

Brand, A., Behrend, O., Marquardt, T., McAlpine, D., & Grothe, B. (2002). Precise inhibition is essential for microsecond interaural time difference coding. *Nature, 417*(6888), 543–547.

Brew, H. M., & Forsythe, I. D. (1995). Two voltage-dependent K+ conductances with complementary functions in postsynaptic integration at a central auditory synapse. *The Journal of Neuroscience, 15*(12), 8011–8022.

Brew, H. M., & Forsythe, I. D. (2005). Systematic variation of potassium current amplitudes across the tonotopic axis of the rat medial nucleus of the trapezoid body. *Hearing Research, 2066*(1–2), 116–132.

Carr, C. E., & Konishi, M. (1988). Axonal delay lines for time measurement in the owl's brainstem. *Proceedings of the National Academy of Sciences of the United States of America, 85*(21), 8311–8315.

Case, D. T., & Gillespie, D. C. (2011). Pre- and postsynaptic properties of glutamatergic transmission in the immature inhibitory MNTB-LSO pathway. *Journal of Neurophysiology, 1066*(5), 2570–2579.

Case, D. T., Zhao, X., & Gillespie, D. C. (2011). Functional refinement in the projection from ventral cochlear nucleus to lateral superior olive precedes hearing onset in rat. *PLoS One, 66*(6), e20756.

Couchman, K., Grothe, B., & Felmy, F. (2010). Medial superior olivary neurons receive surprisingly few excitatory and inhibitory inputs with balanced strength and short-term dynamics. *Journal of Neuroscience, 30*(50), 17111–17121.

Couchman, K., Grothe, B., & Felmy, F. (2012). Functional localization of neurotransmitter receptors and synaptic inputs to mature neurons of the medial superior olive. *Journal of Neurophysiology, 107*(4), 1186–1198.

Darrow, K. N., Maison, S. F., & Liberman, M. C. (2006). Cochlear efferent feedback balances interaural sensitivity. *Nature Neuroscience, 9*(12), 1474–1476.

Dehmel, S., Kopp-Scheinpflug, C., Dorrscheidt, G. J., & Rubsamen, R. (2002). Electrophysiological characterization of the superior paraolivary nucleus in the Mongolian gerbil. *Hearing Research, 172*(1–2), 18–36.

Dondzillo, A., Thompson, J. A., & Klug, A. (2016). Recurrent inhibition to the medial nucleus of the trapezoid body in the mongolian gerbil (*Meriones unguiculatus*). *PLoS One, 11*(8), e0160241.

Felix, R. A., 2nd, Fridberger, A., Leijon, S., Berrebi, A. S., & Magnusson, A. K. (2011). Sound rhythms are encoded by postinhibitory rebound spiking in the superior paraolivary nucleus. *The Journal of Neuroscience, 31*(35), 12566–12578.

Felix, R. A., 2nd, Vonderschen, K., Berrebi, A. S., & Magnusson, A. K. (2013). Development of on-off spiking in superior paraolivary nucleus neurons of the mouse. *Journal of Neurophysiology, 109*(11), 2691–2704.

Fischl, M. J., Burger, R. M., Schmidt-Pauly, M., Alexandrova, O., Sinclair, J. L., Grothe, B., et al. (2016). Physiology and anatomy of neurons in the medial superior olive of the mouse. *Journal of Neurophysiology, 1166*(6), 2676–2688.

Fischl, M. J., Combs, T. D., Klug, A., Grothe, B., & Burger, R. M. (2012). Modulation of synaptic input by $GABA_B$ receptors improves coincidence detection for computation of sound location. *The Journal of Physiology, 590*(13), 3047–3066.

Ford, M. C., Alexandrova, O., Cossell, L., Stange-Marten, A., Sinclair, J., Kopp-Scheinpflug, C., et al. (2015). Tuning of Ranvier node and internode properties in myelinated axons to adjust action potential timing. *Nature Communications, 66*, 8073.

Forsythe, I. D. (1994). Direct patch recording from identified presynaptic terminals mediating glutamatergic EPSCs in the rat CNS, in vitro. *The Journal of Physiology, 479*(3), 381–387.

Friauf, E., & Ostwald, J. (1988). Divergent projections of physiologically characterized rat ventral cochlear nucleus neurons as shown by intra-axonal injection of horseradish peroxidase. *Experimental Brain Research, 73*(2), 263–284.

Fujino, K., Koyano, K., & Ohmori, H. (1997). Lateral and medial olivocochlear neurons have distinct electrophysiological properties in the rat brain slice. *Journal of Neurophysiology, 77*(5), 2788–2804.

Giugovaz-Tropper, B., Gonzalez-Inchauspe, C., Di Guilmi, M. N., Urbano, F. J., Forsythe, I. D., & Uchitel, O. D. (2011). P/Q-type calcium channel ablation in a mice glycinergic synapse mediated by multiple types of $Ca^{(2)+}$ channels alters transmitter release and short term plasticity. *Neuroscience, 192*, 219–230.

Grothe, B. (2000). The evolution of temporal processing in the medial superior olive, an auditory brainstem structure. *Progress in Neurobiology, 661*(6), 581–610.

Grothe, B. (2003). New roles for synaptic inhibition in sound localization. *Nature Reviews. Neuroscience, 4*(7), 540–550.

Grothe, B., Covey, E., & Casseday, J. H. (2001). Medial superior olive of the big brown bat: Neuronal responses to pure tones, amplitude modulations, and pulse trains. *Journal of Neurophysiology, 866*(5), 2219–2230.

Grothe, B., & Pecka, M. (2014). The natural history of sound localization in mammals—A story of neuronal inhibition. *Frontiers in Neural Circuits, 8*, 116. https://doi.org/10.3389/fncir.2014.00116.

Grothe, B., Pecka, M., & McAlpine, D. (2010). Mechanisms of sound localization in mammals. *Physiological Reviews, 90*(3), 983–1012.

Guinan, J. J., Jr., & Li, R. Y. (1990). Signal processing in brainstem auditory neurons which receive giant endings (calyces of Held) in the medial nucleus of the trapezoid body of the cat. *Hearing Research, 49*(1–3), 321–334.

Hamann, M., Billups, B., & Forsythe, I. D. (2003). Non-calyceal excitatory inputs mediate low fidelity synaptic transmission in rat auditory brainstem slices. *European Journal of Neuroscience, 18*(10), 2899–2902.

Hardman, R. M., & Forsythe, I. D. (2009). Ether-á-go-go-related gene K+ channels contribute to threshold excitability of mouse auditory brainstem neurons. *The Journal of Physiology, 587*(11), 2487–2497.

Hassfurth, B., Magnusson, A. K., Grothe, B., & Koch, U. (2009). Sensory deprivation regulates the development of the hyperpolarization-activated current in auditory brainstem neurons. *European Journal of Neuroscience, 30*(7), 1227–1238.

Held, H. (1893). Die centrale Gehörleitung. *Archiv für Anatomie und Physiologie/Anatomische Abteilung, 1893*, 201–248. https://core.ac.uk/download/pdf/14520980.pdf.

Helfert, R. H., & Schwartz, I. R. (1987). Morphological features of five neuronal classes in the gerbil lateral superior olive. *American Journal of Anatomy, 179*(1), 55–69.

Hirtz, J. J., Boesen, M., Braun, N., Deitmer, J. W., Kramer, F., Lohr, C., et al. (2011). Cav1.3 calcium channels are required for normal development of the auditory brainstem. *The Journal of Neuroscience, 31*(22), 8280–8294.

Holcomb, P. S., Deerinck, T. J., Ellisman, M. H., & Spirou, G. A. (2013). Construction of a polarized neuron. *The Journal of Physiology, 591*(13), 3145–3150.

Jalabi, W., Kopp-Scheinpflug, C., Allen, P. D., Schiavon, E., DiGiacomo, R. R., Forsythe, I. D., et al. (2013). Sound localization ability and glycinergic innervation of the superior olivary complex persist after genetic deletion of the medial nucleus of the trapezoid body. *The Journal of Neuroscience, 33*(38), 15044–15049.

Jeffress, L. A. (1948). A place theory of sound localization. *Journal of Comparative and Physiological Psychology, 41*(1), 35–39.

Johnston, J., Forsythe, I. D., & Kopp-Scheinpflug, C. (2010). Going native: Voltage-gated potassium channels controlling neuronal excitability. *The Journal of Physiology, 588*(17), 3187–3200.

Johnston, J., Griffin, S. J., Baker, C., Skrzypiec, A., Chernova, T., & Forsythe, I. D. (2008). Initial segment Kv2.2 channels mediate a slow delayed rectifier and maintain high frequency action potential firing in medial nucleus of the trapezoid body neurons. *The Journal of Physiology, 5866*(14), 3493–3509.

Kadner, A., & Berrebi, A. S. (2008). Encoding of temporal features of auditory stimuli in the medial nucleus of the trapezoid body and superior paraolivary nucleus of the rat. *Neuroscience, 151*(3), 868–887.

Kadner, A., Kulesza, R. J., Jr., & Berrebi, A. S. (2006). Neurons in the medial nucleus of the trapezoid body and superior paraolivary nucleus of the rat may play a role in sound duration coding. *Journal of Neurophysiology, 95*(3), 1499–1508.

Kandler, K., Clause, A., & Noh, J. (2009). Tonotopic reorganization of developing auditory brainstem circuits. *Nature Neuroscience, 12*(6), 711–717.

Kapfer, C., Seidl, A. H., Schweizer, H., & Grothe, B. (2002). Experience-dependent refinement of inhibitory inputs to auditory coincidence-detector neurons. *Nature Neuroscience, 5*(3), 247–253.

Karcz, A., Hennig, M. H., Robbins, C. A., Tempel, B. L., Rübsamen, R., & Kopp-Scheinpflug, C. (2011). Low-voltage activated Kv1.1 subunits are crucial for the processing of sound source location in the lateral superior olive in mice. *The Journal of Physiology, 589*(5), 1143–1157.

Kim, G., & Kandler, K. (2003). Elimination and strengthening of glycinergic/GABAergic connections during tonotopic map formation. *Nature Neuroscience, 66*(3), 282–290.

Kim, J. H., Kushmerick, C., & von Gersdorff, H. (2010). Presynaptic resurgent Na+ currents sculpt the action potential waveform and increase firing reliability at a CNS nerve terminal. *The Journal of Neuroscience, 30*(46), 15479–15490.

Koch, U., Braun, M., Kapfer, C., & Grothe, B. (2004). Distribution of HCN1 and HCN2 in rat auditory brainstem nuclei. *European Journal of Neuroscience, 20*(1), 79–91.

Koike-Tani, M., Saitoh, N., & Takahashi, T. (2005). Mechanisms underlying developmental speeding in AMPA-EPSC decay time at the calyx of Held. *The Journal of Neuroscience, 25*(1), 199–207.

Koka, K., & Tollin, D. J. (2014). Linear coding of complex sound spectra by discharge rate in neurons of the medial nucleus of the trapezoid body (MNTB) and its inputs. *Frontiers in Neural Circuits, 8*, 144.

Kolston, J., Osen, K. K., Hackney, C. M., Ottersen, O. P., & Storm-Mathisen, J. (1992). An atlas of glycine- and GABA-like immunoreactivity and colocalization in the cochlear nuclear complex of the guinea pig. *Anatomy and Embryology (Berlin), 1866*(5), 443–465.

Kopp-Scheinpflug, C., Dehmel, S., Tolnai, S., Dietz, B., Milenkovic, I., & Rübsamen, R. (2008). Glycine-mediated changes of onset reliability at a mammalian central synapse. *Neuroscience, 157*(2), 432–445.

Kopp-Scheinpflug, C., Lippe, W. R., Dorrscheidt, G. J., & Rubsamen, R. (2003). The medial nucleus of the trapezoid body in the gerbil is more than a relay: Comparison of pre- and postsynaptic activity. *Journal of the Association for Research in Otolaryngology, 4*(1), 1–23.

Kopp-Scheinpflug, C., Pigott, B. M., & Forsythe, I. D. (2015). Nitric oxide selectively suppresses IH currents mediated by HCN1-containing channels. *The Journal of Physiology, 593*(7), 1685–1700.

Kopp-Scheinpflug, C., Tozer, A. J., Robinson, S. W., Tempel, B. L., Hennig, M. H., & Forsythe, I. D. (2011). The sound of silence: Ionic mechanisms encoding sound termination. *Neuron, 71*(5), 911–925.

Kulesza, R. J., Jr., & Berrebi, A. S. (2000). Superior paraolivary nucleus of the rat is a GABAergic nucleus. *Journal of the Association for Research in Otolaryngology, 1*(4), 255–269.

Kulesza, R. J., Jr., & Grothe, B. (2015). Yes, there is a medial nucleus of the trapezoid body in humans. *Frontiers in Neuroanatomy, 9*, 35. https://doi.org/10.3389/fnana.2015.00035.

Kulesza, R. J., Jr., Spirou, G. A., & Berrebi, A. S. (2003). Physiological response properties of neurons in the superior paraolivary nucleus of the rat. *Journal of Neurophysiology, 89*(4), 2299–2312.

Kuwabara, N., DiCaprio, R. A., & Zook, J. M. (1991). Afferents to the medial nucleus of the trapezoid body and their collateral projections. *The Journal of Comparative Neurology, 314*(4), 684–706.

Kuwabara, N., & Zook, J. M. (1991). Classification of the principal cells of the medial nucleus of the trapezoid body. *The Journal of Comparative Neurology, 314*(4), 707–720.

Kuwabara, N., & Zook, J. M. (1999). Local collateral projections from the medial superior olive to the superior paraolivary nucleus in the gerbil. *Brain Research, 8466*(1), 59–71.

Leao, R. N., Naves, M. M., Leao, K. E., & Walmsley, B. (2006). Altered sodium currents in auditory neurons of congenitally deaf mice. *European Journal of Neuroscience, 24*(4), 1137–1146.

Leao, R. N., Svahn, K., Berntson, A., & Walmsley, B. (2005). Hyperpolarization-activated (I) currents in auditory brainstem neurons of normal and congenitally deaf mice. *European Journal of Neuroscience, 22*(1), 147–157.

Lehnert, S., Ford, M. C., Alexandrova, O., Hellmundt, F., Felmy, F., Grothe, B., et al. (2014). Action potential generation in an anatomically constrained model of medial superior olive axons. *The Journal of Neuroscience, 34*(15), 5370–5384.

Li, L., & Kelly, J. B. (1992). Binaural responses in rat inferior colliculus following kainic acid lesions of the superior olive: Interaural intensity difference functions. *Hearing Research, 661*(1–2), 73–85.

Li, W., Kaczmarek, L. K., & Perney, T. M. (2001). Localization of two high-threshold potassium channel subunits in the rat central auditory system. *Journal of Comparative Neurology, 437*(2), 196–218.

Magnusson, A. K., Park, T. J., Pecka, M., Grothe, B., & Koch, U. (2008). Retrograde GABA signaling adjusts sound localization by balancing excitation and inhibition in the brainstem. *Neuron, 59*(1), 125–137.

Mathews, P. J., Jercog, P. E., Rinzel, J., Scott, L. L., & Golding, N. L. (2010). Control of submillisecond synaptic timing in binaural coincidence detectors by K(v)1 channels. *Nature Neuroscience, 13*(5), 601–609.

Moore, J. K., & Moore, R. Y. (1987). Glutamic acid decarboxylase-like immunoreactivity in brainstem auditory nuclei of the rat. *The Journal of Comparative Neurology, 2660*(2), 157–174.

Myoga, M. H., Lehnert, S., Leibold, C., Felmy, F., & Grothe, B. (2014). Glycinergic inhibition tunes coincidence detection in the auditory brainstem. *Nature Communications, 5*, 3790. https://doi.org/10.1038/ncomms4790.

Oertel, D. (2009). A team of potassium channels tunes up auditory neurons. *The Journal of Physiology, 587*(11), 2417–2418.

Pecka, M., Brand, A., Behrend, O., & Grothe, B. (2008). Interaural time difference processing in the mammalian medial superior olive: The role of glycinergic inhibition. *The Journal of Neuroscience, 28*(27), 6914–6925.

Pilati, N., Linley, D. M., Selvaskandan, H., Uchitel, O., Hennig, M. H., Kopp-Scheinpflug, C., et al. (2016). Acoustic trauma slows AMPA receptor-mediated EPSCs in the auditory brainstem, reducing GluA4 subunit expression as a mechanism to rescue binaural function. *The Journal of Physiology, 594*(13), 3683–3703.

Roberts, M. T., Seeman, S. C., & Golding, N. L. (2014). The relative contributions of MNTB and LNTB neurons to inhibition in the medial superior olive assessed through single and paired recordings. *Frontiers in Neural Circuits, 8*, 49. https://doi.org/10.3389/fncir.2014.00049.

Roberts, R. C., & Ribak, C. E. (1987). GABAergic neurons and axon terminals in the brainstem auditory nuclei of the gerbil. *The Journal of Comparative Neurology, 258*(2), 267–280.

Ryugo, D. K., Fay, R. R., & Popper, A. N. (Eds.). (2011). *Auditory and vestibular efferents*. New York: Springer.

Schneggenburger, R., & Forsythe, I. D. (2006). The calyx of Held. *Cell and Tissue Research, 3266*(2), 311–337. [Research Support Non-U.S. Gov't Review].

Schofield, B. R. (1991). Superior paraolivary nucleus in the pigmented guinea pig: Separate classes of neurons project to the inferior colliculus and the cochlear nucleus. *The Journal of Comparative Neurology, 312*(1), 68–76.

Schofield, B. R. (1995). Projections from the cochlear nucleus to the superior paraolivary nucleus in guinea pigs. *Journal of Comparative Neurology, 3660*(1), 135–149.
Schofield, B. R., & Cant, N. B. (1991). Organization of the superior olivary complex in the guinea pig. I. Cytoarchitecture, cytochrome oxidase histochemistry, and dendritic morphology. *The Journal of Comparative Neurology, 314*(4), 645–670.
Schofield, B. R., & Cant, N. B. (1992). Organization of the superior olivary complex in the guinea pig: II. Patterns of projection from the periolivary nuclei to the inferior colliculus. *The Journal of Comparative Neurology, 317*(4), 438–455.
Schofield, B. R., Mellott, J. G., & Motts, S. D. (2014). Subcollicular projections to the auditory thalamus and collateral projections to the inferior colliculus. *Frontiers in Neuroanatomy, 8*, 70. https://doi.org/10.3389/fnana.2014.00070.
Scott, L. L., Mathews, P. J., & Golding, N. L. (2005). Posthearing developmental refinement of temporal processing in principal neurons of the medial superior olive. *The Journal of Neuroscience, 25*(35), 7887–7895.
Seidl, A. H., & Rubel, E. W. (2016). Systematic and differential myelination of axon collaterals in the mammalian auditory brainstem. *Glia, 664*(4), 487–494.
Sinclair, J. L., Barnes-Davies, M., Kopp-Scheinpflug, C., & Forsythe, I. D. (2017a). Strain-specific differences in the development of neuronal excitability in the mouse ventral nucleus of the trapezoid body. *Hearing Research, 354*, 28–37.
Sinclair, J. L., Fischl, M. J., Alexandrova, O., Heß, M., Grothe, B., Leibold, C., et al. (2017b). Sound-evoked activity influences myelination of brainstem axons in the trapezoid body. *The Journal of Neuroscience, 37*(34), 8239–8255.
Siveke, I., Pecka, M., Seidl, A. H., Baudoux, S., & Grothe, B. (2006). Binaural response properties of low-frequency neurons in the gerbil dorsal nucleus of the lateral lemniscus. *Journal of Neurophysiology, 966*(3), 1425–1440.
Smith, A. J., Owens, S., & Forsythe, I. D. (2000). Characterization of inhibitory and excitatory postsynaptic currents of the rat medial superior olive. *The Journal of Physiology, 529*(3), 681–698.
Sommer, I., Lingenhohl, K., & Friauf, E. (1993). Principal cells of the rat medial nucleus of the trapezoid body: An intracellular in vivo study of their physiology and morphology. *Experimental Brain Research, 95*(2), 223–239.
Spirou, G. A., & Berrebi, A. S. (1996). Organization of ventrolateral periolivary cells of the cat superior olive as revealed by PEP-19 immunocytochemistry and Nissl stain. *The Journal of Comparative Neurology, 3668*(1), 100–120.
Spirou, G. A., Rowland, K. C., & Berrebi, A. S. (1998). Ultrastructure of neurons and large synaptic terminals in the lateral nucleus of the trapezoid body of the cat. *The Journal of Comparative Neurology, 398*(2), 257–272.
Stange, A., Myoga, M. H., Lingner, A., Ford, M. C., Alexandrova, O., Felmy, F., et al. (2013). Adaptation in sound localization: From GABA(B) receptor-mediated synaptic modulation to perception. *Nature Neuroscience, 166*(12), 1840–1847.
Steinert, J. R., Postlethwaite, M., Jordan, M. D., Chernova, T., Robinson, S. W., & Forsythe, I. D. (2010). NMDAR-mediated EPSCs are maintained and accelerate in time course during maturation of mouse and rat auditory brainstem in vitro. *The Journal of Physiology, 588*(3), 447–463.
Steinert, J. R., Robinson, S. W., Tong, H., Haustein, M. D., Kopp-Scheinpflug, C., & Forsythe, I. D. (2011). Nitric oxide is an activity-dependent regulator of target neuron intrinsic excitability. *Neuron, 71*(2), 291–305.
Sterenborg, J. C., Pilati, N., Sheridan, C. J., Uchitel, O. D., Forsythe, I. D., & Barnes-Davies, M. (2010). Lateral olivocochlear (LOC) neurons of the mouse LSO receive excitatory and inhibitory synaptic inputs with slower kinetics than LSO principal neurons. *Hearing Research, 270*(1–2), 119–126.
Thompson, A. M., & Schofield, B. R. (2000). Afferent projections of the superior olivary complex. *Microscopy Research and Techniques, 51*(4), 330–354.

Thompson, A. M., & Thompson, G. C. (1991). Posteroventral cochlear nucleus projections to olivocochlear neurons. *The Journal of Comparative Neurology, 303*(2), 267–285.
Tollin, D. J. (2003). The lateral superior olive: A functional role in sound source localization. *The Neuroscientist, 9*(2), 127–143.
Tolnai, S., Englitz, B., Kopp-Scheinpflug, C., Dehmel, S., Jost, J., & Rübsamen, R. (2008). Dynamic coupling of excitatory and inhibitory responses in the medial nucleus of the trapezoid body. *European Journal of Neuroscience, 27*(12), 3191–3204.
Tong, H., Kopp-Scheinpflug, C., Pilati, N., Robinson, S. W., Sinclair, J. L., Steinert, J. R., et al. (2013). Protection from noise-induced hearing loss by Kv2.2 potassium currents in the central medial olivocochlear system. *The Journal of Neuroscience, 33*(21), 9113–9121.
Tozer, A. J., Forsythe, I. D., & Steinert, J. R. (2012). Nitric oxide signalling augments neuronal voltage-gated L-type (Ca(v)1) and P/q-type (Ca(v)2.1) channels in the mouse medial nucleus of the trapezoid body. *PLoS One, 7*(2), e32256.
Tsai, J. J., Koka, K., & Tollin, D. J. (2010). Varying overall sound intensity to the two ears impacts interaural level difference discrimination thresholds by single neurons in the lateral superior olive. *Journal of Neurophysiology, 103*(2), 875–886.
Vinuela, A., Aparicio, M. A., Berrebi, A. S., & Saldana, E. (2011). Connections of the superior paraolivary nucleus of the rat: II. Reciprocal connections with the tectal longitudinal column. *Frontiers in Neuroanatomy, 5*, 1. https://doi.org/10.3389/fnana.2011.00001.
von Gersdorff, H., & Borst, J. G. (2002). Short-term plasticity at the calyx of Held. *Nature Reviews Neuroscience, 3*(1), 53–64.
Wang, H., Kunkel, D. D., Martin, T. M., Schwartzkroin, P. A., & Tempel, B. L. (1993). Heteromultimeric K+ channels in terminal and juxtaparanodal regions of neurons. *Nature, 3665*(6441), 75–79.
Wang, L. Y., Gan, L., Forsythe, I. D., & Kaczmarek, L. K. (1998). Contribution of the Kv3.1 potassium channel to high-frequency firing in mouse auditory neurones. *The Journal of Physiology, 509*(1), 183–194.
Warr, W. B., & Beck, J. E. (1996). Multiple projections from the ventral nucleus of the trapezoid body in the rat. *Hearing Research, 93*(1–2), 83–101.
Warr, W. B., Boche, J. B., & Neely, S. T. (1997). Efferent innervation of the inner hair cell region: Origins and terminations of two lateral olivocochlear systems. *Hearing Research, 108*(1–2), 89–111.
Weisz, C. J., Rubio, M. E., Givens, R. S., & Kandler, K. (2016). Excitation by axon terminal GABA spillover in a sound localization circuit. *The Journal of Neuroscience, 366*(3), 911–925.
Whitley, J. M., & Henkel, C. K. (1984). Topographical organization of the inferior collicular projection and other connections of the ventral nucleus of the lateral lemniscus in the cat. *The Journal of Comparative Neurology, 229*(2), 257–270.
Yang, B., Desai, R., & Kaczmarek, L. K. (2007). Slack and Slick K(Na) channels regulate the accuracy of timing of auditory neurons. *The Journal of Neuroscience, 27*(10), 2617–2627.
Yassin, L., Radtke-Schuller, S., Asraf, H., Grothe, B., Hershfinkel, M., Forsythe, I. D., et al. (2014). Nitric oxide signaling modulates synaptic inhibition in the superior paraolivary nucleus (SPN) via cGMP-dependent suppression of KCC2. *Frontiers in Neural Circuits, 8*, 65. https://doi.org/10.3389/fncir.2014.00065.

Chapter 6
Neurons, Connections, and Microcircuits of the Inferior Colliculus

Tetsufumi Ito and Manuel S. Malmierca

Abstract In this chapter, the neural circuitry of the inferior colliculus (IC) is described at a cellular level. The IC is subdivided into the central nucleus (ICC) and surrounding cortices that receive specific combinations of inputs and generate diverse outputs. Neuronal types in the IC can be distinguished by their dendritic arborization patterns, neurochemical profiles, and physiological properties. Based on these properties, neuronal types organizing the ICC appear to be different from those organizing the IC cortex. The IC receives ascending inputs from the cochlear nuclei, superior olivary complex, and nuclei of the lateral lemniscus, and the IC receives descending inputs from the auditory cortex and nonauditory inputs from various nuclei. Each input source forms terminal fields in particular zones of the IC. Massive commissural and local inputs are also present. One well-characterized cell type, the large GABAergic IC neuron, receives convergent input from all of these sources. The IC neurons project to the auditory thalamus as well as to lower brain stem nuclei. Anatomical and physiological features suggest that ICC acts as an auditory integration center, and IC cortex acts as a multimodal integration center and novelty detector. These data support the view that the lemniscal and nonlemniscal pathways emerge at the midbrain level.

T. Ito (✉)
Department of Anatomy, School of Medicine, Kanazawa Medical University, Uchinada, Ishikawa, Japan
e-mail: itot@kanazawa-med.ac.jp

M. S. Malmierca
Auditory Neuroscience Laboratory, Institute for Neuroscience of Castilla y León, University of Salamanca, Salamanca, Spain

Department of Cell Biology and Pathology, Faculty of Medicine, University of Salamanca, Salamanca, Spain

The Salamanca Institute for Biomedical Research (IBSAL), Salamanca, Spain
e-mail: msm@usal.es

© Springer International Publishing AG 2018

D. L. Oliver et al. (eds.), *The Mammalian Auditory Pathways*,
Springer Handbook of Auditory Research 65,
https://doi.org/10.1007/978-3-319-71798-2_6

Keywords Ascending fibers · Commissural fibers · Descending fibers · GABA · Glutamate · Intrinsic membrane properties · Lemniscal auditory pathway · Local circuit · Neurotransmitter · Nonlemniscal auditory pathway · Sound-evoked activity · Synaptic domain

6.1 Introduction

The inferior colliculus (IC) is a midbrain auditory structure that receives inputs from virtually all lower brainstem auditory nuclei and sends information to the thalamus. In addition to the convergence of inputs from these nuclei, IC neurons, which can be subdivided into various types, interconnect with each other. This unique local circuitry allows the emergence of new coding strategies for auditory information. This chapter highlights the anatomical organization of the IC and its physiological correlates. For most of the chapter, the focus is on the differences between the core area (central nucleus, lemniscal pathway) and shell areas (IC cortex, nonlemniscal pathways). All abbreviations are explained in Table 6.1.

6.2 Cytoarchitecture of the Inferior Colliculus

Numerous cytoarchitectonic studies of the IC have accumulated over the past 50 years. These have focused on its organization in a variety of species, including nonmammalian clades. There is a general consensus that the IC consists of a central nucleus that is surrounded by collicular cortices. The main goal here is to provide the reader with a brief introduction as background for the discussion of the intrinsic organization of the IC. A detailed description of interspecies variations is outside the scope of this chapter and is described elsewhere (e.g., Covey and Carr 2005). Here, descriptions are focused on the IC of rats and mice for which there is now a great deal of molecular, electrophysiological, and behavioral data.

The IC is a large, oval-shaped structure located in the dorsocaudal midbrain. The left and right ICs are interconnected by a large, heavily myelinated commissure (the commissure of the IC, CoIC). The central nucleus of the IC (ICC) is surrounded by dorsal, lateral, and rostral cortices. In coronal sections at levels including the central nucleus, the dorsal cortex (ICD) and lateral cortex (ICL) are located dorsomedial and lateral to the central nucleus, respectively (Fig. 6.1) (Loftus et al. 2008). In more caudal sections, the central nucleus disappears and the dorsal and lateral cortices merge together (Faye-Lund and Osen 1985); this merged zone is called the caudal cortex (Morest and Oliver 1984). Rostral to the central nucleus, forming the rostromedial boundary of the IC is a region that, in rats, is now commonly referred to as the rostral cortex (ICR) (Malmierca et al. 2011). Together with the ICL, the ICR was included as a part of the external cortex by Faye-Lund and Osen (1985) in

Table 6.1 Abbreviations

A1	Primary auditory cortex
AC	Auditory cortex
CF	Characteristic frequency
CN	Cochlear nucleus
CoIC	Commissure of the inferior colliculus
DCN	Dorsal cochlear nucleus
DNLL	Dorsal nucleus of the lateral lemniscus
EPSP	Excitatory postsynaptic potential
F	Flat or disc-shaped neurons in ICC
FD-BDA	Fluorescein dextran and biotinylated dextran amine
FM	Frequency modulation
GABAergic	Gamma-aminobutyric acid neurotransmitter
GAD	Glutamic acid decarboxylase
GFP	Green fluorescent protein
IC	Inferior colliculus
ICC	Central nucleus of the IC
ICD	Dorsal cortex of the IC
ICL	Lateral cortex of the IC
ICR	Rostral cortex of the IC
INLL	Intermediate nucleus of the lateral lemniscus
IPSP	Inhibitory postsynaptic potential
ITD	Interaural time difference
LF	Less flat or stellate neurons in ICC
LG	Large GABAergic
LSO	Lateral superior olive
MGB	Medial geniculate body
MGD	Dorsal division of the MGB
MGM	Medial division of the MGB
MGV	Ventral division of the MGB
MSO	Medial superior olive
NLL	Nucleus of the lateral lemniscus
SG	Small GABAergic
SOC	Superior olivary complex
SSA	Stimulus-specific adaptation
VCN	Ventral cochlear nucleus
VGLUT	Vesicular glutamate transporter
VLN	Ventrolateral nucleus of the IC
VNLL	Ventral NLL

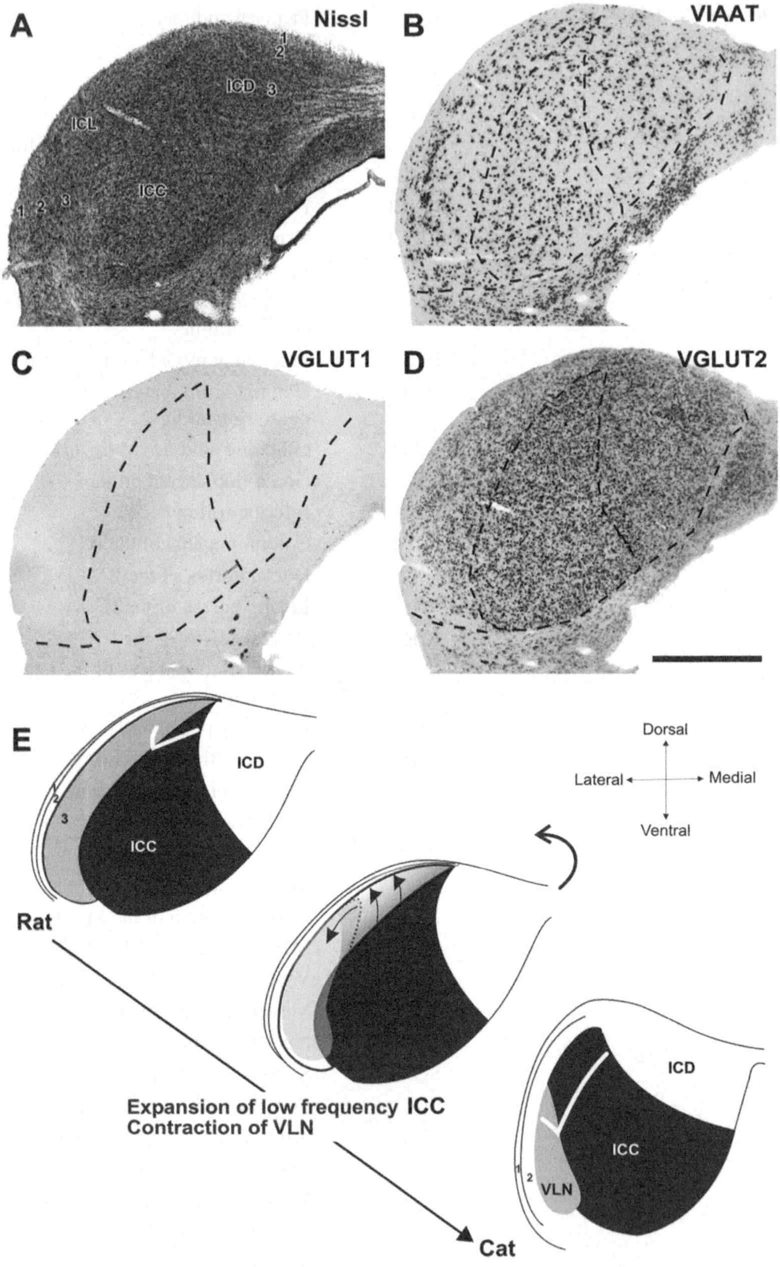

Fig. 6.1 Cytoarchitecture of the inferior colliculus (IC). **(A–D)** Labeling of cell bodies in the rat IC with several markers. Serial IC sections were processed for each staining. **(A)** Three subdivisions and layers in the cortices are discriminated with Nissl staining: central nucleus, *ICC*; dorsal cortex, *ICD*; lateral cortex, *ICL*. Layers *1, 2, 3* are also illustrated. **(B)** In the IC, VIAAT-positive neurons represent the GABAergic phenotype. Density of VIAAT-positive neurons is higher in the

their study of subdivisions in the rat, but now the ICR is known to have cell types and physiological properties different from those of the lateral (external) cortex. At the ventral extreme of the IC, a rostral extension of the central nucleus is sometimes referred to as the *rostral pole nucleus*. More laterally, layer 1 of the lateral cortex thickens and is continuous with the brachium of the IC.

6.2.1 Central Nucleus

The ICC is defined by the presence of fibrodendritic laminae that are composed of bundles of axons and dendritic arbors (Oliver and Morest 1984). The thickness of one lamina is about 50–70 μm (Malmierca et al. 1993). Within the laminae, numerous flat cell bodies are identified by Nissl staining and in situ hybridization (Fig. 6.1A, D) and correspond to disc-shaped or flat neurons. The main origin of axons in the fibrodendritic laminae is lemniscal, consisting of ascending fibers (Oliver and Morest 1984), but they also contain local collaterals of intrinsic IC axons (Sect. 6.5.2). The presence of fibrodendritic laminae is the basis of the tonotopic organization found in the ICC (Fig. 6.2A, B).

6.2.2 Dorsal Cortex

In cats, the ICD is subdivided into two outer layers and two deeper layers (Morest and Oliver 1984). A difference between layers 3 and 4 is less clear in rodents (Fig. 6.1A), and the dorsal cortex is subdivided into three layers (Faye-Lund and Osen 1985). Layer 1 is the continuum of that in the ICL. The deepest layer forms the largest part of the ICD. The ICD receives more descending inputs and fewer lemniscal ascending inputs than the ICC (Coleman and Clerici 1987; Herbert et al. 1991). However, defining a distinct border between the ICC and ICD based on Golgi or fiber staining is challenging, if not impossible. Rather, it is more likely that there is a loose transition between these subdivisions.

Fig. 6.1 (continued) ICC than cortices except for *layer 2* of the ICL, which represents GABA modules. *Dashed lines* represent approximate boundaries for regions labeled in **A.** **(C)** Absence of VGLUT1 expression in the IC. Note that ventral to the ICC, mesencephalic trigeminal neurons express strong VGLUT1 mRNA. **(D)** Numerous IC neurons express VGLUT2. It is clear that neurons expressing VGLUT2 outnumber cells expressing VIAAT. (*Scale bar*: 1 mm for **A–D**) **(E)** Schematic diagram of the IC subdivisions in the rat and cat. The low frequency region in the ICC is contracted in rats while the size of the ventrolateral nucleus, *VLN*, (corresponding roughly to *layer 3* of ICL) is expanded in rats. [**A–D** modified from Ito et al. (2011) with permission of Wiley; **E** redrawn from Loftus et al. (2008) with permission of Elsevier]

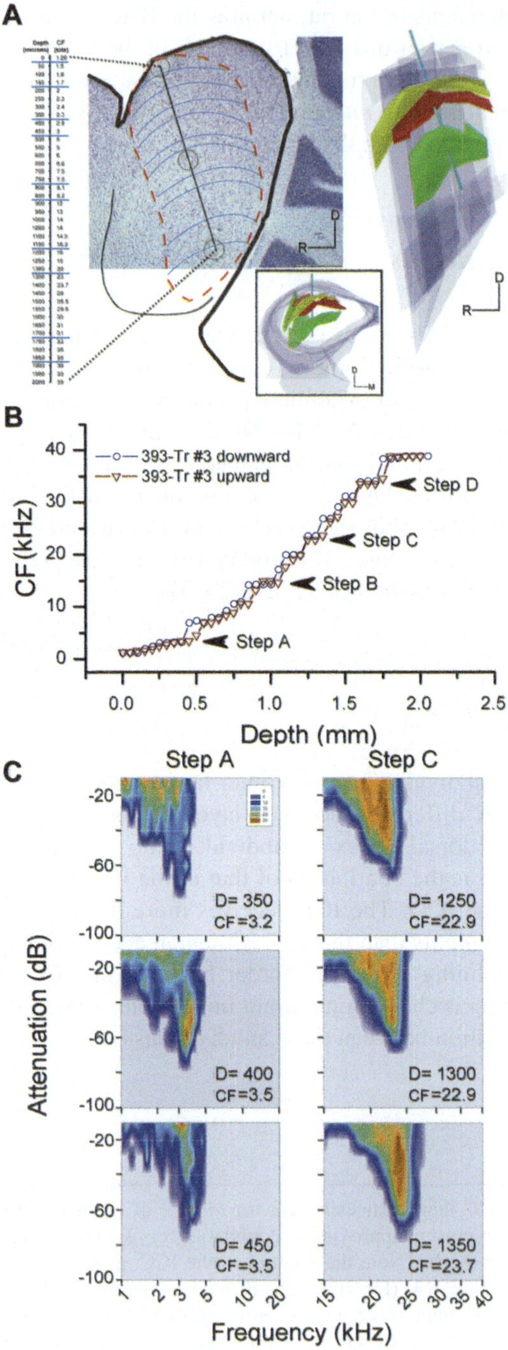

Fig. 6.2 Characteristic frequencies (CF) in the inferior colliculus. **(A)** ***Top left***: Sagittal section after Nissl staining; *dashed lines* show the boundaries of the fibrodendritic lamina in ICC. Frequency representation in the ICC along a 2000 μm electrode track (*black line*) identified

6.2.3 Lateral Cortex

The ICL is subdivided into three layers (Fig. 6.1). Since each layer has different types of inputs and outputs, they are likely to serve different functional roles. Ascending lemniscal inputs to the ICL, especially the peripheral layers, are scanty (Tokunaga et al. 1984). Instead, the auditory input originates mostly from the ICC. *Layer 1* is a thin, myelin-rich fibrous layer that is continuous with layer 1 of the dorsal cortex and brachium of the IC (Faye-Lund and Osen 1985; Oliver 2005). It contains numerous ascending fibers to the thalamus. *Layer 2* is composed of densely packed small neurons that form several clusters within a myelin-rich neuropil (Faye-Lund and Osen 1985). These two layers together constitute about one third of the maximum thickness of the ICL. *Layer 3* is larger than the other layers and may correspond to the ventrolateral nucleus of cats (VLN, Fig. 6.1E) (Faye-Lund and Osen 1985; Loftus et al. 2008). Layer 3 receives tonotopically organized axons that form band-like axonal plexuses almost orthogonal to the plexuses extending into the ICC (Saldaña and Merchán 1992; Malmierca et al. 1995).

6.2.4 Rostral Cortex

The ICR covers the rostral part of the IC. In coronal sections the ICR is easily recognized by the presence of many commissural fibers that interconnect both ICs and cross the ICR. The ICR may be homologous to a part of the intercollicular tegmentum described in the cat (Morest and Oliver 1984). Cytoarchitectonic features of the ICR or intercollicular tegmentum were revealed by Golgi preparations as very different from other IC regions (Morest and Oliver 1984). The neurons that populate this region are mostly multipolar cells of different sizes, with the largest ones being the most conspicuous (Malmierca et al. 2011). The function of the ICR is speculated to be multimodal integration (Oliver 2005).

Fig. 6.2 (continued) with three electrolytic lesions (indicated by *circles*); CF recorded at 50 μm intervals revealed frequency increases as a function of depth. Curves roughly perpendicular to the electrode track show the approximate orientations of the fibrodendritic laminae in ICC. CFs remain the same for roughly 150 μm and then change abruptly. ***Top right***: Three-dimensional reconstruction of three axonal laminae. The recording electrode entered the ICC at a 10° angle with respect to the main dorsoventral axis of the ICC. Three different anatomical laminae formed by afferent axons are illustrated (*top*: 1.7 kHz lamina; *middle*: 1.8 kHz; *bottom*: 4.5 kHz). ***Bottom***: Frontal view of the same three-dimensional reconstruction after 90° rotation. (*D*, dorsal; *M*, medial; *R*, rostral) (**B**) A single electrode penetration downward (*circles*) and upward (*triangles*) along the same electrode track (*Tr*) through the ICC in which frequency response areas were obtained from multiunit clusters. (*CF*, characteristic frequency) (**C**) Frequency response areas from Step A and C shown in **B** recorded at 50 μm intervals; *Color scale* represents neuronal firing rate. *D*, Depth. [**A** anatomical data from Malmierca et al. (2005, 2008); **B** redrawn from Malmierca et al. (2008) with permission from the Society for Neuroscience; **C** redrawn from Oliver et al. (2011) with permission from Elsevier]

6.3 Neuron Types

There have been many attempts to classify neuron types in the IC. Classically, neurons in the IC were classified by dendritic morphology in Golgi impregnated material (Fig. 6.3). More recent studies utilize molecular profiles or electrophysiological properties to classify neuron types. These attempts have been only partially successful as the data indicate that there is no simple relationship between neuron classes defined with different methods. Most probably, a combination of several methods will be needed to draw a clearer portrait of IC neuron types.

6.3.1 Dendritic Morphologies of Neurons in the Inferior Colliculus

6.3.1.1 Dendritic Arborizations of Central Nucleus Neurons

Studies based on computer-assisted three-dimensional reconstructions of Golgi-impregnated material have shown that two main types of neurons populate the ICC (Fig. 6.3): disc-shaped or flat (F) neurons that have flattened dendritic arbors, and stellate cells or less flat (LF) neurons (Fig. 6.3B, C) with dendritic arbors that often extend across the laminae (Oliver and Morest 1984; Malmierca et al. 1993). The majority of ICC neurons (78%, cat; 67%, rat) are classified as disc-shaped or flat neurons.

The F and LF neurons differ in several respects, including dendritic arbor thickness, dendritic branching pattern, orientation, and location with regard to the laminae (Figs. 6.3, and 6.4B). In rats, the dendritic arbor of the F neurons lies parallel to a lamina that is about 50 μm wide. The laminae are separated by 100 μm-thick interlaminar compartments that are populated by the LF neurons. The dendritic arbor of these LF neurons is 100 μm wide and is less dense than the dendritic arbor of the F neurons (Malmierca et al. 1993). The dendritic arbors of most F and LF neurons are elongated and oriented in parallel with the ventrolaterally to dorsomedially oriented long axis of the laminae, although a few F neurons are oriented rostrocaudally (Fig. 6.4). A gradual shift takes place so that the orientation is more horizontal in the dorsolateral (low frequency) part of the nucleus and more vertical in the medial (high frequency) part (Malmierca et al. 1993).

6.3.1.2 Dendritic Morphologies of Cortex Neurons

Studies on the dendritic architecture in cortical regions of the IC are limited (Malmierca et al. 2011). Layer 1 of the ICD and ICL is made of small flattened neurons that are sparsely distributed (Faye-Lund and Osen 1985) and extend their dendrites parallel to the surface of the IC (Ito, unpublished observations). The deeper, slightly thicker layer 2 of the ICD consists of small and medium-sized, mostly multipolar neurons (Th+ cells in Fig. 6.7) (Ono et al. 2005). Layer 2 of the ICL also

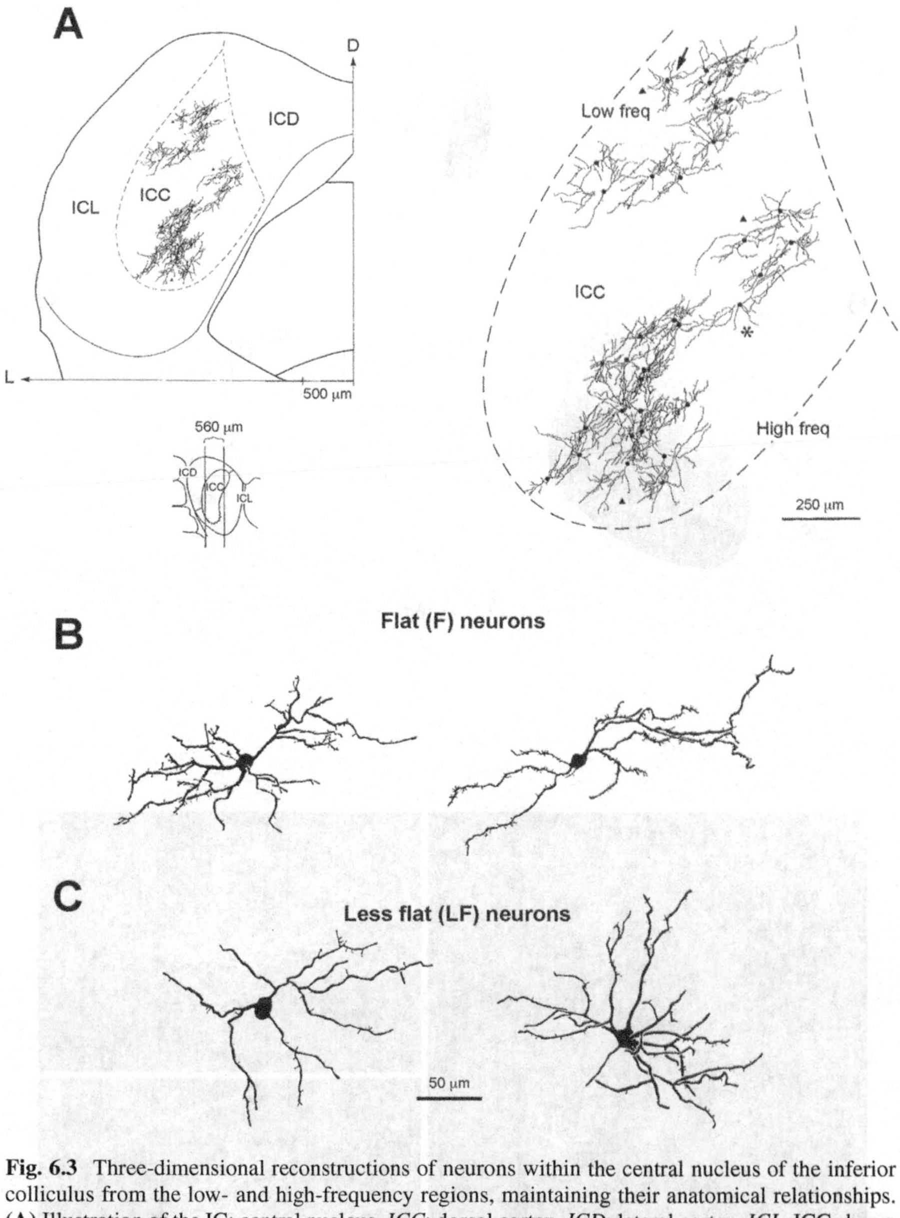

Fig. 6.3 Three-dimensional reconstructions of neurons within the central nucleus of the inferior colliculus from the low- and high-frequency regions, maintaining their anatomical relationships. **(A)** Illustration of the IC: central nucleus, *ICC*; dorsal cortex, *ICD*; lateral cortex, *ICL*. ICC shown enlarged on the right. *Inset* on left below panel indicates a sagittal section from where neurons were drawn. *Arrow* in the enlargement points to a perpendicular F neuron, *asterisk* is next to the longest cell, and *triangles* are adjacent to large cells in the outskirts of the ICC. **(B, C)** Camera lucida drawings of two flat neurons (*F*) and two less flat neurons (*LF*). (*Scale bars*: **A** enlargement 250 μm; **B, C** 50 μm) [Redrawn from Malmierca et al. (1993) with permission from Wiley]

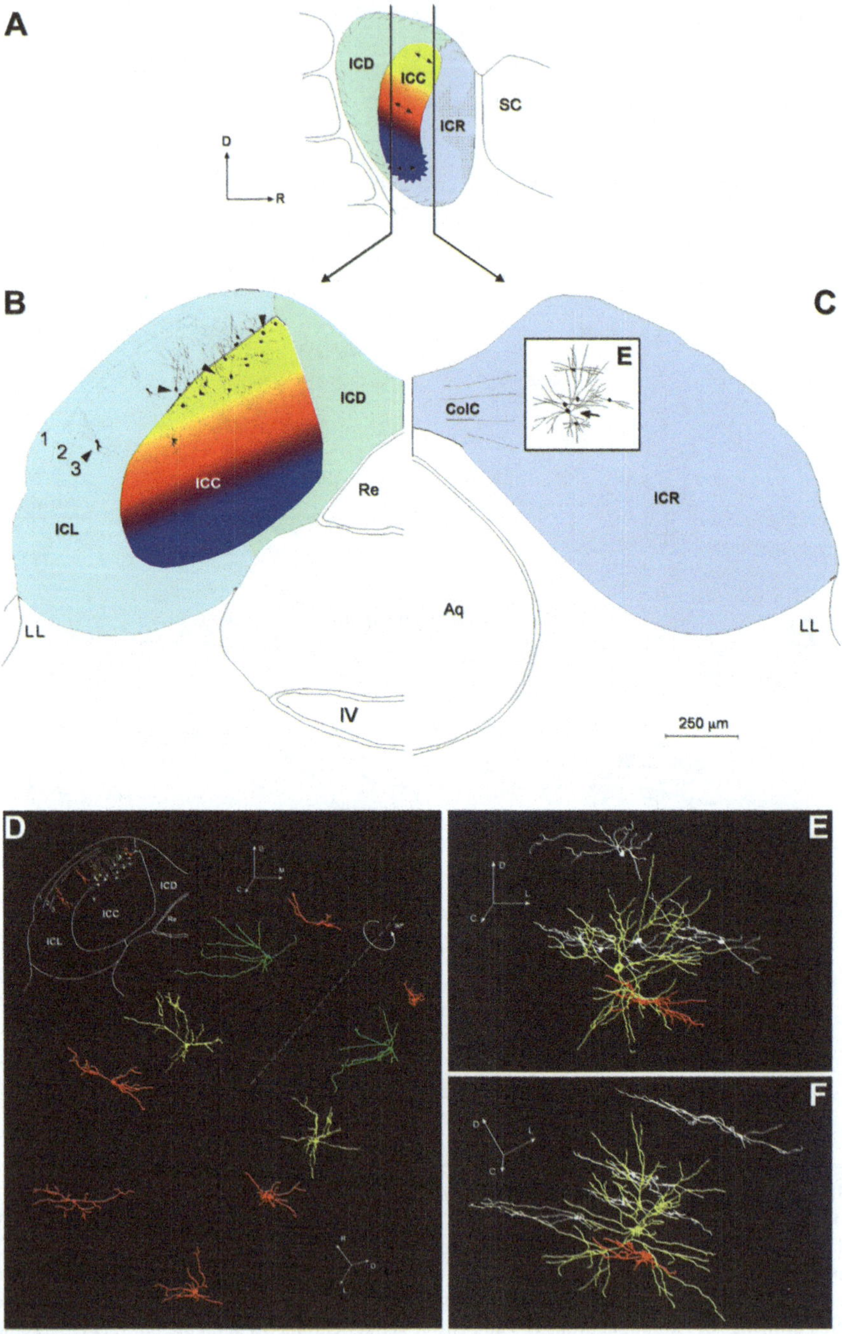

Fig. 6.4 Camera lucida drawings of a transverse section through the inferior colliculus (IC) of an adult rat. **(A)** Diagram illustrating the divisions of the IC. **(B)** Left IC is from a more caudal plane as indicated in the sagittal diagram in **A**. The left IC shows 7 flat neurons from the low frequency

contains small cells, but they form distinct clusters. Layer 3 of the ICD contains at least two types of neurons: large multipolar neurons and transitional neurons with elongated dendritic arbors that parallel the orientation of the ICC laminae and are located at the border between the ICC and the ICD (Malmierca et al. 2011).

Three-dimensional reconstructions of ICL neurons have shown that in addition to small and medium-sized cells, layer 3 contains large multipolar cells, especially ventromedially and rostrally, which makes the border with the ICC appear blurred (Fig. 6.1) (Loftus et al. 2008). In contrast, the dorsolateral portion has three morphologically distinct neuron types: bitufted, pyramidal-like, and chandelier. The dendritic orientations of these neurons mark a border with the ICC (Fig. 6.4B, D) (Faye-Lund and Osen 1985; Malmierca et al. 2011). Classical studies in the mouse by Willard and Ryugo (1983) described "large stellate cells with elongated dendritic arbors whose axes are oriented perpendicular to the pial surface." These types of neurons and distinct inputs to each layer suggest that the ICL possesses a cortical-like neuronal architecture, as originally described by Ramón y Cajal in 1902 (Malmierca et al. 2011).

Finally, neurons in the ICR lack a distinct dendritic orientation, so there is no indication of layering or anisotropy in the tissue (Malmierca et al. 2011). The neurons that populate this region are mostly multipolar cells of different sizes. The dendritic trees of large multipolar neurons are made of four or more thick dendritic processes that branch repeatedly without any preferred orientation and exhibit a few pedunculated spines. These neurons have the largest dendritic arbors in the entire IC (Fig. 6.4E, F).

6.3.2 *Neurotransmitters in the Inferior Colliculus*

Most IC neurons are considered to use GABA (gamma-aminobutyric acid) or glutamate as their neurotransmitter (see Fig. 6.1B–D). The use of glycine as a neurotransmitter is unlikely in the IC neurons since GLYT2 (necessary for the glycinergic

Fig. 6.4 (continued) region of the ICC and 16 neurons in the adjacent ICL. *Arrowheads* indicate some bitufted neurons. Some of the neurons are reconstructed in **D** (*scale bar* for B & C only). **(C)** The right IC showing neurons at the border zone between the rostral part of the ICC and the ICR with one typical large multipolar cell (*arrow*) and some flat neurons of the ICC. The same group of neurons is reconstructed in **E** and **F**. **(D)** Three-dimensional reconstruction of pyramidal (*yellow*), bitufted (*red*), and chandelier (*green*) neurons selected from the group of ICL neurons shown in the left IC in **B** and reproduced in the *inset at the upper-left* corner. The neurons are shown as drawn in the transverse section and then rotated about 90° until seen from a dorsolateral view with the neurons on end. Note that the cross section of the cells is cylindrical and not flattened. **(E, F)** Three-dimensional reconstructions of a selected group of neurons shown in **C** (right IC) from the transversely cut section at the border zone between the ICC and the ICR. The *yellow neuron* is a typical large multipolar neuron from the ICR while the *white neurons* are flat (*F*) neurons and the *red* is a less-flat (*LF*) neuron from the ICC (**E** as drawn in **B**; **F** after being rotated until seen on edge). The caudal part of the dendritic tree of the ICR neuron interdigitates with the F and LF neurons in this border zone. (*Aq*, aqueduct; *C*, caudal; *CoIC*, commissure; *D*, dorsal; *ICC*, central nucleus; *ICD*, dorsal IC; *ICL*, lateral IC; *ICR*, rostral IC; *LL*, lateral lemniscus; *SC*, superior colliculus; Re, recesus IV ventricle; *IV*, trochlear cranial nerve) [Redrawn from Malmierca et al. (2011) with permission from Elsevier]

phenotype) (Aubrey et al. 2007) is not expressed in the IC (Tanaka and Ezure 2004). The following sections include a description of IC neurons that use one of the two classical neurotransmitters and then a brief discussion of other neurotransmitters/neuromodulators follows.

6.3.2.1 Glutamate and GABA are Expressed in the Entire Inferior Colliculus

Numerous glutamatergic neurons are present in the IC. Electrical stimulation of the brachium of the IC elicited glutamate-receptor-mediated excitatory responses in medial geniculate neurons (Hu et al. 1994), suggesting that many IC ascending fibers are glutamatergic. Uptake of D-[^{3}H]aspartate by synaptic terminals of IC axons also suggested the presence of glutamatergic neurons in the IC (Saint Marie 1996). Indeed, numerous IC neurons express mRNA for vesicular glutamate transporter 2 (VGLUT2) (Fig. 6.1D) (Ito et al. 2011), which takes up cytoplasmic glutamate into synaptic vesicles (Fremeau et al. 2001). Messengers for the other isoforms of VGLUT (VGLUT1 and VGLUT3) are not expressed in adult IC (Fig. 6.1C) (Ito et al. 2009, 2011).

The presence of GABAergic neurons in the IC is well established, and they make up 20–30% of all IC neurons (Oliver et al. 1994; Merchán et al. 2005). The mean diameter of GABAergic neurons is larger than that of non-GABAergic neurons (Oliver et al. 1994; Merchán et al. 2005) and, indeed, the largest IC neurons are GABAergic (Oliver et al. 1994; Ito et al. 2009). IC GABAergic neurons show large diversity in size and are heterogeneously distributed (Fig. 6.1B). GABAergic neurons are more abundant and have higher density in the ICC than in the cortices (Merchán et al. 2005) except for the GABA modules in layer 2 of the ICL (Chernock et al. 2004), which are strongly positive for glutamic acid decarboxylase (GAD), a key enzyme in GABA synthesis. The GABAergic neurons in the GABA modules are likely to be a distinct neuron type (see Sect. 6.3.4). On the other hand, GABAergic neurons in the other part of the IC can be classified into at least 2 categories: large and small GABAergic (LG and SG) neurons (Fig. 6.5) (Ito et al. 2009; Ito and Oliver 2012). These two types receive different inputs and target different regions (Sect. 6.5.3). GABAergic neurons can be either disc-shaped or stellate (Ono et al. 2005); however, it seems that LG neurons are mostly stellate (Oliver et al. 1994; Ito and Oliver 2012). Most GABAergic neurons are likely to colocalize parvalbumin (Chernock et al. 2004).

6.3.2.2 Neuromodulators are Common in the Cortex

Several molecules that act as neuromodulators are expressed in the cortices of the IC. Nitric oxide, which is considered to be a retrograde messenger, is produced by ICD and ICL neurons (Coote and Rees 2008; Loftus et al. 2008). Enkephalin-expressing neurons are found in the ICD and in layers 1 and 3 of the ICL, with fewer in the ICC (Nakagawa et al. 1995). Since IC neurons express opioid receptors (Tongjaroenbuangam et al. 2006), the enkephalin may act locally. Layer 2 receives

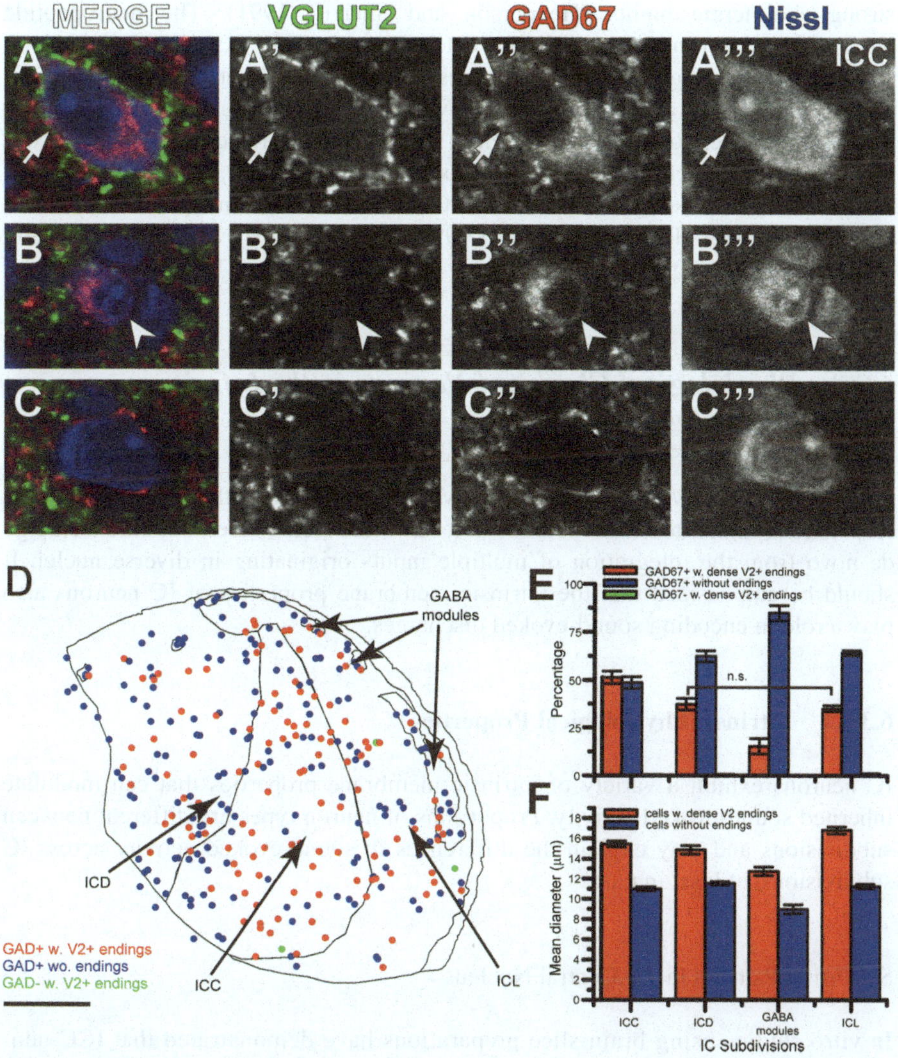

Fig. 6.5 Two classes of GABAergic neurons in the inferior colliculus (IC). **(A–C)** Confocal images of large GABAergic neurons (*arrows in* **A**), small GABAergic neurons (*arrowheads in* **B**), and nonGABAergic neurons (**C**) in the central nucleus (*ICC*). The large GABAergic (LG) neuron had a GAD67-positive (*3rd column*) large soma that was encircled by dense VGLUT2-positive (*2nd column*) terminals. The small GABAergic (SG) and GAD67-negative Nissl-stained (*4th column*) cell bodies lacked the dense VGLUT2-positive axosomatic terminals. **(D)** Distribution of GABAergic neurons. *Red dots* (GAD+ with V2 endings) indicate LG cell bodies; *blue dots* (GAD+ without endings) indicate SG cell bodies. *Green dots* are GAD67-negative cell bodies with dense VGLUT2-positive axosomatic terminals (GAD− with V2+ endings), which were rare. **(E)** Histogram illustrating relative percentage of the three neuron types shown in **D**; LG neurons were more commonly found in the ICC and were less common in the GABA modules (*scale bar:* 500 µm). **(F)** Histogram illustrating mean diameter of GABAergic somata: in all subdivisions, LG neurons had larger mean diameters than SG neurons. (*GABA*, gamma-aminobutyric acid; *GAD*, glutamic acid decarboxylase; *ICD*, dorsal IC; *ICL*, lateral IC; *VGLUT*, vesicular glutamate transporter) [Modified from Ito et al. (2009) with permission from the Society for Neuroscience]

strong cholinergic inputs (Henderson and Sherriff 1991). The neuropeptide Y-expressing neurons are mainly present in layer 2 of the ICL (Nakagawa et al. 1995). It appears that other neuropeptides, such as somatostatin, substance P, and cholecystokinin, are also expressed mainly in the ICD and ICL (Wynne et al. 1995; Wynne and Robertson 1997). These data suggest that many neuromodulators control layer-specific activity of the IC cortex. It should be noted, however, that some neuromodulators, like acetylcholine, serotonin, or noradrenaline, also innervate the ICC diffusely (Henderson and Sherriff 1991; Klepper and Herbert 1991; Schofield and Hurley, Chap. 9).

6.3.3 Physiological Classes of Neurons in the Inferior Colliculus

There is a large diversity of auditory-evoked responses in the IC. Some of the responses are inherited from neurons in lower brain stem nuclei, but others emerge de novo from the integration of multiple inputs originating in diverse nuclei. It should be emphasized that the intrinsic membrane properties of IC neurons also play a role in encoding sound-evoked discharges.

6.3.3.1 Intrinsic Physiological Properties

IC neurons exhibit a variety of intrinsic membrane properties that can modulate inherited sound-evoked activity. Proportions of neuron types are different between subdivisions and may explain the differences in sound-evoked activity across IC subdivisions, at least in part.

Six Firing Patterns in the Central Nucleus

In vitro studies using brain slice preparations have demonstrated that ICC neurons show one of six firing patterns in response to depolarizing current injection (Fig. 6.6) (Sivaramakrishnan and Oliver 2001). Onset neurons (8.6% of ICC neurons; Fig. 6.6B) can be quickly repolarized after both single and summed depolarizing currents so that they are well-suited for temporal coding (Peruzzi et al. 2000) but cannot encode the duration or intensity of sounds (Sivaramakrishnan and Oliver 2001). Sustained neurons exhibit a heterogeneous array of responses that can be grouped into five types: (1) sustained regular (19.2% of ICC neurons; Fig. 6.6A); (2) pause/build (12.5%; Fig. 6.6C); (3) rebound regular (10.6%; Fig. 6.6D); (4) rebound adapting (25%; Fig. 6.6E); and (5) rebound transient (21.1%; Fig. 6.6F) (Sivaramakrishnan and Oliver 2001). Sustained-regular, pause/build, and rebound-regular neurons show sustained firing with constant interspike intervals throughout injection of depolarizing current (Fig. 6.6). Since they have a linear voltage response to a wide range of input currents, they may have linear rate-level functions to sound. The relatively simple properties of onset neurons and sustained-regular neurons imply that these neuron types may serve to relay intensity and temporal information. On the other hand, the remaining neuron types, which make up about 70% of ICC neurons, substantially modify inputs and create de novo responses.

Pause/build neurons have an A-current. Hyperpolarization that precedes depolarization, such as that produced by a previous after-hyperpolarization or an inhibitory postsynaptic potential (IPSP), removes inactivation of the A-current and modifies the initial part of the neural response. This produces a buildup response or a pause in firing after an onset spike. The pause duration is proportional to the magnitude of hyperpolarization. Therefore, the pause/build cells may show temporal interaction between two stimuli with a time lag smaller than 100 ms.

Rebound neurons fire just after the cessation of hyperpolarization and may code offset of inhibition. Each of the three types of rebound neurons show different degrees of adaptation of firing to depolarization. Reboundregular neurons do not exhibit adaptation, whereas reboundtransient neurons show only phasic activity. Reboundadaptation neurons start to show adaptation for stimuli longer than 200 ms.

Tan et al. (2007) demonstrated firing properties of rodent IC neurons to current injections in vivo. In living animals, spontaneous excitatory postsynaptic potentials (EPSP) are common and affect firing regularity. Although responses to depolarizing current were generally similar to the slice studies, there were several differences. These authors found accelerated firing in 11% of cells, but they did not observe onset neurons that fired only a single spike. Burst-firing neurons were seen more commonly (15%) in the in vivo preparations than in vitro.

Firing Patterns in the Cortex

In contrast to the six firing patterns of ICC neurons described in Sect. 6.3.3.1.1, the IC cortex appears to be composed of fewer neuron types. In the ICL, the majority of neurons studied (66%) showed regular firing, and transient neurons were never observed (Ahuja and Wu 2007). ICL neurons have a lower threshold of firing than ICC and ICD neurons (Li et al. 1998). The relatively uniform membrane properties and the low thresholds of ICL neurons imply that the ICL may relay auditory information from the ICC (Ahuja and Wu 2007).

In the ICD, almost all neurons studied (91%) showed sustained firing; the rest showed a buildup response (Sun and Wu 2008). Half of the sustained neurons exhibited a rebound response after hyperpolarizing current injection. Interestingly, preceding hyperpolarization modifies the response to depolarization in more than half of the non-rebound neurons. These findings suggest that most ICD neurons modify their firing activity based on preceding inhibition.

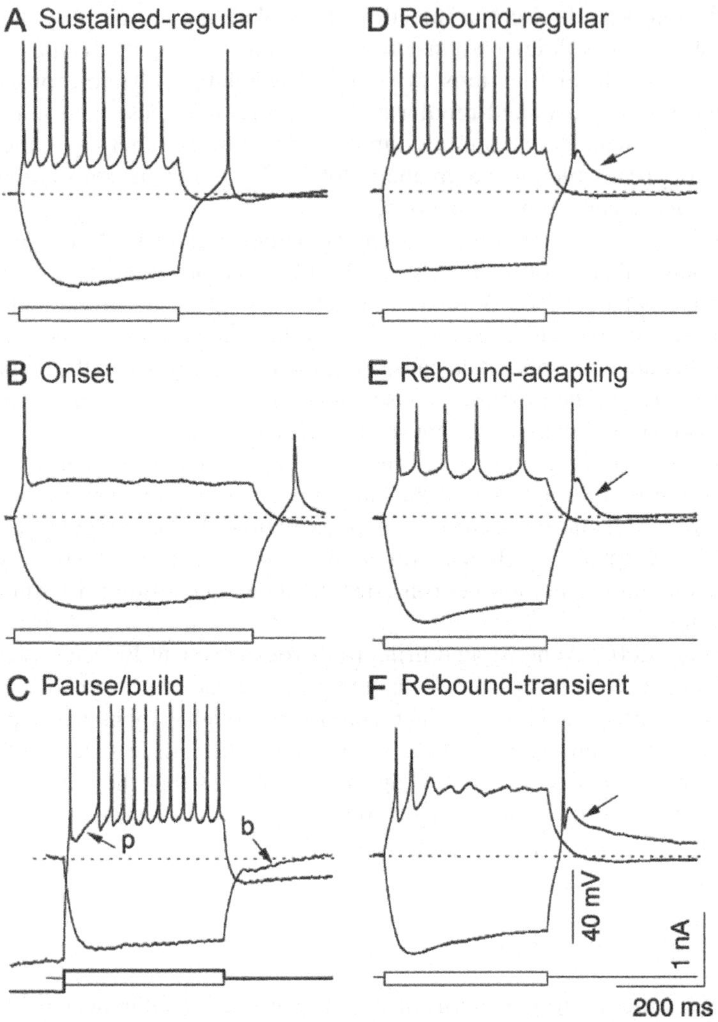

Fig. 6.6 The six firing patterns found in the inferior colliculus after injection of depolarizing and hyperpolarizing current pulses. In each panel, the *top two traces* are the voltage response to each current pulse (*bottom traces*). **(A)** Sustained-regular and **(B)** onset firing to depolarization with an anode-break spike and no calcium rebound following hyperpolarization. **(C)** Pause/build response to depolarization after prehyperpolarization and no anode-break spike after hyperopolarization. This cell shows a pause in firing (*p*) after the first spike; the response to a hyperpolarizing current shows a buildup response (*b*). **(D)** Sustained regular firing to depolarization with an anode-break sodium spike and a calcium rebound following the hyperpolarization. **(E)** Sustained firing with adaptation and calcium-sodium rebound activity. **(F)** Transient response to depolarization with calcium-sodium rebound activity. [Figure kindly provided by Dr. Douglas L. Oliver; redrawn from Sivaramakrishnan and Oliver (2001) with permission from the Society for Neuroscience]

6.3.3.2 Responses to Sound

Tuning of Cortex Neurons

A fundamental physiological feature of the ICC is its tonotopic organization. Neurons located in dorsolateral laminae respond to low frequency sounds, and neurons located more ventrally respond to progressively higher frequency sounds. However, this representation does not appear to reflect a continuous mapping—fine grained tonotopic mapping studies have demonstrated a distinct stepwise organization in the progression of characteristic frequencies (CF) (Schreiner and Langner 1997; Malmierca et al. 2008). Moreover, independent data from single neurons showed that CFs at octave intervals of approximately one-third are more prevalent than others (Malmierca et al. 2008). Thus, a narrow range of CFs is represented within each isofrequency lamina (Fig. 6.2B, C). Since the physiological step width is similar to the laminar thickness, and the frequency increase is comparable to one critical band width, the isofrequency laminae have been postulated to correlate with critical bands (Schreiner and Langner 1997).

An elegant study recently explored the possible relationships between the microstructure of the cochlear map and the tiered tonotopy observed in the IC (Shera 2015). The stepwise tonotopy may be an emergent property arising from wave reflection and interference within the cochlea, a mechanism similar to that responsible for the microstructure of the hearing threshold (Shera 2015). Each lamina of the ICC also shows a highly organized representation of both spectral and temporal features of the acoustic signal. An axis that maps the best modulation frequency appears to be oriented orthogonal to the frequency or tonotopic axis and within the plane of the laminae (reviewed by Schreiner and Langner 1988, 1997).

Single-unit recording studies of ICC neurons in response to pure tone stimulation have revealed that IC neurons show a large variety of frequency response areas that include both V-shaped (Fig. 6.2C) and non-V-shaped ones (Palmer et al. 2013). The non-V-shaped response types form a heterogeneous group that includes closed, narrow, low- and high-tilt, and multipeaked response areas. In addition, ICC neuron responses produce different types of peristimulus time histograms including onset, on-sustained, pauser, sustained, and regular responses (Duque et al. 2012). The ICC neurons are also sensitive to the duration of the sound stimuli (Pérez-González et al. 2006) and to binaural stimulation. Thus ICC binaural responses can be suppressive, summative, or mixed (Zhang and Kelly 2010). Binaural suppression responses are more numerous at high frequencies and summation is more common at low frequencies. These characteristics of binaural interaction likely reflect the time course of converging excitatory and inhibitory synaptic inputs to ICC neurons as well as the intrinsic membrane properties of those neurons. Studies based on the iontophoretic application of GABA and glycine antagonists have shown that neural inhibition contributes to the binaural responses of neurons in the IC (Faingold et al. 1989, 1991). The ICC neurons show responses to amplitude modulation and frequency modulation (FM) as a function of the modulation frequency and other parameters (Moller and Rees 1986; Rees and Moller 1987).

Novelty Detection in Cortex Neurons

Studies on the neuronal responses of cortical regions of the IC are still sparse, but there is evidence that these neurons exhibit multisensory integration that mirrors a similar integration at higher levels of the nonlemniscal auditory pathway (reviewed by Malmierca and Ryugo 2011). Another prominent feature of neurons in the nonlemniscal IC regions is the phenomenon of stimulus-specific adaptation (SSA). This property is a specific adaptation of neuronal responses to a repeated stimulus that does not fully generalize to other stimuli. It provides a mechanism for emphasizing rare and potentially interesting sensory events. The SSA phenomenon has been linked to deviance detection, auditory memory, recognition of acoustic objects, auditory scene analysis, and behavioral habituation (Netser et al. 2011; Ayala and Malmierca 2012). Moreover, SSA may be a component of short-term memory in neuronal networks since it creates a nontrivial dependence of neuronal responses on the history of the stimulation. A series of recent studies of cortical IC neurons (i.e., nonlemniscal neurons of the IC) demonstrated that the great majority exhibit SSA, display transient onset responses, and have broad frequency response areas (Pérez-González et al. 2005, 2012). As described in Sect. 6.3.1.2, neurons in the ICD and ICR regions possess widespread dendritic arbors (Malmierca et al. 2011) and broader frequency tuning than those in the ICC (Duque et al. 2012). Although SSA was originally described in the cat primary auditory cortex and was suggested to be an emergent property of the auditory cortex (AC) (Ulanovsky et al. 2003), IC neurons do not inherit their SSA sensitivity from the AC (Anderson and Malmierca 2013). However, they may be strongly modulated in a gain control fashion (Anderson and Malmierca 2013) directly through the corticofugal projections (Ayala et al. 2015) or indirectly via cholinergic inputs (Ayala and Malmierca 2015).

6.3.4 Dendritic Morphologies, Neurotransmitters, and Physiological Properties

As the numbers of morphological and physiological classes are both relatively small, one might expect a simple one-to-one relationship between dendritic morphology and firing properties. However, the dichotomy of dendritic morphology (disc-shaped or stellate) does not generally predict differences in physiological properties. Both disc-shaped and stellate cells appear to be subdivided into several physiological classes.

Correlations between intrinsic membrane properties and dendritic arborization was addressed in slice studies (Reetz and Ehret 1999; Peruzzi et al. 2000). Although both transient and sustained responses were observed, the ratio of each response differed across studies. Interestingly, the ratio of neuron types based on dendritic arborization also differed among the studies, and it is likely that the higher the percentage of neurons with unoriented dendrites, the higher the percentage of onset responses. Indeed, an in vivo juxtacellular study demonstrated that the onset response was characteristic of stellate neurons but not of those with oriented dendrites (Wallace et al. 2012). Further, rebound neurons may have larger somata with more complex dendritic trees, whereas buildup neurons have smaller somata with a simpler dendritic tree (Peruzzi et al. 2000).

Ono et al. (2005) studied GABAergic neurons in the IC by using GAD67-GFP mice in which GABAergic neurons express GFP (green fluorescent protein). They demonstrated that GABAergic neurons can be subdivided into several classes by their dendritic arborizations and intrinsic membrane properties (Fig. 6.7).

A common feature of GABAergic neurons is the presence of a depolarizing sag, which may enhance temporal precision in response to synaptic inputs. The GABA neurons in layer 2 of the ICL (Th+ in Fig. 6.7), in other words, the GABA modules, form a single distinct class of IC neurons since they share a small, oriented dendritic morphology and a transient response with a hump in response to current injections, which is only expressed in these neurons. Regular-sustained neurons (RS in Fig. 6.7)

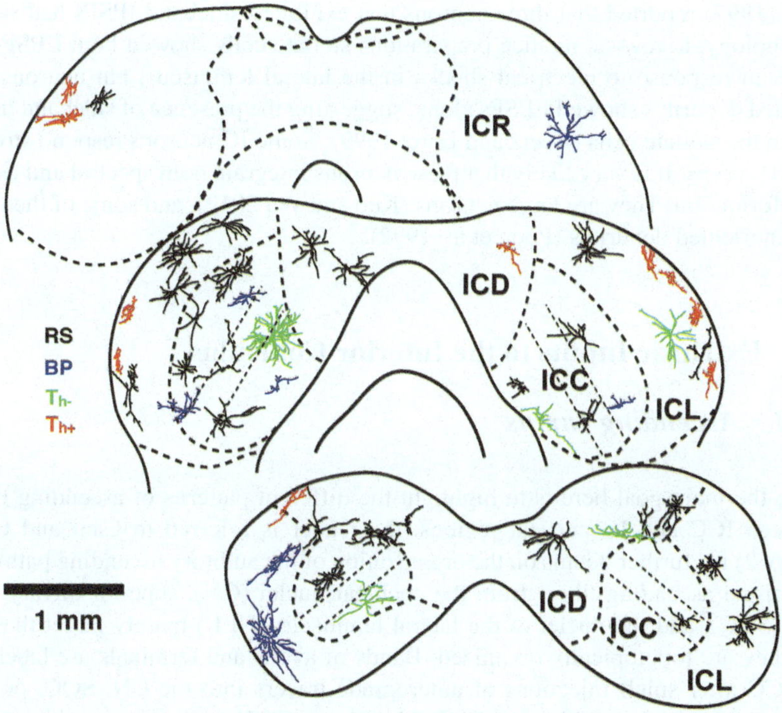

Fig. 6.7 Dendritic arborization and distribution of GABAergic neurons in the inferior colliculus (IC) for which intrinsic properties were identified and classified into regular-sustained (*RS*, *black*), buildup-pauser (*BP*, *blue*), transient with hump (*Th*−, *green*), or transient without hump (*Th*+, *red*). *Thick dashed lines* indicate the border of IC subdivisions, and *thin dashed lines* indicate laminae in the central nucleus. (*ICC*, central nucleus of IC; *ICD*, dorsal nucleus of IC; *ICL*, lateral nucleus of the IC; *ICR*, rostral nucleus of the IC) [Figure kindly provided by Dr. Munenori Ono; modified from Ono et al. (2005) with permission from Elsevier]

are commonly found in every subdivision. Buildup-pauser neurons (BP in Fig. 6.7) are likewise common in ICC and ICL but almost absent in the ICD. These two neuron types do not correlate with specific dendritic morphologies. A minority of GABAergic neurons in the ICC and ICD express a transient response without a hump (Th– in Fig. 6.7) and are large stellate neurons. Some of these GABAergic neurons may correspond to those classified as LG neurons that are contacted by dense excitatory axosomatic synapses and project to the thalamus (Ito et al. 2009; Ito and Oliver 2012). Geis and Borst (2013) demonstrated that LG cells in mouse ICD show an EPSP with a short delay at the onset of sound, have low input resistance, and exhibit a depolarizing sag in response to hyperpolarizing current. These characteristics imply that activation of the LG cells requires the occurrence of many EPSPs within a short period of time (Ito and Oliver 2014). Geis and Borst (2013) also noted that adaption to depolarization current is more commonly seen in neurons with larger cell bodies.

Since stellate neurons extend dendrites into neighboring isofrequency laminae, they may be associated with spectral integration. In an intracellular study, Kuwada et al. (1997) reported that three neurons that exhibited sideband IPSPs had stellate morphology. Likewise, in slice preparations stellate cells showed both EPSPs and IPSPs in response to electrical shocks in the lateral lemniscus, but neurons with oriented dendrites showed EPSPs alone, suggesting the presence of sideband inhibition in the stellate cells (Reetz and Ehret 1999). Some IC neurons respond strongly to FM sweeps. It is very likely that these neurons integrate both spectral and temporal information. They are large neurons (Kuo and Wu 2012), and some of them also had unoriented dendrites (Poon et al. 1992).

6.4 Extrinsic Inputs to the Inferior Colliculus

6.4.1 Ascending Inputs

Since the main goal here is to highlight the different patterns of ascending inputs between ICC and IC cortical regions, the reader is referred to Cant and Oliver (Chap. 2) for further details on the organization of the auditory ascending pathways. Lemniscal ascending fibers from the cochlear nuclei (CN), superior olivary complex (SOC), and the nuclei of the lateral lemniscus (NLL) mainly target the ICC, and they are tonotopically organized. Bands of axons and terminals are labeled in the ICC after small injections of anterograde tracers into the CN, SOC, or NLL (reviewed by Oliver and Huerta 1992; Malmierca 2015).

A portion of the IC cortex, mainly the deeper layers, also receives tonotopically organized afferent inputs (details in Oliver and Huerta 1992). Band-like ascending lemniscal fibers arising from the CN and NLL invade the deep layers of the ICD. In the cat, ventral CN (VCN) axons are found in layers 3 and 4, while dorsal CN (DCN) axons are restricted to layer 4, which is adjacent to the ICC. Layer 3 of the ICL also receives inputs from the same nuclei as the deeper layers of the

ICD. Ventral NLL (VNLL) axons appear to send fewer projections into the collicular cortices than dorsal or intermediate NLL (DNLL or INLL) axons (Whitley and Henkel 1984).

In contrast to ICC, the IC cortex receives inputs from various nonauditory subcortical nuclei (reviewed by Gruters and Groh 2012). Laminar organization of the IC cortex reflects different combinations of nonauditory input sources. Layer 1 of the ICD and ICL receives visual inputs from the contralateral retina (Itaya and Van Hoesen 1982). Layer 2 receives dense fibers from the dorsal column nuclei (Wiberg et al. 1987). Terminals of these axons form several patches that seem to correspond to the GABA modules (Lesicko and Llano 2015). Therefore, neurons in GABA modules may receive somatosensory information and transfer onset inhibition to other neurons. All layers receive nondopaminergic inputs from substantia nigra pars lateralis (Yasui et al. 1991). Neurons in the globus pallidus appear to project to upper layers of the ICD and ICL (Shammah-Lagnado et al. 1996). Cholinergic fibers that originate from midbrain tegmental nuclei (Motts and Schofield 2009) terminate densely in layer 2 and sparsely in other layers (Henderson and Sherriff 1991). Noradrenergic fibers that arise from locus coeruleus are dense in layers 2 and 3 while serotonergic fibers from the dorsal raphe nucleus are dense in layer 1 (Klepper and Herbert 1991). Considering the dendritic orientation and morphology of neurons in the deeper layers of the IC cortex, these neurons are likely to integrate visual (layer 1), somatosensory (layer 2), and auditory (layer 3) information as well as behavioral context via the substantia nigra and globus pallidus. The layer preference of monoaminergic and cholinergic innervation suggests that the modes of multimodal integration in the IC cortex can be altered by behavioral and mental states.

6.4.2 Descending Inputs

The descending pathways from the AC are described by Cant and Oliver (Chap. 2). Therefore, only the major differences in the patterns of descending inputs between ICC and IC cortex are summarized here. Glutamatergic pyramidal neurons located mainly in layer V but also a small number in layer VI (Games and Winer 1988) project to both ICs, although the ipsilateral projection is heavier than the contralateral one (Herbert et al. 1991; Saldaña et al. 1996). The ICC receives tonotopically organized fibers from the primary auditory cortex (A1) (Saldaña et al. 1996): the caudal low-frequency regions of A1 project to the dorsal laminae, and the rostral high-frequency regions project to the ventral laminae in the ICC (Saldaña et al. 1996).

The IC cortex receives inputs from multiple areas of the forebrain. The A1 neurons form tonotopically organized terminal areas in the ICD and ICL. The caudal low-frequency regions of A1 project to the dorsolateral laminae in the ICL and dorsal laminae in the ICD, which is the continuum of those in the ICC. The rostral high-frequency region projects to the ventrolateral laminae in the ICL and ventral laminae in the ICD (Saldaña et al. 1996). Secondary, multimodal auditory cortical areas also project to the IC cortex. Area Te2 projects to layer 1, whereas Area Te3 projects to layers 2 and 3 of the ICD and ICL (Herbert et al. 1991). It is interesting to note that Areas Te2 and Te3 seem to correspond to the cat's posterior auditory field and the suprarhinal auditory field, respectively (Malmierca 2015), which constitute high-order cortical fields linked to the nonlemniscal auditory pathway.

Although the corticollicular projection is purely excitatory, electrical stimulation of the AC in cat (Mitani et al. 1983) or its inactivation in rat, using either tetrodotoxin (Nwabueze-Ogbo et al. 2002) or a cooling technique (Anderson and Malmierca 2013), produce not only excitatory but also inhibitory responses in IC neurons. Those studies indicate that the AC modulates IC responses through the activation of local inhibitory connections.

6.5 Microcircuits

Knowing the details of organization of the IC microcircuits is essential for gaining a better understanding of how auditory information is encoded in the IC. Thus far, most studies of IC microcircuitry have been restricted to the ICC and very little is known about the microcircuitry of the cortical areas.

6.5.1 Synaptic Domain Hypothesis

As described in Sect. 6.3.1.1, the ICC is composed of 150–200 μm sheets that form isofrequency laminae and interlaminar components. Neurons in one lamina share similar CFs within a range of about 0.3 octaves (Schreiner and Langner 1997; Malmierca et al. 2008). Other characteristics of sound are represented in a nonuniform fashion, although adjacent neurons tend to have similar responsiveness to sound (Chen et al. 2012). In an in vitro study in laminar slices, repetitive shocks to the lateral lemniscus revealed several zones showing different activities (Chandrasekaran et al. 2013). These observations are consistent with the concept that the laminae contain distinct functional zones. These functional zones are referred to as synaptic domains because they are created by specific shared inputs, at least in part (Casseday et al. 2002; Oliver 2005). Neurons in the same synaptic domain may share the same terminal field in the medial geniculate body (MGB), which is described later in this section (Cant and Benson 2007).

Earlier anterogradetracing studies had already revealed that projections from lower brainstem nuclei produce segregated terminal fields in the ICC but with various degrees of overlap (details in Oliver and Huerta 1992). These terminal zones have been examined in detail in the projections from CN and SOC. The projections from the two sides of the lateral superior olive (LSO) interdigitate, forming bands suggestive of a sublaminar organization within the isofrequency laminae (Shneiderman and Henkel 1987). Terminal fields of axons from DCN and LSO on the same side overlap in the lateral part of the contralateral ICC, but the inputs from the DCN extend more medially than do those from the LSO (Oliver et al. 1997).

In more recent studies, focal injections of anterograde tracers were made in functionally defined zones of the medial superior olive (MSO) and LSO to reveal the organization of inputs to specific synaptic domains (Oliver et al. 2003; Loftus et al. 2004). Axons from two different regions of MSO with similar CFs but presumably different interaural time difference (ITD) sensitivities gave rise to overlapping terminal fields within a lamina, suggesting convergence of ITD information in the ICC (Oliver et al. 2003). Within the laminae representing low frequencies, axons from ipsilateral MSO terminate in the centraltocaudal part of the ICC, while those from the contralateral LSO terminate in the rostral ICC. Therefore, the main excitatory drives from the SOC appear to be largely segregated. On the other hand, axons from the ipsilateral LSO, which are mainly inhibitory, target the same domain as the ipsilateral MSO (Loftus et al. 2004). Since the MSO showed maximal discharge rates at the best ITD (peak-type responses) and the LSO showed minimal responses at best ITD (trough-type responses) (Joris and Yin 1995), the results suggested that ITD coding is different in the sublaminar domains.

Segregation of the terminal fields of ascending axons from different origins suggests that different synaptic domains receive unique combinations of inputs. By making medium-sized tracer deposits, Cant and Benson (2006) demonstrated patterns of ascending inputs to 74 injection sites that covered most parts of the gerbil IC. Three combinations of ascending inputs were identified: group 1 received substantial inputs from the MSO and LSO (>5% of total labeled cells), group 2 almost lacked inputs from these nuclei, and group 3 had few ascending inputs. Moreover, and importantly, the group categorization was a good predictor of the injection site: group 1 sites fell into the rostrolateral zone of the ICC while group 2 sites fell into the caudal ventromedial zone. Sites of both groups received substantial inputs from the CN, VNLL, and superior paraolivary nucleus of the SOC. Group 3 sites corresponded to the ICL and ICD. Group 1 might be further subdivided into dorsal and ventral parts with the dorsal part receiving more input from the MSO than the ventral part, which is dominated by inputs from the LSO. Once again, these data are consistent with the hypothesis that the IC is made of synaptic domains that receive unique combinations of inputs.

The unique combinations of inputs into each synaptic domain may create domain-specific responses to features of sound. By making small injections of retrograde tracers into physiologically characterized sites, Loftus et al. (2010) correlated anatomical and physiological features of synaptic domains of cat ICC. Based on the proportions of each input, three types of domains were classified. The first type (confined mainly to the caudal ICC) was characterized by inputs from the contralateral CN and ipsilateral VNLL, and physiologically the neurons exhibited monaural responses. The second type (confined to the dorsal ICC) was characterized by inputs from the MSO and ipsilateral LSO, and the neurons showed ITD sensitivity to low frequencies. The third type of domain (confined to the rostral ICC) was relatively heterogeneous but had in common inputs that arose bilaterally from the SOC nuclei, and the neurons showed ITD sensitivity to complex sounds.

Each synaptic domain has a characteristic ascending projection. Binaural and monaural domains in the gerbil ICC appear to target different regions in the ventral division of the MGB (the MGV) with some overlap (Cant and Benson 2007). The lateral zone of the ICC, which receives substantial inputs from the MSO and LSO, targets the rostral part of the MGV. In contrast, the medial zone, which lacks inputs from the MSO and LSO, targets the caudal part of the MGV. Overlap of inputs from the two zones is obvious in the ventral part of the MGV. These findings suggest the existence of parallel pathways from midbrain to thalamus such that each functional domain created in the IC projects in parallel to a functional domain of the medial geniculate.

6.5.2 Contributions of the Local Circuitry

The studies discussed in Sect. 6.5.1 emphasized the organization of ascending inputs within a synaptic domain. However, it was not possible to analyze the contributions of local axon collaterals in those tracer injection studies. Within the IC, a massive system of local collaterals provides a substrate for interactions among neurons (Fig. 6.8). The axonal arborization patterns of single neurons give the most detailed information about local circuitry, although the number of studies is limited due to the difficulty of filling individual IC neurons. Oliver et al. (1991) filled eighteen cat IC neurons with HRP and demonstrated that nearly all labeled neurons (except for one in the VLN) had local collaterals. Two ICC neurons with rostrocaudally oriented dendritic fields (disc-shaped cells) gave rise to axonal terminations confined to the fibrodendritic laminae.

Neurons that have dendrites with mediolateral orientation or no orientation (stellate cells) gave rise to axons with no orientation (Fig. 6.8A). Some cells had both a thick branch entering the brachium of the IC (an ascending projection branch; arrowheads in Fig. 6.8) and local collaterals. Wallace et al. (2012) examined eleven juxtacellularly labeled guinea pig ICC cells that had oriented dendrites and formed terminations within fibrodendritic laminae (Fig. 6.8B). The axonal plexuses of neurons with the CF higher than 1 kHz had a laminar thickness of about 200 μm, while those with lower CFs were less extensive. In concert with the extensive axonal plexuses in the low CF area (dorsal ICC), the volume of the dendritic arbor is reduced in the dorsal ICC neurons of mice (Reetz and Ehret 1999), suggesting that dorsal, low CF neurons form a less complex local circuitry than do ventral, high CF neurons.

Ito and Oliver (2014) used a viral tracer to investigate local and projection axons (e.g., Ito et al. 2007; Matsuda et al. 2009) and reconstructed axons of three glutamatergic neurons in the ICC and ICL. All of them had a laminar axon (Fig. 6.8C). One ICC cell formed two axonal plexuses extending from the ICC to ICD and had one plexus in the ICL. Although the virus has been shown to label very long projection axons (e.g., sympathetic preganglionic neurons; Ito et al. 2007) or highly arborized terminal bushes (e.g., nigrostriatal fibers; Matsuda et al. 2009), labeled axons outside the IC were not found in these three infected neurons, suggesting that they were

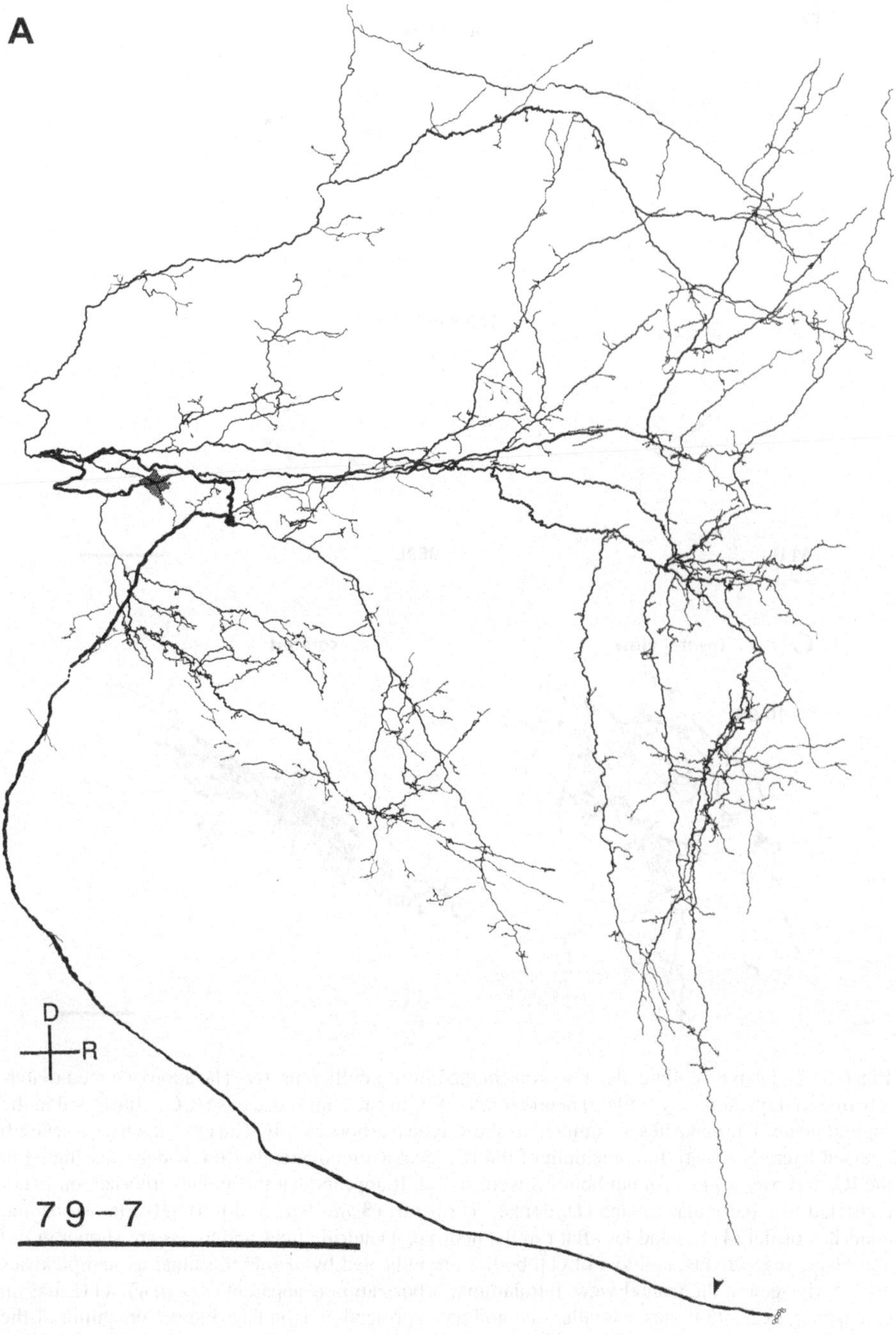

Fig. 6.8 (continued)

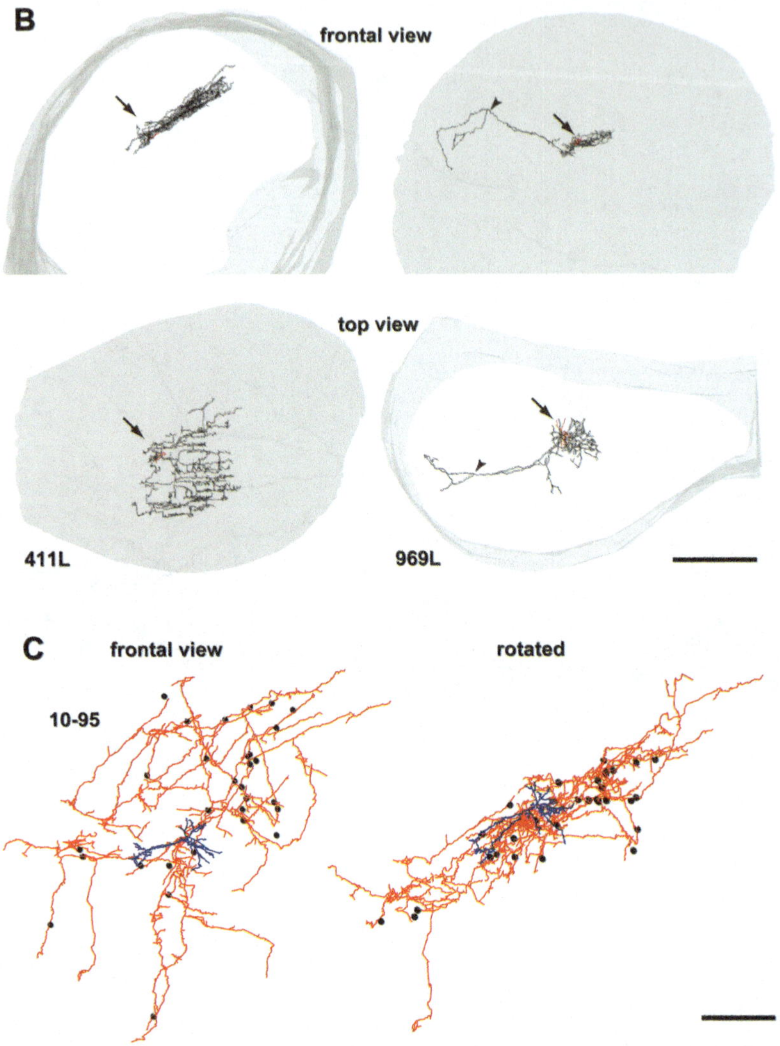

Fig. 6.8 Local axon collaterals of neurons in the inferior colliculus. **(A)** The axon of a mediolaterally oriented (presumably stellate) neuron (cell 79-7) in cat central nucleus (ICC), illustrated in the sagittal plane. The dendrites are omitted to show axonal arbors clearly. The main axon (*arrowhead*) coursed laterally toward the brachium of the IC, local axon collaterals were widely distributed in the IC, and over 2000 terminal boutons were found. It appears that the axonal arborization is not restricted to a particular lamina (*D*, dorsal; *R*, rostral) (*Scale bar:* 500 μm). **(B)** The axons and dendrites of flat (411L) and less-flat (969L) neurons. Dendritic trees and axons are shown in *red* and *black*, respectively, and loci of cell bodies are indicated by *arrows*. Laminar axonal plexuses are clearly seen in the frontal view. Intralaminar arborization is apparent (*top view*): 411L has an extensive plexus, 969L has a smaller one and has a projection axon entering the brachium of the IC (*arrowheads*). The CFs of 411L and 969L are 4.9 and 5.1 kHz, respectively (*scale bar:* 1 mm). **(C)** The axon of a glutamatergic neuron (cell 10-95) with oriented dendrites. This neuron makes axosomatic contacts on 30 large GABAergic neurons (*dots*) (*scale bar:* 250 μm). [**A** was kindly provided by Dr. Douglas L. Oliver, from Oliver et al. (1991) with permission from Wiley; **B** was modified from Wallace et al. (2012), kindly provided by Dr. Mark N. Wallace; **C** was modified from Ito and Oliver (2014) with permission from Wiley]

purely interneurons. In a juxtacellular study, the authors also suggested the presence of interneurons in the ICC (e.g., cell 411L in Fig. 6.8B) (Wallace et al. 2012). These studies clearly showed that many IC neurons give rise to local collaterals with numerous boutons, and, therefore, they may influence many other adjacent neurons.

Since IC neurons outnumber the total number of all lower brainstem auditory nuclei (Kulesza et al. 2002), it is likely that the local collaterals are among the largest sources of inputs to the IC, and they may act as input amplifiers. Such massive local inputs undoubtedly exert a key influence in shaping de novo responses to sound. A wide variety of arborization patterns of axon collaterals have been reported (Fig. 6.8A–C), so the role of local neurons within a synaptic domain may also be very diverse. For example, large stellate neurons with extensive, unoriented axonal terminations and numerous boutons (Fig. 6.8A) may influence multiple synaptic domains in multiple isofrequency laminae. Laminar neurons (Fig. 6.8B, C) are more likely to exert their influence in the single lamina in which their cell bodies lie, although the area under influence may differ among neurons. Indeed, a functional imaging study on IC slices (Chandrasekaran et al. 2013) showed both cross-laminar and intra-laminar activity elicited by electrical stimulation. Unoriented axonal terminations in multiple laminae are ideal for causing the cross-laminar activity, which may affect spectral integration. These authors also demonstrated that repetitive stimulation in laminar slices (which in theory preserves local circuitry in a single lamina) resulted in the emergence of domains of high activity, which may reflect anatomical synaptic domains. Focal application of concentrated divalent cations, which causes inhibition of local circuit activity (Sivaramakrishnan et al. 2013), narrowed the dynamic range for sound intensity, suggesting that local circuits also affect encoding sound intensity (Grimsley et al. 2013).

6.5.3 Roles of GABAergic Cells

Neurotransmission of GABA plays an important role in IC function, both in the intrinsic circuitry and in the projections to the MGB. Two types of GABAergic neurons have been identified (Ito et al. 2009). The large and small GABAergic neurons differ in their synaptic organization, and both types are randomly distributed within the ICC (Fig. 6.5D) such that each type of synaptic domain may contain these two types of neurons in addition to glutamatergic neurons (Ito and Oliver 2012). The LG neurons receive dense excitatory axosomatic inputs (Ito et al. 2009). The excitatory terminals are positive for VGLUT2 but not VGLUT1 (Ito et al. 2009) and originate from the IC, INLL, SOC, and DCN (Ito and Oliver 2014; Ito et al. 2015). In all of those sites, the majority of IC-projecting glutamatergic neurons express VGLUT2 but not VGLUT1 (Ito and Oliver 2010; Ito et al. 2011). Since a single axon makes only a few (1–7) axosomatic terminals, many excitatory neurons may converge on the cell body of a single LG neuron (Ito and Oliver 2014; Ito et al. 2015). In addition, intrinsic glutamatergic neurons make axosomatic contacts on many nearby LG neurons (Fig. 6.8C).

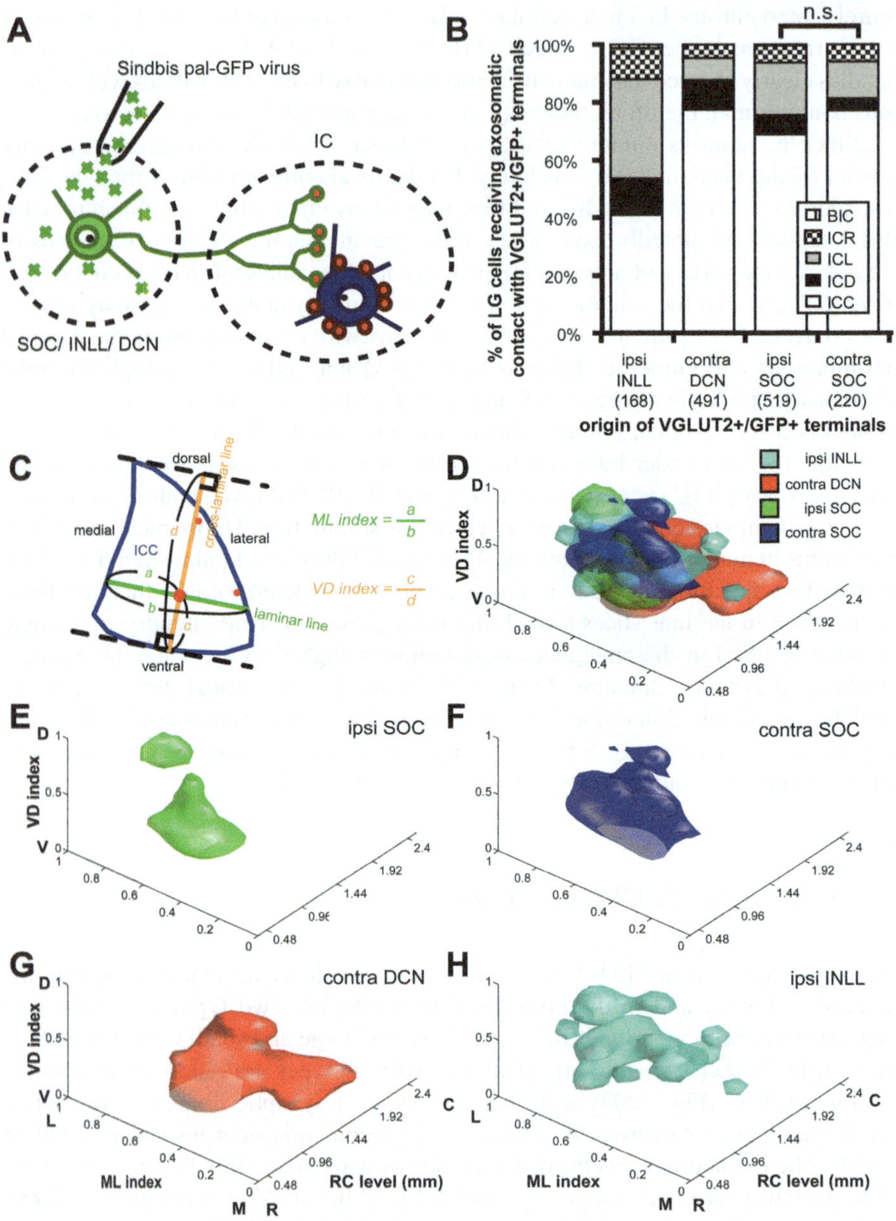

Fig. 6.9 Neurons in a particular lower auditory brain stem nucleus make axosomatic contacts on large GABAergic (LG) neurons in a particular region of the inferior colliculus (IC). (**A**) Schematic drawing of the experiment (Ito and Oliver 2014; Ito et al. 2015): an injection of Sindbis pal-GFP virus was made in either the superior olivary complex (*SOC*), dorsal cochlear nucleus (*DCN*), or intermediate nucleus of the lateral lemniscus (*INLL*), and the distribution of LG neurons that received axosomatic contacts with axons from GFP-expressing excitatory (GFP+/VGLUT2) neurons was examined. (**B**) The proportion of LG neurons receiving axosomatic contact with GFP+/VGLUT2+ terminals in each IC subdivision is different among the origins of GFP+ axons

Within each IC subdivision, the distribution of excitatory axosomatic terminals on LG neurons differs, depending on which brain stem nucleus is the source of input (Ito et al. 2015). Excitatory neurons in the DCN and SOC mainly target LG cells in the ICC, whereas those in the INLL target cortices of the IC as well as the ICC. After an injection of Sindbis pal-GFP virus in either DCN or SOC (Fig. 6.9A), the majority of LG neurons that receive labeled axosomatic contacts is located in the ICC, but less than half is located in the ICC in the INLL injection cases (Fig. 6.9B). The percentage of LG neurons receiving labeled axosomatic contacts in each subdivision was significantly different among the source nuclei except for the ipsilateral versus contralateral SOCs (Fig. 6.9B).

Even within the ICC, the spatial distribution of LG cells that receive axosomatic contacts from identified sources differs according to the nucleus of origin. SOC glutamatergic axons made contacts on LG cells mainly in the lateral part of the ICC, while DCN axons made contacts on LG cells mostly in the caudal and ventral parts. LG cells receiving axosomatic contacts with INLL terminals were sparsely distributed within the ICC. To quantify the distribution of LG neurons in ICC that received inputs from a particular source, the mediolateral (ML) and ventrodorsal (VD) position was calculated for all LG cells receiving particular inputs (Fig. 6.9C). The quantitative results are consistent with the conclusion that LG cells in different IC locations tend to receive axosomatic inputs from different sources. The spatial distribution of these LG neurons is illustrated in Fig. 6.9D–H in three-dimensional surface plots. In the plots, three-dimensional histograms of the LG neuron distribution were calculated (dimension was $9 \times 11 \times 11$ on ML, VD indices, and rostrocaudal (RC) level axes, respectively) and histogram blocks that contain more LG neurons (>0.5% of all LG cells counted) were visualized. Each plot represents an ICC region that contains more LG neurons receiving axosomatic contact from a given source than the region outside. The axons from the contralateral DCN mainly made axosomatic contacts on LG cells located in the ventral and caudal part of the ICC (Fig. 6.9D, G), while SOC axons mainly terminated on LG cells in the rostral and lateral part (Fig. 6.9D–F). The region receiving INLL axons extended to the dorsal and rostral part of the ICC (Fig. 6.9D, H). Although these regions overlapped, the shape of the distributions was significantly different with the exception of the contralateral SOC versus the ipsilateral INLL and the ipsilateral versus contralateral

Fig. 6.9 (continued) (DCN, SOC, or INLL). *Numbers in parentheses* indicate the number of LG neurons with contact. Statistical significance (contralateral DCN vs. contralateral SOC, $P = 0.019$; other pairs, $P < 0.001$) was detected between nuclei of origin except for ipsilateral SOC versus contralateral SOC ($P = 0.146$; pairwise comparisons using Fisher's exact test). **(C)** A schematic diagram of ML and VD indices. For each LG neuron in the ICC, a line passing through the cell body and parallel to the most ventral lamina of the ICC (*laminar line segment*) is drawn. The *ML index* reflects mediolateral location of the LG cell. The *VD index* reflects relative ventrodorsal location of an LG cell in ICC. This is the position of an LG cell on a cross-laminar line orthogonal to the laminar line. The RC position of an LG cell was measured from the rostral end of the ICC. **(D–H)** Three-dimensional surface plots showing ICC zones of LG cells receiving axosomatic contact from ipsilateral SOC (**D, E**), contralateral SOC (**D, F**), contralateral DCN (**D, G**), and ipsilateral INLL (**D, H**). (*C*, caudal; *L*, lateral; *M*, medial; *R*, rostral)

SOCs. Thus, the LG neurons in an ICC synaptic domain mix particular excitatory ascending inputs with local IC inputs. The local inputs could be from another domain on the same lamina that received a different ascending input. Further, since LG neurons most likely have a stellate dendritic morphology (Oliver et al. 1994), they may also integrate spectral information across laminae.

Both GABAergic and glutamatergic IC neurons project to the MGB (Bartlett and Smith 1999). In rodents, feedforward inhibition by IC GABAergic neurons may be especially important because GABAergic interneurons are virtually absent in the medial geniculate (Winer et al. 1996; Ito et al. 2011). It is highly likely that the main source of feedforward GABAergic inputs in the MGV is LG neurons because most of the GABAergic neurons in the ICC that project to the MGB are LG neurons (Ito et al. 2009), but SG neurons in the IC cortices may also project to the thalamus since neurons in layer 2, which are made of GABA modules, project to the suprageniculate nucleus (Linke 1999).

LG neurons terminate on specific neuron types of the MGB. Inhibitory synapses of IC axons are small and terminate both on cell bodies and higher order dendritic branches of MGB neurons, while excitatory synapses of IC axons vary in size and terminate more widely on dendrites (Bartlett et al. 2000). In vitro studies in slice preparations have shown that electrical shock on the brachium of the IC stimulates ascending fibers from the IC and produces postsynaptic potentials in MGB neurons. An IPSP is evoked in stellate neurons in the dorsal division of the MGB (the MGD) and a subpopulation of tufted neurons in MGV and MGD (Bartlett and Smith 1999). During development, the IPSPs elicited by similar electrical shocks of the brachium show increasing intensity, shortened latencies, and reduced depression of responses (Venkataraman and Bartlett 2013). Mature LG neurons may make secure synapses that control the rising phase of excitation. Indeed, GABAergic axons have the largest diameters among brachial axons and are especially suitable for rapid transmission of action potentials (Saint Marie et al. 1997). Furthermore, at postnatal day 27, 40% of MGB neurons exhibit purely inhibitory responses to shock of the brachium (Venkataraman and Bartlett 2013). It is possible that the main drivers of these MGB neurons are corticothalamic descending fibers, while ascending sensory inputs from LG neurons block their activity.

6.6 Efferent Projections of Neurons in the Inferior Colliculus

6.6.1 Ascending Projections

Since the details of organization of the tectothalamic projection are described by Cant and Oliver (Chap. 2), the following section summarizes only the major differences in the main targets of the ICC and IC cortex. Most ICC neurons are projection neurons (at least 80% of cat ICC neurons) (Oliver 1984), and their main targets are the MGV and the medial division of MGB (the MGM) (LeDoux et al. 1985).

In the IC cortex, the outer layers (layers 1 and 2) in ICL and ICD share similar inputs and outputs (Herbert et al. 1991; Linke 1999), suggesting that they serve a similar function. Layer 1 neurons project to the posterior intralaminar and peripeduncular nucleus (Linke 1999). Layer 2 neurons project to the suprageniculate nucleus in the thalamus (Linke 1999). In rats, ICL layer 3 neurons project to the MGM (Linke 1999). In cats, the ICD deep layer projects to the MGD (Kudo and Niimi 1980). All the targets of the IC cortex are considered to be multimodal nuclei.

6.6.2 Commissural Connections

A large number of neurons interconnect the two ICs through the commissure of the IC (the CoIC) (Okoyama et al. 2006; Malmierca et al. 2009). Neurons that give rise to uniquely commissural projections may be rare since González-Hernández et al. (1986) have shown that the majority of cells that project to the ipsilateral medial geniculate body also send collaterals to the contralateral IC.

Small injections of tracers into an isofrequency lamina of the ICC label fibers in the corresponding lamina on the contralateral side (Saldaña and Merchán 1992), indicating connections between the homotopic isofrequency lamina on the two sides. However, the density of these projections is biased toward a point that matches the position of the tracer injection (Fig. 6.10), which is consistent with a point-to-point emphasis in the wiring pattern of commissural fibers (Malmierca et al. 2009). At least some ICD neurons make laminar terminations in the contralateral ICC and ICD (Fig. 6.10), whereas ICL neurons do not form laminar plexuses and terminate mainly in the contralateral ICD and ICL (Saldaña and Merchán 2005). Recent electrophysiological studies demonstrated that commissural connections modulate the spectral, temporal, and binaural properties of many IC neurons (Malmierca et al. 2003, 2005). More recent experiments using the cooling technique have demonstrated that mostly non-Vshaped frequency response areas in the IC are modulated by the contralateral IC. Moreover, the supra-threshold sensitivity of rate-level functions decreases during deactivation, and the ability to signal changes in sound level is diminished. This commissural enhancement suggests the ICs should be viewed as a single entity in which the representation of sound in each is governed by the other (Orton and Rees 2014). A large proportion of neurons contributing to the commissural projection may be glutamatergic, as in guinea pig (Saint Marie 1996), and about 20% of the neurons in rat are GABAergic (Hernández et al. 2006). These neurochemical properties are consistent with physiological studies that show that the commissural inputs may have either an excitatory or inhibitory influence on neurons in the contralateral IC (in vitro, guinea pig: Smith 1992; Li et al. 1998; in vivo, rat: Malmierca et al. 2003, 2005).

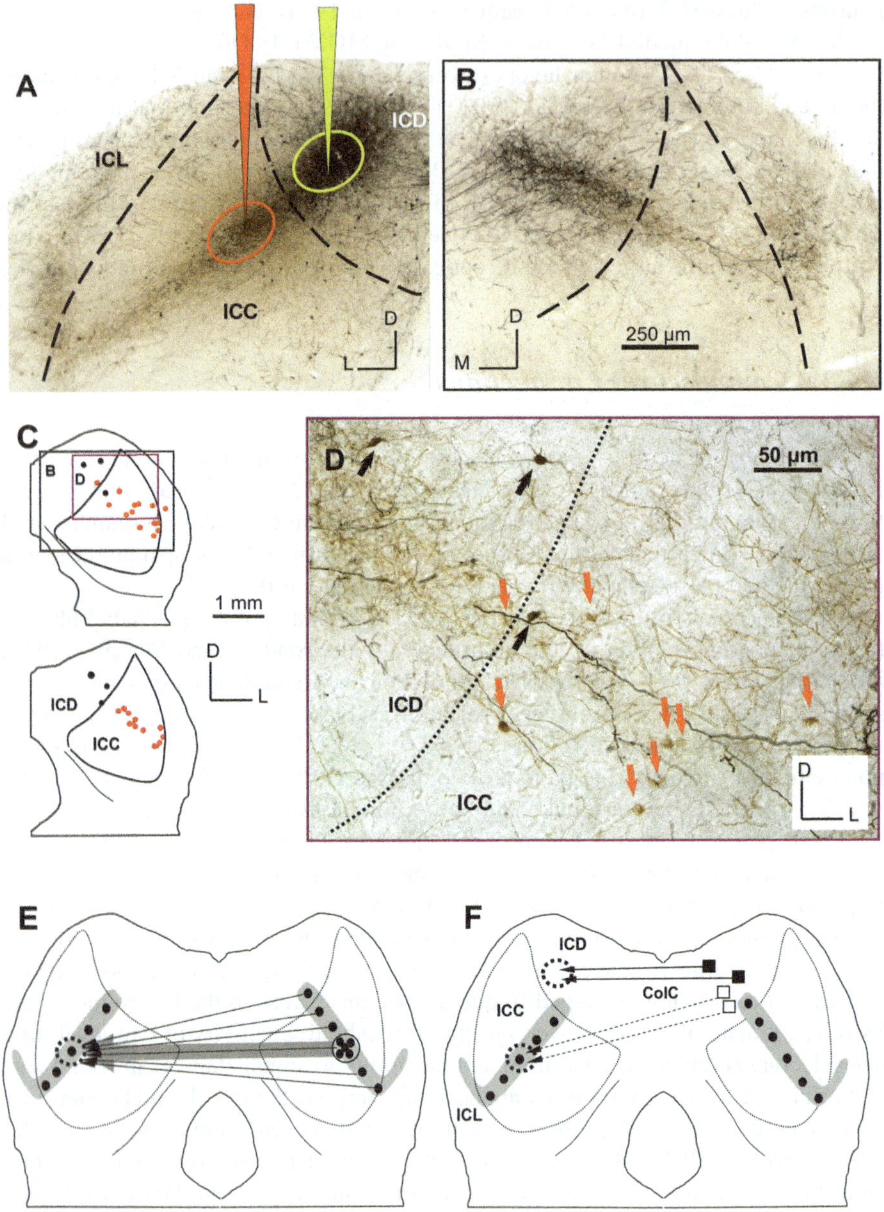

Fig. 6.10 **(A)** A photomicrograph showing two injections into the same lamina of the inferior colliculus (IC) (case # 279 in Malmierca et al. 2005). An injection (FD-BDA, *yellow circle*) was confined to the dorsal IC (*ICD*); another injection (tetramethylrhodamine dextran, TRD, *orange circle*) was located at the central nucleus (*ICC*). The characteristic frequency (CF) at the injection sites was similar and within the same laminae (10–10.5 kHz). Note the typical V-shaped or wing-like plexus of intrinsic axons with a central wing that extends into the ICC (described by Saldaña and Merchán 1992) and a lateral wing located in the lateral IC (*ICL*). The vertex of the plexus

6.6.3 Descending Projections

In addition to the thalamic projections, the IC also gives rise to colliculo-lemniscal, colliculo-olivary, and colliculo-cochlear projections (Caicedo and Herbert 1993). The colliculo-lemniscal projections are largely restricted to the DNLL, the sagulum, the horizontal cell group, and the perilemniscal zone (Caicedo and Herbert 1993). The colliculo-olivary projections originate in the ICC, ICL, and ICR and terminate as bands of terminals that target the rostral and medioventral periolivary nuclei (Caicedo and Herbert 1993). This projection is tonotopic and the axonal fields terminate at the site of origin of the medial olivocochlear system (White and Warr 1983), suggesting that the IC modulates cochlear activity. The classical electrophysiological studies by Dolan and Nuttall (1988) in guinea pigs and by Rajan (1990) in cats support this notion. The colliculo-cochlear projection originates in the ICC and ICL and targets the DCN and granule cell domain of the VCN (Caicedo and Herbert 1993), but the functional roles of these projections are still unknown.

Finally, it is interesting to note that the IC also projects to nonauditory areas, including the pontine nuclei, the lateral paragigantocellular nucleus, the gigantocellular reticular nucleus, the ventrolateral tegmental nucleus, and the caudal pontine reticular nucleus (Caicedo and Herbert 1993). These nuclei also receive strong projections from the CN (Kandler and Herbert 1991). These nonauditory areas may play a role in acoustically elicited autonomic responses or long- and short-latency auditory–motor behaviors (Kandler and Herbert 1991; Caicedo and Herbert 1993).

Fig. 6.10 (continued) marks the border between the ICC and the ICL. **(B)** Anterogradely labeled axons and retrogradely labeled neurons on the side contralateral to the injections in **A**. **(C)** Camera lucida drawings of the retrogradely labeled neurons originating from the TRD (*orange*) and mixture of fluorescein dextran and biotinylated dextran amine (FD-BDA) (*black*) injections at two rostrocaudal levels of the IC. *Black box* in the most caudal section highlights area shown in **B** and the *purple box* highlights area shown in **D** at higher magnification. **(D)** Area in *purple box* in **C**; the *orange arrows* point to TRD-labeled neurons and *black arrows* FD-BDA-labeled neurons. **(E, F)** Schematic wiring diagrams of the commissural connections; *filled circles* are somata and *dotted circles* indicate that an injection into one point on the lamina (e.g., dotted circle, left IC) retrogradely labels neurons over the whole extent of the contralateral lamina, consistent with a divergent pattern of connections (*thin arrows*). **(E)** In the ICC, anterograde labeling and the retrograde labeling of neurons also support a convergent projection. In addition, however, the density of the projection is centered on a point matching the position of the tracer injection. This result is consistent with a point-to-point weighted-wiring pattern (*thick arrow*). The coexistence of convergent and point-to-point projections suggests that there are two functionally different systems of commissural connections (*CoIC*). **(F)** In the dorsal IC (*ICD*), two populations of neurons seem to contribute to the commissural projections to the contralateral IC: one ICD population projects to the frequency-band laminae in the ICC in a tonotopic manner (*white squares, dashed lines*), while the other ICD population projects diffusely to its counterpart (*black squares, solid lines*). (*C*, caudal; *D*, dorsal; *L*, lateral; *R*, rostral) (*Scale bars* as labeled in **B**, **C**, **D**)

6.7 Summary

The IC is subdivided into the ICC and surrounding cortices. The ICC is characterized by the presence of fibrodendritic laminae that are composed of oriented axons from lemniscal nuclei and the dendritic arbors of flat neurons. IC cortices, especially ICL, are organized in several layers. Since different layers receive different combinations of sensory modal inputs, the IC cortex may be involved in multimodal integration.

Various neuron types have been identified in the IC. Different types of dendritic arborizations may reflect different degrees of frequency integration in the ICC and multisensory integration in the IC cortex. Some GABAergic neurons can be linked to specific functions. Although there is diversity in the intrinsic properties of IC neurons, the proportion of each neuron type classified with intrinsic properties is different among subdivisions. There is also wide diversity of sound-evoked activity in the IC. The IC cortex can be characterized by the presence of neurons showing SSA. These neuron types successfully distinguish IC subdivisions.

Lower brain stem auditory nuclei (e.g., CN, SOC, and NLL) mainly target the ICC, while AC axons mainly target the IC cortex (although A1 also targets the ICC). The ICC and ICD neurons form a laminar axonal plexus in both ICs. The ICL neurons terminate bilaterally in ICD and ICL in a nonlaminar pattern. These data clearly indicate the different combinations of inputs in different subdivisions of the IC.

Within a subdivision, there are synaptic domains that receive different combinations of inputs. In the ICC, LG neurons may mix information from multiple synaptic domains via collaterals of excitatory neurons located in neighboring domains. Taken together, it is plausible to conclude that microcircuits in the ICC are especially suited to integrate ascending auditory information and create de novo coding properties, whereas microcircuits of the IC cortex are more suitable for integrating multimodal information and detecting novel features of a stimulus.

Acknowledgments The authors are grateful to Drs. Douglas L. Oliver, Munenori Ono, and Mark N. Wallace for providing excellent images. Financial support was provided by the Spanish Ministerio de Economía y Competitividad (SAF2016-75803-P) and Consejeria de Educación JCYL grant (SA343U14) to MSM; support was provided by Japan Society for the Promotion of Science (KAKENHI Grant numbers: 16H01501, 16K07026, 22700365, and 25430034), the Uehara Memorial Foundation, the Ichiro Kanehara Foundation, Novartis Foundation of Japan, Takahashi Industrial and Economic Research Foundation, and Research and Education Program for Life Science of University of Fukui to TI.

Compliance with Ethics Requirements Tetsufumi Ito declares no conflicts of interest.

Manuel S. Malmierca declares no conflicts of interest.

References

Ahuja, T. K., & Wu, S. H. (2007). Intrinsic membrane properties and synaptic response characteristics of neurons in the rat's external cortex of the inferior colliculus. *Neuroscience, 145*(3), 851–865.

Anderson, L. A., & Malmierca, M. S. (2013). The effect of auditory cortex deactivation on stimulus-specific adaptation in the inferior colliculus of the rat. *European Journal of Neuroscience, 37*(1), 52–62.

Aubrey, K. R., Rossi, F. M., Ruivo, R., Alboni, S., Bellenchi, G. C., Le Goff, A., et al. (2007). The transporters GlyT2 and VIAAT cooperate to determine the vesicular glycinergic phenotype. *The Journal of Neuroscience, 27*(23), 6273–6281.

Ayala, Y. A., & Malmierca, M. S. (2012). Stimulus-specific adaptation and deviance detection in the inferior colliculus. *Frontiers in Neural Circuits, 6*, 89. https://doi.org/10.3389/fncir.2012.00089.

Ayala, Y. A., & Malmierca, M. S. (2015). Cholinergic modulation of stimulus-specific adaptation in the inferior colliculus. *The Journal of Neuroscience, 35*(35), 12261–12272.

Ayala, Y. A., Udeh, A., Dutta, K., Bishop, D., Malmierca, M. S., & Oliver, D. L. (2015). Differences in the strength of cortical and brainstem inputs to SSA and non-SSA neurons in the inferior colliculus. *Scientific Reports, 5*, 10383. https://doi.org/10.1038/srep10383.

Bartlett, E. L., & Smith, P. H. (1999). Anatomic, intrinsic, and synaptic properties of dorsal and ventral division neurons in rat medial geniculate body. *Journal of Neurophysiology, 81*(5), 1999–2016.

Bartlett, E. L., Stark, J. M., Guillery, R. W., & Smith, P. H. (2000). Comparison of the fine structure of cortical and collicular terminals in the rat medial geniculate body. *Neuroscience, 100*(4), 811–828.

Caicedo, A., & Herbert, H. (1993). Topography of descending projections from the inferior colliculus to auditory brainstem nuclei in the rat. *The Journal of Comparative Neurology, 328*(3), 377–392.

Cant, N. B., & Benson, C. G. (2006). Organization of the inferior colliculus of the gerbil (*Meriones unguiculatus*): Differences in distribution of projections from the cochlear nuclei and the superior olivary complex. *The Journal of Comparative Neurology, 495*(5), 511–528.

Cant, N. B., & Benson, C. G. (2007). Multiple topographically organized projections connect the central nucleus of the inferior colliculus to the ventral division of the medial geniculate nucleus in the gerbil, *Meriones unguiculatus*. *The Journal of Comparative Neurology, 503*(3), 432–453.

Casseday, J. H., Fremouw, T., & Covey, E. (2002). The inferior colliculus: A hub for the central auditory system. In D. Oertel, R. R. Fay, & A. N. Popper (Eds.), *Integrative functions in the mammalian auditory pathway* (pp. 238–318). New York: Springer.

Chandrasekaran, L., Xiao, Y., & Sivaramakrishnan, S. (2013). Functional architecture of the inferior colliculus revealed with voltage-sensitive dyes. *Frontiers in Neural Circuits, 7*, 41. https://doi.org/10.3389/fncir.2013.00041.

Chen, C., Rodriguez, F. C., Read, H. L., & Escabi, M. A. (2012). Spectrotemporal sound preferences of neighboring inferior colliculus neurons: Implications for local circuitry and processing. *Frontiers in Neural Circuits, 6*, 62. https://doi.org/10.3389/fncir.2012.00062.

Chernock, M. L., Larue, D. T., & Winer, J. A. (2004). A periodic network of neurochemical modules in the inferior colliculus. *Hearing Research, 188*(1-2), 12–20.

Coleman, J. R., & Clerici, W. J. (1987). Sources of projections to subdivisions of the inferior colliculus in the rat. *The Journal of Comparative Neurology, 262*(2), 215–226.

Coote, E. J., & Rees, A. (2008). The distribution of nitric oxide synthase in the inferior colliculus of guinea pig. *Neuroscience, 154*(1), 218–225.

Covey, E., & Carr, C. E. (2005). The auditory midbrain in bats and birds. In J. A. Winer & C. E. Schreiner (Eds.), *The inferior colliculus* (pp. 493–536). New York: Springer.

Dolan, D. F., & Nuttall, A. L. (1988). Masked cochlear whole-nerve response intensity functions altered by electrical stimulation of the crossed olivocochlear bundle. *Journal of the Acoustical Society of America, 83*(3), 1081–1086.

Duque, D., Perez-Gonzalez, D., Ayala, Y. A., Palmer, A. R., & Malmierca, M. S. (2012). Topographic distribution, frequency, and intensity dependence of stimulus-specific adaptation in the inferior colliculus of the rat. *The Journal of Neuroscience, 32*(49), 17762–17774.

Faingold, C. L., Gehlbach, G., & Caspary, D. M. (1989). On the role of GABA as an inhibitory neurotransmitter in inferior colliculus neurons: Iontophoretic studies. *Brain Research, 500*(1-2), 302–312.

Faingold, C. L., Boersma Anderson, C. A., & Caspary, D. M. (1991). Involvement of GABA in acoustically evoked inhibition in inferior colliculus neurons. *Hearing Research, 52*(1), 201–216.

Faye-Lund, H., & Osen, K. K. (1985). Anatomy of the inferior colliculus in rat. *Anatomy and Embryology (Berlin), 171*(1), 1–20.

Fremeau, R. T., Jr., Troyer, M. D., Pahner, I., Nygaard, G. O., Tran, C. H., Reimer, R. J., Bellocchio, E. E., Fortin, D., Storm-Mathisen, J., & Edwards, R. H. (2001). The expression of vesicular glutamate transporters defines two classes of excitatory synapse. *Neuron, 31*(2), 247–260.

Games, K. D., & Winer, J. A. (1988). Layer V in rat auditory cortex: Projections to the inferior colliculus and contralateral cortex. *Hearing Research, 34*(1), 1–25.

Geis, H. R., & Borst, J. G. (2013). Large GABAergic neurons form a distinct subclass within the mouse dorsal cortex of the inferior colliculus with respect to intrinsic properties, synaptic inputs, sound responses, and projections. *The Journal of Comparative Neurology, 521*(1), 189–202.

González-Hernández, T. H., Meyer, G., & Ferres-Torres, R. (1986). The commissural interconnections of the inferior colliculus in the albino mouse. *Brain Research, 368*(2), 268–276.

Grimsley, C. A., Sanchez, J. T., & Sivaramakrishnan, S. (2013). Midbrain local circuits shape sound intensity codes. *Frontiers in Neural Circuits, 7*, 174. https://doi.org/10.3389/fncir.2013.00174.

Gruters, K. G., & Groh, J. M. (2012). Sounds and beyond: Multisensory and other non-auditory signals in the inferior colliculus. *Frontiers in Neural Circuits, 6*, 96. https://doi.org/10.3389/fncir.2012.00096.

Henderson, Z., & Sherriff, F. E. (1991). Distribution of choline acetyltransferase immunoreactive axons and terminals in the rat and ferret brainstem. *The Journal of Comparative Neurology, 314*(1), 147–163.

Herbert, H., Aschoff, A., & Ostwald, J. (1991). Topography of projections from the auditory cortex to the inferior colliculus in the rat. *The Journal of Comparative Neurology, 304*(1), 103–122.

Hernández, O., Rees, A., & Malmierca, M. S. (2006). A GABAergic component in the commissure of the inferior colliculus in rat. *Neuroreport, 17*(15), 1611–1614.

Hu, B., Senatorov, V., & Mooney, D. (1994). Lemniscal and non-lemniscal synaptic transmission in rat auditory thalamus. *The Journal of Physiology, 479*(2), 217–231.

Itaya, S. K., & Van Hoesen, G. W. (1982). Retinal innervation of the inferior colliculus in rat and monkey. *Brain Research, 233*(1), 45–52.

Ito, T., & Oliver, D. L. (2010). Origins of glutamatergic terminals in the inferior colliculus identified by retrograde transport and expression of VGLUT1 and VGLUT2 genes. *Frontiers in Neuroanatomy, 4*, 135. https://doi.org/10.3389/fnana.2010.00135.

Ito, T., & Oliver, D. L. (2012). The basic circuit of the IC: Tectothalamic neurons with different patterns of synaptic organization send different messages to the thalamus. *Frontiers in Neural Circuits, 6*, 48. https://doi.org/10.3389/fncir.2012.00048.

Ito, T., & Oliver, D. L. (2014). Local and commissural IC neurons make axosomatic inputs on large GABAergic tectothalamic neurons. *The Journal of Comparative Neurology, 522*(15), 3539–3554.

Ito, T., Hioki, H., Nakamura, K., Tanaka, Y., Nakade, H., Kaneko, T., Iino, S., & Nojyo, Y. (2007). Gamma-aminobutyric acid-containing sympathetic preganglionic neurons in rat thoracic spinal

cord send their axons to the superior cervical ganglion. *The Journal of Comparative Neurology, 502*(1), 113–125.
Ito, T., Bishop, D. C., & Oliver, D. L. (2009). Two classes of GABAergic neurons in the inferior colliculus. *The Journal of Neuroscience, 29*(44), 13860–13869.
Ito, T., Bishop, D. C., & Oliver, D. L. (2011). Expression of glutamate and inhibitory amino acid vesicular transporters in the rodent auditory brainstem. *The Journal of Comparative Neurology, 519*(2), 316–340.
Ito, T., Hioki, H., Sohn, J., Okamoto, S., Kaneko, T., Iino, S., & Oliver, D. L. (2015). Convergence of lemniscal and local excitatory inputs on large GABAergic tectothalamic neurons. *The Journal of Comparative Neurology, 523*, 2277–2296.
Joris, P. X., & Yin, T. C. (1995). Envelope coding in the lateral superior olive. I. Sensitivity to interaural time differences. *Journal of Neurophysiology, 73*(3), 1043–1062.
Kandler, K., & Herbert, H. (1991). Auditory projections from the cochlear nucleus to pontine and mesencephalic reticular nuclei in the rat. *Brain Research, 562*(2), 230–242.
Klepper, A., & Herbert, H. (1991). Distribution and origin of noradrenergic and serotonergic fibers in the cochlear nucleus and inferior colliculus of the rat. *Brain Research, 557*(1-2), 190–201.
Kudo, M., & Niimi, K. (1980). Ascending projections of the inferior colliculus in the cat: An autoradiographic study. *The Journal of Comparative Neurology, 191*(4), 545–556.
Kulesza, R. J., Vinuela, A., Saldana, E., & Berrebi, A. S. (2002). Unbiased stereological estimates of neuron number in subcortical auditory nuclei of the rat. *Hearing Research, 168*(1-2), 12–24.
Kuo, R. I., & Wu, G. K. (2012). The generation of direction selectivity in the auditory system. *Neuron, 73*(5), 1016–1027.
Kuwada, S., Batra, R., Yin, T. C. T., Oliver, D. L., Haberly, L. B., & Stanford, T. R. (1997). Intracellular recordings in response to monaural and binaural stimulation of neurons in the inferior colliculus of the cat. *The Journal of Neuroscience, 17*(19), 7565–7581.
LeDoux, J. E., Ruggiero, D. A., & Reis, D. J. (1985). Projections to the subcortical forebrain from anatomically defined regions of the medial geniculate body in the rat. *The Journal of Comparative Neurology, 242*(2), 182–213.
Lesicko, A., & Llano, D. (2015). Connectional and neurochemical modularity of the mouse inferior colliculus. Abstract PS-564, Association for Research in Otolaryngology Midwinter Meeting, February 21–25, Baltimore, MD.
Li, Y., Evans, M. S., & Faingold, C. L. (1998). In vitro electrophysiology of neurons in subnuclei of rat inferior colliculus. *Hearing Research, 121*(1-2), 1–10.
Linke, R. (1999). Differential projection patterns of superior and inferior collicular neurons onto posterior paralaminar nuclei of the thalamus surrounding the medial geniculate body in the rat. *European Journal of Neuroscience, 11*(1), 187–203.
Loftus, W. C., Bishop, D. C., Saint Marie, R. L., & Oliver, D. L. (2004). Organization of binaural excitatory and inhibitory inputs to the inferior colliculus from the superior olive. *The Journal of Comparative Neurology, 472*(3), 330–344.
Loftus, W. C., Bishop, D. C., & Oliver, D. L. (2010). Differential patterns of inputs create functional zones in central nucleus of inferior colliculus. *The Journal of Neuroscience, 30*(40), 13396–13408.
Loftus, W. C., Malmierca, M. S., Bishop, D. C., & Oliver, D. L. (2008). The cytoarchitecture of the inferior colliculus revisited: A common organization of the lateral cortex in rat and cat. *Neuroscience, 154*(1), 196–205.
Malmierca, M. S. (2015). Auditory system. In G. Paxinos (Ed.), *The rat nervous system* (4th ed., pp. 865–946). Amsterdam: Academic Press.
Malmierca, M. S., & Ryugo, D. K. (2011). Descending connections of auditory cortex to the midbrain and brainstem. In J. A. Winer & C. E. Schreiner (Eds.), *The auditory cortex* (pp. 189–208). New York: Springer.
Malmierca, M. S., Blackstad, T. W., Osen, K. K., Karagulle, T., & Molowny, R. L. (1993). The central nucleus of the inferior colliculus in rat: A Golgi and computer reconstruction study of neuronal and laminar structure. *The Journal of Comparative Neurology, 333*(1), 1–27.

Malmierca, M. S., Rees, A., Le Beau, F. E., & Bjaalie, J. G. (1995). Laminar organization of frequency-defined local axons within and between the inferior colliculi of the guinea pig. *The Journal of Comparative Neurology, 357*(1), 124–144.
Malmierca, M. S., Hernández, O., Falconi, A., Lopez-Poveda, E. A., Merchán, M., & Rees, A. (2003). The commissure of the inferior colliculus shapes frequency response areas in rat: An in vivo study using reversible blockade with microinjection of kynurenic acid. *Experimental Brain Research, 153*(4), 522–529.
Malmierca, M. S., Hernández, O., & Rees, A. (2005). Intercollicular commissural projections modulate neuronal responses in the inferior colliculus. *European Journal of Neuroscience, 21*(10), 2701–2710.
Malmierca, M. S., Izquierdo, M. A., Cristaudo, S., Hernandez, O., Pérez-González, D., Covey, E., et al. (2008). A discontinuous tonotopic organization in the inferior colliculus of the rat. *The Journal of Neuroscience, 28*(18), 4767–4776.
Malmierca, M. S., Hernandez, O., Antunes, F. M., & Rees, A. (2009). Divergent and point-to-point connections in the commissural pathway between the inferior colliculi. *The Journal of Comparative Neurology, 514*(3), 226–239.
Malmierca, M. S., Blackstad, T. W., & Osen, K. K. (2011). Computer-assisted 3-D reconstructions of Golgi-impregnated neurons in the cortical regions of the inferior colliculus of rat. *Hearing Research, 274*(1-2), 13–26.
Matsuda, W., Furuta, T., Nakamura, K. C., Hioki, H., Fujiyama, F., Arai, R., et al. (2009). Single nigrostriatal dopaminergic neurons form widely spread and highly dense axonal arborizations in the neostriatum. *The Journal of Neuroscience, 29*(2), 444–453.
Merchán, M., Aguilar, L. A., Lopez-Poveda, E. A., & Malmierca, M. S. (2005). The inferior colliculus of the rat: Quantitative immunocytochemical study of GABA and glycine. *Neuroscience, 136*(3), 907–925.
Mitani, A., Shimokouchi, M., & Nomura, S. (1983). Effects of stimulation of the primary auditory cortex upon colliculogeniculate neurons in the inferior colliculus of the cat. *Neuroscience Letters, 42*(2), 185–189.
Moller, A. R., & Rees, A. (1986). Dynamic properties of the responses of single neurons in the inferior colliculus of the rat. *Hearing Research, 24*(3), 203–215.
Morest, D. K., & Oliver, D. L. (1984). The neuronal architecture of the inferior colliculus in the cat: Defining the functional anatomy of the auditory midbrain. *The Journal of Comparative Neurology, 222*(2), 209–236.
Motts, S. D., & Schofield, B. R. (2009). Sources of cholinergic input to the inferior colliculus. *Neuroscience, 160*(1), 103–114.
Nakagawa, H., Ikeda, M., Houtani, T., Ueyama, T., Baba, K., Kondoh, A., et al. (1995). Immunohistochemical evidence for enkephalin and neuropeptide Y in rat inferior colliculus neurons that provide ascending or commissural fibers. *Brain Research, 690*(2), 236–240.
Netser, S., Zahar, Y., & Gutfreund, Y. (2011). Stimulus-specific adaptation: Can it be a neural correlate of behavioral habituation? *The Journal of Neuroscience, 31*(49), 17811–17820.
Nwabueze-Ogbo, F. C., Popelar, J., & Syka, J. (2002). Changes in the acoustically evoked activity in the inferior colliculus of the rat after functional ablation of the auditory cortex. *Physiological Research, 51*(Supplement 1), 95–104.
Okoyama, S., Ohbayashi, M., Ito, M., & Harada, S. (2006). Neuronal organization of the rat inferior colliculus participating in four major auditory pathways. *Hearing Research, 218*(1-2), 72–80.
Oliver, D. L. (1984). Neuron types in the central nucleus of the inferior colliculus that project to the medial geniculate body. *Neuroscience, 11*(2), 409–424.
Oliver, D. L. (2005). Neuronal organization in the inferior colliculus. In J. A. Winer & C. E. Schreiner (Eds.), *The inferior colliculus* (pp. 69–114). New York: Springer.
Oliver, D. L., & Morest, D. K. (1984). The central nucleus of the inferior colliculus in the cat. *The Journal of Comparative Neurology, 222*(2), 237–264.

Oliver, D. L., & Huerta, M. F. (1992). Inferior and superior colliculi. In D. B. Webster, A. N. Popper, & R. R. Fay (Eds.), *The mammalian auditory pathway: Neuroanatomy* (pp. 168–221). New York: Springer.
Oliver, D. L., Kuwada, S., Yin, T. C., Haberly, L. B., & Henkel, C. K. (1991). Dendritic and axonal morphology of HRP-injected neurons in the inferior colliculus of the cat. *The Journal of Comparative Neurology, 303*(1), 75–100.
Oliver, D. L., Winer, J. A., Beckius, G. E., & Saint Marie, R. L. (1994). Morphology of GABAergic neurons in the inferior colliculus of the cat. *The Journal of Comparative Neurology, 340*(1), 27–42.
Oliver, D. L., Beckius, G. E., Bishop, D. C., & Kuwada, S. (1997). Simultaneous anterograde labeling of axonal layers from lateral superior olive and dorsal cochlear nucleus in the inferior colliculus of the cat. *The Journal of Comparative Neurology, 382*(2), 215–229.
Oliver, D. L., Beckius, G. E., Bishop, D. C., Loftus, W. C., & Batra, R. (2003). Topography of interaural temporal disparity coding in projections of medial superior olive to inferior colliculus. *The Journal of Neuroscience, 23*(19), 7438–7449.
Oliver, D. L., Izquierdo, M. A., & Malmierca, M. S. (2011). Persistent effects of early augmented acoustic environment on the auditory brainstem. *Neuroscience, 184*, 75–87.
Ono, M., Yanagawa, Y., & Koyano, K. (2005). GABAergic neurons in inferior colliculus of the GAD67-GFP knock-in mouse: Electrophysiological and morphological properties. *Neuroscience Research, 51*(4), 475–492.
Orton, L. D., & Rees, A. (2014). Intercollicular commissural connections refine the representation of sound frequency and level in the auditory midbrain. *eLife, 3*, e03764.
Palmer, A. R., Shackleton, T. M., Sumner, C. J., Zobay, O., & Rees, A. (2013). Classification of frequency response areas in the inferior colliculus reveals continua not discrete classes. *The Journal of Physiology, 591*(16), 4003–4025.
Pérez-González, D., Malmierca, M. S., & Covey, E. (2005). Novelty detector neurons in the mammalian auditory midbrain. *European Journal of Neuroscience, 22*(11), 2879–2885.
Pérez-González, D., Malmierca, M. S., Moore, J. M., Hernández, O., & Covey, E. (2006). Duration selective neurons in the inferior colliculus of the rat: Topographic distribution and relation of duration sensitivity to other response properties. *Journal of Neurophysiology, 95*(2), 823–836.
Pérez-González, D., Hernández, O., Covey, E., & Malmierca, M. S. (2012). $GABA_A$-mediated inhibition modulates stimulus-specific adaptation in the inferior colliculus. *PLoS One, 7*(3), e34297. https://doi.org/10.1371/journal.pone.0034297.
Peruzzi, D., Sivaramakrishnan, S., & Oliver, D. L. (2000). Identification of cell types in brain slices of the inferior colliculus. *Neuroscience, 101*(2), 403–416.
Poon, P. W., Chen, X., & Cheung, Y. M. (1992). Differences in FM response correlate with morphology of neurons in the rat inferior colliculus. *Experimental Brain Research, 91*(1), 94–104.
Rajan, R. (1990). Electrical stimulation of the inferior colliculus at low rates protects the cochlea from auditory desensitization. *Brain Research, 506*(2), 192–204.
Rees, A., & Moller, A. R. (1987). Stimulus properties influencing the responses of inferior colliculus neurons to amplitude-modulated sounds. *Hearing Research, 27*(2), 129–143.
Reetz, G., & Ehret, G. (1999). Inputs from three brainstem sources to identified neurons of the mouse inferior colliculus slice. *Brain Research, 816*(2), 527–543.
Saint Marie, R. L. (1996). Glutamatergic connections of the auditory midbrain: Selective uptake and axonal transport of D-[3H]aspartate. *The Journal of Comparative Neurology, 373*(2), 255–270.
Saint Marie, R. L., Stanforth, D. A., & Jubelier, E. M. (1997). Substrate for rapid feedforward inhibition of the auditory forebrain. *Brain Research, 765*(1), 173–176.
Saldaña, E., & Merchán, M. A. (1992). Intrinsic and commissural connections of the rat inferior colliculus. *The Journal of Comparative Neurology, 319*(3), 417–437.
Saldaña, E., & Merchán, M. A. (2005). Intrinsic and commissural connections of the inferior colliculus. In J. A. Winer & C. E. Schreiner (Eds.), *The inferior colliculus* (pp. 155–181). New York: Springer.

Saldaña, E., Feliciano, M., & Mugnaini, E. (1996). Distribution of descending projections from primary auditory neocortex to inferior colliculus mimics the topography of intracollicular projections. *The Journal of Comparative Neurology, 371*(1), 15–40.
Schreiner, C. E., & Langner, G. (1988). Periodicity coding in the inferior colliculus of the cat. II. Topographical organization. *Journal of Neurophysiology, 60*(6), 1823–1840.
Schreiner, C. E., & Langner, G. (1997). Laminar fine structure of frequency organization in auditory midbrain. *Nature, 388*(6640), 383–386.
Shammah-Lagnado, S. J., Alheid, G. F., & Heimer, L. (1996). Efferent connections of the caudal part of the globus pallidus in the rat. *The Journal of Comparative Neurology, 376*(3), 489–507.
Shera, C. A. (2015). The spiral staircase: Tonotopic microstructure and cochlear tuning. *The Journal of Neuroscience, 35*(11), 4683–4690.
Shneiderman, A., & Henkel, C. K. (1987). Banding of lateral superior olivary nucleus afferents in the inferior colliculus: A possible substrate for sensory integration. *The Journal of Comparative Neurology, 266*(4), 519–534.
Sivaramakrishnan, S., & Oliver, D. L. (2001). Distinct K currents result in physiologically distinct cell types in the inferior colliculus of the rat. *The Journal of Neuroscience, 21*(8), 2861–2877.
Sivaramakrishnan, S., Sanchez, J. T., & Grimsley, C. A. (2013). High concentrations of divalent cations isolate monosynaptic inputs from local circuits in the auditory midbrain. *Frontiers in Neural Circuits, 7*, 175. https://doi.org/10.3389/fncir.2013.00175.
Smith, P. H. (1992). Anatomy and physiology of multipolar cells in the rat inferior collicular cortex using the in vitro brain slice technique. *The Journal of Neuroscience, 12*(9), 3700–3715.
Sun, H., & Wu, S. H. (2008). Modification of membrane excitability of neurons in the rat's dorsal cortex of the inferior colliculus by preceding hyperpolarization. *Neuroscience, 154*(1), 257–272.
Tan, M. L., Theeuwes, H. P., Feenstra, L., & Borst, J. G. (2007). Membrane properties and firing patterns of inferior colliculus neurons: An in vivo patch-clamp study in rodents. *Journal of Neurophysiology, 98*(1), 443–453.
Tanaka, I., & Ezure, K. (2004). Overall distribution of GLYT2 mRNA-containing versus GAD67 mRNA-containing neurons and colocalization of both mRNAs in midbrain, pons, and cerebellum in rats. *Neuroscience Research, 49*(2), 165–178.
Tokunaga, A., Sugita, S., & Otani, K. (1984). Auditory and non-auditory subcortical afferents to the inferior colliculus in the rat. *Journal für Hirnforschung, 25*(4), 461–472.
Tongjaroenbuangam, W., Jongkamonwiwat, N., Phansuwan-Pujito, P., Casalotti, S. O., Forge, A., Dodson, H., et al. (2006). Relationship of opioid receptors with GABAergic neurons in the rat inferior colliculus. *European Journal of Neuroscience, 24*(7), 1987–1994.
Ulanovsky, N., Las, L., & Nelken, I. (2003). Processing of low-probability sounds by cortical neurons. *Nature Neuroscience, 6*(4), 391–398.
Venkataraman, Y., & Bartlett, E. L. (2013). Post-natal development of synaptic properties of the gabaergic projection from inferior colliculus to auditory thalamus. *Journal of Neurophysiology, 109*(12), 2866–2882.
Wallace, M. N., Shackleton, T. M., & Palmer, A. R. (2012). Morphological and physiological characteristics of laminar cells in the central nucleus of the inferior colliculus. *Frontiers in Neural Circuits, 6*, 55. https://doi.org/10.3389/fncir.2012.00055.
White, J. S., & Warr, W. B. (1983). The dual origins of the olivocochlear bundle in the albino rat. *The Journal of Comparative Neurology, 219*(2), 203–214.
Whitley, J. M., & Henkel, C. K. (1984). Topographical organization of the inferior collicular projection and other connections of the ventral nucleus of the lateral lemniscus in the cat. *The Journal of Comparative Neurology, 229*(2), 257–270.
Wiberg, M., Westman, J., & Blomqvist, A. (1987). Somatosensory projection to the mesencephalon: An anatomical study in the monkey. *The Journal of Comparative Neurology, 264*(1), 92–117.
Willard, F. H., & Ryugo, D. (1983). Anatomy of the central auditory system. In J. F. Willot (Ed.), *The auditory psychobiology of the mouse* (pp. 201–304). Springfield, IL: Charles C. Thomas.

Winer, J. A., Saint Marie, R. L., Larue, D. T., & Oliver, D. L. (1996). GABAergic feedforward projections from the inferior colliculus to the medial geniculate body. *Proceedings of the National Academy of Sciences of the United States of America, 93*(15), 8005–8010.
Wynne, B., & Robertson, D. (1997). Somatostatin and substance P-like immunoreactivity in the auditory brainstem of the adult rat. *Journal of Chemical Neuroanatomy, 12*(4), 259–266.
Wynne, B., Harvey, A. R., Robertson, D., & Sirinathsinghji, D. J. (1995). Neurotransmitter and neuromodulator systems of the rat inferior colliculus and auditory brainstem studied by in situ hybridization. *Journal of Chemical Neuroanatomy, 9*(4), 289–300.
Yasui, Y., Nakano, K., Kayahara, T., & Mizuno, N. (1991). Non-dopaminergic projections from the substantia nigra pars lateralis to the inferior colliculus in the rat. *Brain Research, 559*(1), 139–144.
Zhang, H., & Kelly, J. B. (2010). Time dependence of binaural responses in the rat's central nucleus of the inferior colliculus. *Hearing Research, 268*(1-2), 271–280.

Chapter 7
Sensing Sound Through Thalamocortical Afferent Architecture and Cortical Microcircuits

Heather L. Read and Alex D. Reyes

Abstract Mammals have multiple cortical fields with neurons responding to the full range of sound frequencies sensed by the sensory epithelium of the ear, the cochlea. At first glance, sensitivities to sound frequency seem highly similar or even redundant across different auditory cortical fields; however, closer inspection reveals marked differences in the architecture of underlying thalamocortical pathways, afferents, and sound sensitivities within each cortical field. Here we summarize the differences in the architecture of the thalamocortical pathway and the termination patterns across auditory cortices. A conceptual framework is presented for how pathway architecture in combination with intrinsic cortical microcircuits can account for observed differences in sound sensitivity. This organization has the capacity to create parallel sound processing streams with diverse and even dynamic sound processing properties.

Keywords Ascending auditory pathway · Cochleotopy · Comparative physiology · Cortical fields · Cortico-cortical networks · Neural coding · Sensory processing · Spectral bandwidth · Sound localization cues · Spectral resolution · Thalamocortical · Wide-field imaging

H. L. Read (✉)
Department of Psychological Sciences and Department of Biomedical Engineering, University of Connecticut, Storrs, CT, USA

Kavli Institute for Theoretical Physics, University of California, Santa Barbara, CA, USA
e-mail: heather.read@uconn.edu

A. D. Reyes
Center for Neural Science, New York University, New York, NY, USA
e-mail: reyes@cns.nyu.edu

© Springer International Publishing AG 2018
D. L. Oliver et al. (eds.), *The Mammalian Auditory Pathways*, Springer Handbook of Auditory Research 65,
https://doi.org/10.1007/978-3-319-71798-2_7

7.1 *Introduction*

In all mammals, there is a remarkable transformation from one neural representation of sound at the sensory epithelium of the ear to many neural representations as one ascends the auditory pathway to its highest levels in the cortex. Initially, sounds are decomposed into tonal frequency components in the cochlea, where maximal sensitivities to the highest and lowest audible frequencies occur at its base and apex, respectively. This topographic organization, which has been dubbed "tonotopy" or "cochleotopy," is conserved in the spatial organization of the auditory nerves, downstream brainstem nuclei, and ultimately cortex, thereby producing a direct correspondence between sound frequency and the location of activated neurons in each brain structure. Accordingly, *cochlear place* is an initial framework for encoding sound frequency and is the basis for the highly successful cochlear implant prosthetics that ultimately enable deaf patients to discriminate speech, tonal frequency, and pitch (Moore 2003; Carlyon et al. 2010). Though a single cochleotopy is observed subcortically in the central nucleus of the inferior colliculus (Ito and Oliver 2012; Cant and Oliver, Chap. 2), multiple distinct cochleotopic areas emerge at the levels of the thalamus and cortex.

At first glance, the existence of multiple cochleotopies seems redundant, but on close examination it is clear that each cortical field responds differently to sound. As elaborated in this chapter, these differences can be attributed in part to differences in spatial resolution of the corresponding ascending thalamic pathway to each cortical field. Furthermore, computer simulations that capture observed thalamic pathway architecture and cortical microstructure indicate that differences in thalamocortical pathway architecture can ultimately determine what networks of excitatory and inhibitory neurons within cortex are activated by sound. Consequently, having multiple parallel ascending pathway architectures allows for diverse and dynamic ways of sensing sound.

7.1.1 Multiple Cochleotopic Cortical Fields in Mammals

All mammals examined to date have multiple cochleotopic cortical fields. Human and nonhuman primates, such as the macaque monkey (*Mucaca mulatta*), have six or more cochleotopic cortical fields as mapped with spike rate or metabolic responses (e.g., Fig. 7.1A, human) (Formisano et al. 2003; Baumann et al. 2015). Similarly, wide-field response mapping reveals that cat (*Felis silvestris catus*) (Imig and Reale 1980; Hall and Lomber 2015) and ferret (*Mustela putorius furo*) (Nelken et al. 2008; Bizley and King 2009) have about six cochleotopic cortical fields. Up to seven cochleotopic fields can be distinguished in mouse (*Mus musculus*) and rat (*Rattus norvegicus*) (Fig. 7.1B and Fig. 7.2), as determined from mapping layer 4 neuron spike rate (Guo et al. 2012) and wide-field metabolic (Kalatsky et al. 2005; Higgins et al. 2010), calcium (Issa et al. 2014), and surface potential responses to sound (Escabi et al. 2014).

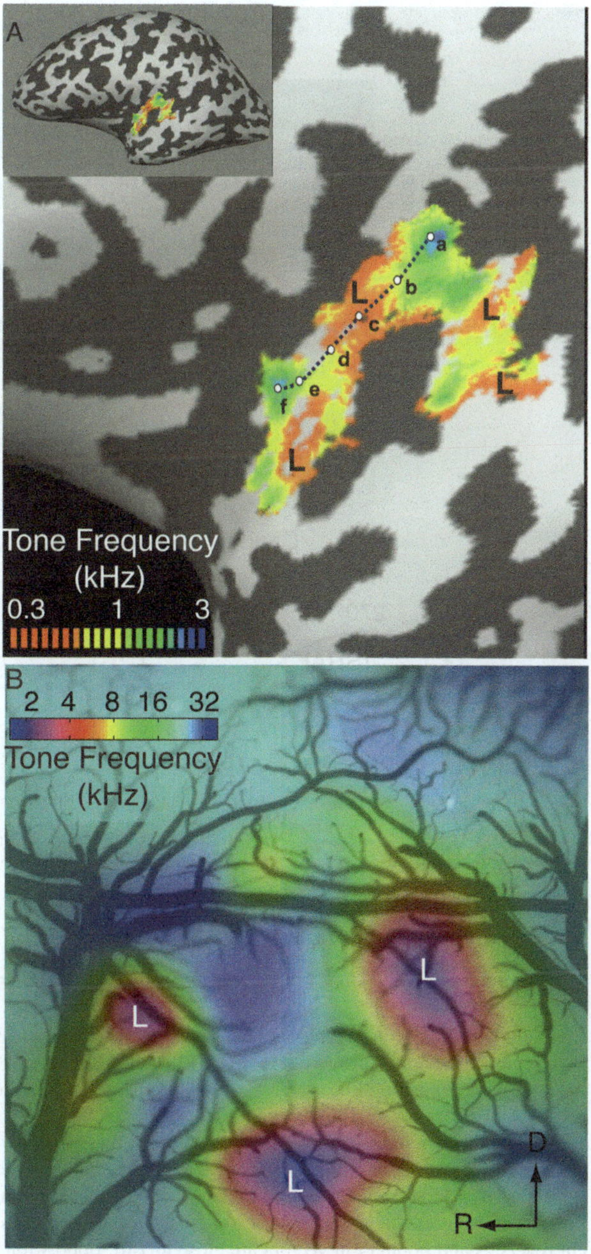

Fig. 7.1 Functional imaging of responses to tones reveals multiple cochleotopic cortical fields in human and rat temporal cortex. With the Fourier imaging approach, ascending and descending tone sequences are played continuously while acquiring metabolic responses to generate high spatial resolution, wide-field cortical response maps. **(A)** Pseudo-color rendering, functional magnetic resonance image (fMRI) reveals four cortical areas with low-tone frequency selectivity in human (labeled *L*, *red regions*). Multiple mirror reversals in cochleotopy are evident (e.g., follow color spectrum changes between areas labeled *a* and *f*). **(B)** Pseudo-color rendering of intrinsic optical image reveals three cortical areas responding to low-frequency tones in rat temporal cortex (labeled *L*, *blue regions*). Multiple mirror reversals in cochleotopy are evident. Note same data here as in Fig. 7.2A. (Orientation arrows refer to both figures: *D*, dorsal; *R*, rostral) [**A** modified from Formisano et al. (2003); **B** adapted from Higgins et al. (2010)]

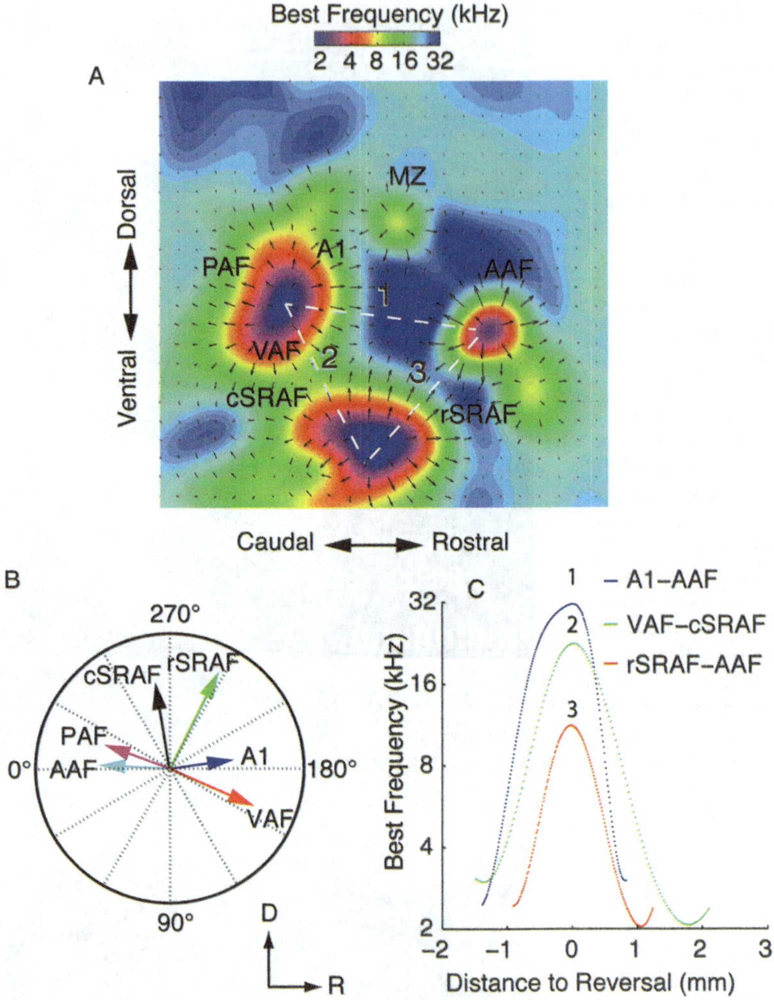

Fig. 7.2 Differences in magnitude and direction of tone best frequency (BF) response gradients between cochleotopic cortical fields in rat. **(A)** Local changes in BF are determined and rendered by a map of corresponding vectors (*overlaid arrows*) with the wide field intrinsic metabolic response (for details see Higgins et al. 2010). *Scale*: x-axis is 4.6 mm. **(B)** *Population BF gradient vectors* are the sum of local BF gradient vectors within each of six cortical fields including primary (*A1*), anterior (*AAF*), ventral (*VAF*), posterior (*PAF*) auditory fields, caudal (*cSRAF*) and rostral (*rSRAF*) suprarhinal auditory fields. Each cortical field has a unique magnitude and direction of the population BF gradient as shown in the circular plot. Dotted lines indicate 30° intervals. BF gradient reversals and population vectors are close to 180° opposition for VAF versus PAF and A1 versus AAF. Opposing gradient reversals (<180°) are observed at borders between VAF and cSRAF, AAF and rSRAF. Cortical field boundaries determined as detailed in original source. **(C)** Scatter plots of BF variation along single anatomical axes indicated with *white dotted lines (1, 2, 3)* in **A**. Note that the data shown here highlight cochleotopy and BFs ranging from 2 to 32 kHz, as illustrated by the color bar in **A**, though responses are mapped up to 48 kHz. Image in **A** covers a 4.6 mm by 4.6 mm area in temporal cortex. (*D*, dorsal; *R*, rostral)[Figures **A** and **B** and data plotted in **C** are adapted from Higgins et al. (2010)]

Table 7.1 Abbreviations

A1	Primary auditory field
A2	Area 2
AAF	Anterior auditory field
BF	Best frequency
CM	Caudal medial area
CMF	Cortical magnification factor
cSRAF	Caudal suprarhinal auditory field
DP	Dorsal posterior field
g_{syn}	Excitatory synaptic conductance
IAF	Insular auditory field
ILD	Interaural level difference
MGB	Medial geniculate body
MZ	Multimodal zone
PAF	Posterior auditory field
Q	Quality factor for spectral resolution
rSRAF	Rostral suprarhinal auditory field
SRAF	Suprarhinal auditory field
VAF	Ventral auditory field
VGLUT	Vesicular glutamate transporter

Cross-modal sensory sensitivities exist in cochleotopic cortical fields outside of the primary auditory cortex (A1; all abbreviations listed in Table 7.1), indicating distinct functionality. This includes visuospatial sensitivities in the posterior auditory field in the cat (Updyke 1986), caudal auditory field in macaque monkey (Miller and Recanzone 2009) and posterior suprasylvian field in ferret (Bizley and King 2009). In addition, cross-sensory modal (auditory and somatosensory) sensitivities are observed in the cochleotopic multimodal zone (MZ) in rat (Brett-Green et al. 2003). Hence, cochleotopy is not an exclusive property of unimodal primary auditory cortex.

The surface area of A1 increases in mammals as one ascends the phylogenetic tree. For example, the length of the rat A1 (1.5 mm) in the caudal–rostral or cochleotopic dimension is about two times larger than that of the mouse (0.75 mm) for the same range of sound frequency sensitivities (e.g., Figs. 7.2C, and 7.3C) (Hackett et al. 2011; Guo et al. 2012). An approximate doubling of A1 surface area is also evident between ferret and cat (Merzenich et al. 1975; Mrsic-Flogel et al. 2006) as well as between Macaque monkey (8 mm) and human (15 mm) (Petkov et al. 2006; Moerel et al. 2014). This phylogenetic expansion of A1 surface area cannot be exclusively attributed to an expansion of the audible range of sound frequencies. For example, the A1 surface area of humans is almost 4-fold greater than that of the cat, though both have highly overlapped audible sound frequency ranges, and the cat has hearing sensitivities that extend to higher sound frequencies that are associated with expanded cortical area representation (details in Sect. 7.3). Though the structure-function relationships are not well understood, the phylogenetic expansion of A1

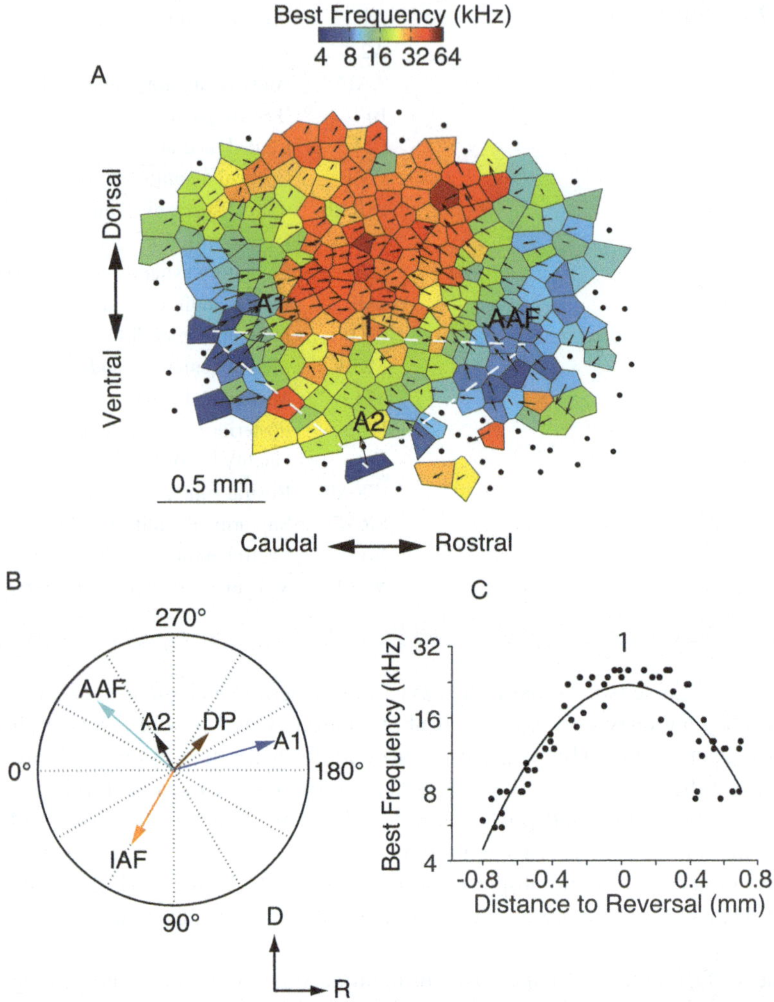

Fig. 7.3 Differences in magnitude and direction of best-frequency (BF) response gradients between cochleotopic cortical fields in mouse. BF responses were estimated from multi-unit spike rate responses to tones, recorded in layers 4 and 3B. **(A)** Local changes in BF were determined and rendered by a map of corresponding vectors overlaid with the wide-field map, multi-unit responses. Auditory fields shown: *A1, A2*, and anterior, *AAF*. BF variations along *white dotted line 1* are shown in **C**. Filled black circles outside polygons indicate unresponsive recording sites. **(B)** Population BF gradient vectors are the sum of local BF gradient vectors within each of five cortical fields, including primary (*A1*), anterior (*AAF*), dorsal medial-posterior (*DP*), area 2 (*A2*), and insular (*IAF*) auditory fields. Each cortical field has a unique magnitude and direction of the population BF gradient as shown in the circular plot. Dotted lines indicate 30° intervals. Cortical field boundaries determined as detailed in original source, cited below. BF gradient reversals and population vectors are reversed with angles greater than 100° for *A1* versus *AAF* and *DP* versus *IAF*. *D*, dorsal; *R*, rostral. **(C)** Scatter plots of BF variation along single anatomical axes approximated by *white dotted line 1* in panel **A**. Note that data shown here include cochleotopy and BFs ranging from 2 to 64 kHz (see color bar in panel **A**). [**A** and **B** adapted from Guo et al. (2012); **C** adapted from Hackett et al. (2011)]

surface area is likely critical for the ability to discriminate pitch (Jenkins and Merzenich 1984; Heffner 2004), intensity, and spatial location of sounds (Polley et al. 2006; Lomber and Malhotra 2008).

7.2 Cochleotopic Field Boundaries Marked by Differences

Outside of A1, the auditory cortical fields have quantifiable differences in cochleotopy. These can be measured by mapping the topography of the best frequency (BF) response to sound, that is, the sound frequency eliciting the largest response at each cortical site. The BF responses and cochleotopy can be measured and cross-validated in several ways, including measures of activity from multi-neuron clusters (Hackett et al. 2011; Storace et al. 2011), calcium imaging of individual neurons (Kanold et al. 2014), surface microelectrocorticographic potentials (micro-ECoG) (Escabi et al. 2014), and intrinsic metabolic responses (Storace et al. 2011). The magnitude and direction of change in BF can be computed locally and represented with local vector maps (e.g., overlaid arrows in Figs. 7.2A and 7.3A). The *population vector* (the mean of local BF gradient vectors) is a metric of the population BF gradient for a given cortical field (Figs. 7.2B and 7.3B).

Using the population vector approach, BF gradient directions and magnitudes can be compared across cortical fields within a cerebral hemisphere. For example, unique population BF vectors may be identified in the rat for fields, including primary (A1), anterior (AAF), ventral (VAF), and posterior (PAF) auditory fields and caudal (cSRAF) and rostral (rSRAF) suprarhinal auditory fields (Fig. 7.2A) (Higgins et al. 2010). Similarly, unique local and population BF gradients identify multiple cochleotopic fields in the mouse, including dorsal posterior (DP), A1, AAF, Area 2 (A2,) and insular auditory fields (IAF) (Fig. 7.3A) (Guo et al. 2012).

Neighboring cortical fields often have opposite or nearly opposite directions of BF gradient. For example, A1 and AAF best frequency gradients oppose one another, forming mirror-reversed BF gradients (e.g., Fig. 7.2A). Accordingly, A1 and AAF population vectors are rotated by about 130° and 180° in the mouse and rat, respectively (e.g., Figs. 7.2B, and 7.3B). In addition, along the rostral-to-caudal direction of rodent cortex, scatterplots illustrate that BF increases to a peak that marks the end of A1 and the start of AAF (e.g., blue and black plots, respectively in Fig. 7.2C and Fig. 7.3C). Similar local and population frequency gradient transitions are evident for A1–AAF in the cat (Imaizumi et al. 2010).

With current technologies, it is challenging and time consuming to map multi-unit BF responses across multiple cortical fields with sufficient density to compute local BF gradients in two dimensions. A more practical approach is to determine BF as a function of cortical position in one dimension and to fit the data with an appropriate mathematical function (e.g., Fig. 7.3C). The BF gradient in this case is the slope of the corresponding function for a given cortical field. The latter approach reveals distinct A1 and anterior cochleotopic fields in cat (Merzenich et al. 1975; Imaizumi and Schreiner 2007), ferret (Mrsic-Flogel et al. 2006), rat

(Storace et al. 2011), guinea pig (*Cavia porcellus*) (Nishimura and Song 2014), and mouse (Guo et al. 2012).

In addition to A1 and AAF, many mammals have one or more ventral cochleotopic fields with unique anatomical directions for BF gradients. In the rat, there is a ventral field called the suprarhinal (SRAF) auditory field as it borders the rhinal fissure. Caudal (cSRAF) and rostral (rSRAF) halves of SRAF have BF gradients that are mirror reversed with VAF and AAF, respectively (Fig. 7.2C). There is an analogous PAF in the cat (Reale and Imig 1980; Butler et al. 2016) and rat (Doron et al. 2002; Pandya et al. 2008) that can be revealed with wide-field tone response mapping. As elaborated in Sect. 7.4, response maps for many nonprimary fields in rodent, such as PAF, VAF, and SRAF, are best obtained when sounds are presented to both ears (binaurally) at relatively low sound levels (e.g., 45 dB, Fig. 7.2A, B). Binaural sensitivities are also characteristic of cat PAF (Harrington et al. 2008).

Neighboring cochleotopic fields can share high-frequency or low-frequency sensitive borders. For example, high BF's are observed at the reversal border for the A1–AAF, VAF–cSRAF, and AAF–rSRAF in the rat (Fig. 7.2C). Conversely, BF gradients between A1 and PAF reverse at a shared border that responds maximally to low frequency sounds (Fig. 7.2A). The *A Laplace transform* of the BF response map can be used to locate such maxima and minima in the BF gradients and identify the borders between cortical fields (Higgins et al. 2010). Borders between fields are also associated with lower response magnitudes with metabolic imaging (Kalatsky et al. 2005) or calcium imaging techniques (Issa et al. 2014). Whether cochleotopic fields outside of A1 are functionally analogous across different mammals remains unclear. Together these studies indicate that cochleotopic fields are readily identified according to unique BF gradients and corresponding reversals at boundaries between fields.

7.3 Cochleotopic Resolution Differences Across Cortical Fields

Cortical fields differ in the cortical area and the number of neurons responding to a given range of sound frequency. The *cortical magnification factor* (CMF) can be calculated as the inverse BF gradient or as the length of a segment of cortex responding to an octave range of sound frequency. For example, a four-octave range of sound frequency activates about a 1.25 mm diameter area of A1 cortex, resulting in a CMF of about 0.312 mm per octave or 312 μm/octave (Fig. 7.2C). In contrast, AAF spans about 0.8 mm per 4 octaves with CMF of 200 μm/octave (Fig. 7.2C). Assuming that the neuron density is the same for both areas, the CMF also provides a relative estimate of the number of neurons that represents an octave of sound frequency. Accordingly, for this example, A1 has 1.5-fold more neurons representing each octave of sound than AAF (Fig. 7.2).

Local BF gradient slopes and CMFs can vary within a cortical field, indicating that some frequencies are represented by more neurons than others. For example, along the caudal–rostral axis in cat A1, the BF gradient or slope is 2-fold greater for an octave of sound between 10 and 20 kHz than it is for an octave between 2 and 4 kHz (Merzenich et al. 1973, 1975). The cortical distance or CMF responding to an octave between 10 and 20 kHz is 1.25 mm/octave and that for 2 to 4 kHz is half as much or 0.625 mm/octave in the cat. Similarly, in rat A1, a 1 mm segment of cortex represents an octave between 32 and 64 Hz; whereas about 0.300 mm of cortex along the same anatomical axis responds to an octave between 10 and 20 kHz (Kim and Bao 2013). This inhomogeneity in the neural representation likely reflects an adaptation allowing mammals to discriminate higher frequency sounds with greater resolution (LePage 2003).

BF gradients and CMFs also differ across cortical fields. As described for rat, cat BF gradients in AAF (Imaizumi et al. 2010) are steeper than those in A1 (Merzenich et al. 1975; Greenwood 1990; LePage 2003). Moreover, the BF values in AAF vary more than those in A1. Similarly in the mouse, BF gradients in A2 are steeper and more variable than those of A1 and AAF (Guo et al. 2012). The origins of these CMF and BF gradient differences are still poorly understood but may arise from differences in underlying thalamocortical pathway architectures and the associated populations of activated neurons (see Sect. 7.8).

7.4 Sound Sensing Differences Across Cortical Fields

Cochleotopic cortical fields and subfields differ in their ability to resolve frequency or spectral cues in sound. The *spectral bandwidth* is a measure of the range of sound frequencies sensed by a neuron or multi-unit cluster of neurons. Spectral bandwidth can be estimated as the standard deviation of the tone-evoked spike rate response across different sound frequencies at a given sound level (e.g., see horizontal bars indicating spectral bandwidth at each sound level, Fig. 7.4A, C). The quality factor (Q = BF/bandwidth) describes how well a neuron resolves sound frequency, that is, its *spectral resolution*. Spectral bandwidth and Q factor can also be estimated from the spectral temporal response field elicited with dynamically modulated sounds (Miller et al. 2002; Escabi et al. 2014). In A1, spectral bandwidths generally increase with increasing sound intensity, resulting in a V–shaped response function (e.g., 40 dB versus 80 dB in Fig. 7.4A). Accordingly, spectral bandwidths can span the entire hearing range of an animal when sounds are presented at high decibel levels (e.g., 5 octaves in rat at 80 dB; Fig. 7.4A). However, this is not the case in regions that are sensitive to input from both ears, such as VAF and cSRAF. When tones are played at 40 to 80 decibels to both ears, VAF neurons respond to a narrow range of sound frequencies (e.g., Fig. 7.4C). For many VAF neurons, spike rate changes non-monotonically with sound intensity, which results in lower spike rates as sound levels are increased above 50 dB (e.g., Fig. 7.4C). Morever, spike rate responses can be greatly increased or facilitated when using binaural as opposed to monaural sound presentation (Higgins et al., 2010).Therefore, it is important to use mid-range (45–65 dB) sound levels and binaural sound presentation for simultaneous wide-field mapping of cochleotopy across multiple cortical fields.

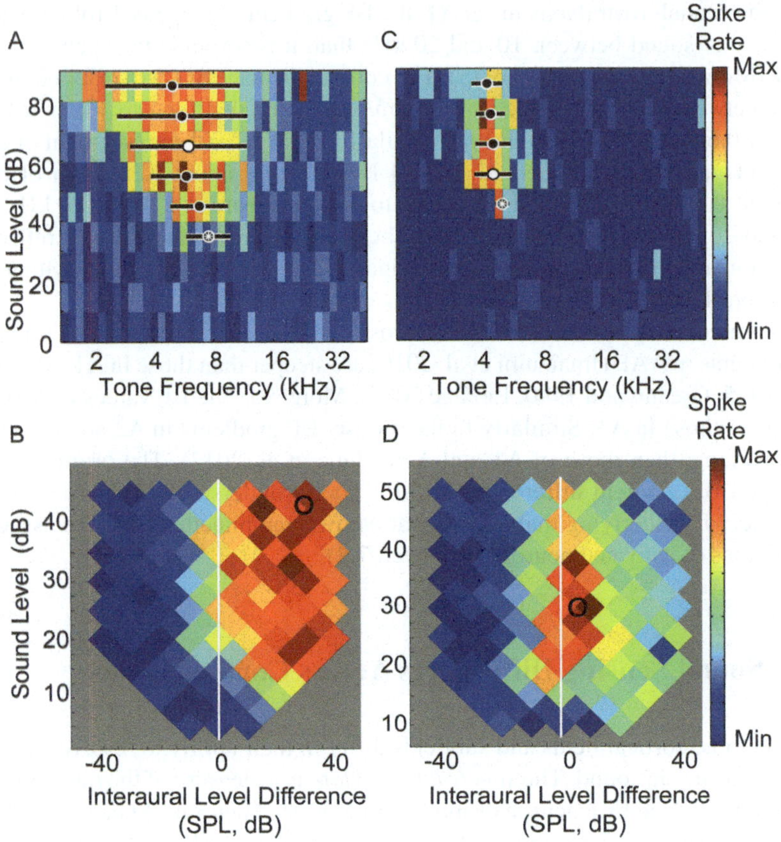

Fig. 7.4 Cortical field differences in resolution of responses to spectral and spatial sound cues in rat primary (*A1*) and nonprimary (*VAF* and *cSRAF*) cochleotopic cortical fields. **(A, C)** Multi-unit spike rate responses vary with tone frequency (*x-axis*) and level (*y-axis*) for two example units from A1 and VAF, respectively. Sound are presented binaurally and levels were matched such that interaural level difference (*ILD*) is zero for all tone responses. *Black filled circles* indicate the best frequency (BF) computed as the center of mass for the spike rate response across all tone frequencies at a given sound level. *White filled circle* indicates BF at the sound level eliciting maximum spike rate. *Circle with asterisk* indicates the BF at sound level threshold for significant spiking, typically called the characteristic frequency. *Bars* indicate the spectral bandwidth (i.e., standard deviation) of the spike rate response for the same conditions. As is typical of A1, the unit in **A** has increasingly broader bandwidths with higher sound levels. As is typical of VAF, the example unit in **C** has narrow bandwidths at all sound levels. **(B, D)** Multi-unit spike rate responses vary with interaural level difference (ILD) cues for virtual horizontal sound position (*x-axis*). A typical A1 unit in **B** responds near maximally to a broad range of ILD cues for sound positions on the contralateral acoustic hemifield or opposite side of the body (i.e., positive values on x-axis). For **B,** the best ILD (*open circle*) associated with the greatest spike rate is for ILD cues far from the body or acoustic hemi-field midline (ILD = 0, *white line*). A typical cSRAF unit shown in **D** responds near maximally to a narrow range of ILD cues for sound positions near the body or acoustic hemi-field midline (ILD = 0, *white line*). In the cSRAF unit, the best ILD (*open circle*) is associated with ILD cues near the acoustic midline. [**A, C** data from Storace et al. (2011); **B, D** data from Higgins et al. (2010)]

Wide-field maps indicate that the spectral resolution of responses to sound changes systematically across cortical fields. Tone response maps obtained either with metabolic (Fig. 7.2A) or multiunit recordings (Fig. 7.5A) show no mirror reversal between cochleotopies in A1 and VAF. However, a clear border is evident with intrinsic metabolic response magnitude maps that find lower magnitude responses at the borders between fields, including A1 and VAF (Kalatsky et al. 2005). Furthermore, the spectral resolution (Q factor) for tone frequency increases between A1 and VAF along the dorsoventral anatomical axis (Fig. 7.5B, C). Accordingly, the population-average spectral resolution increases between A1 and VAF (Polley et al. 2007; Storace et al. 2011). In this and other respects, the rat A1 and VAF fields are analogous to the cat dorsocentral and central subfields of A1, respectively (Read et al. 2001; Imaizumi and Schreiner 2007). These observations indicate that, unlike the sensory epithelium of the ear, the cortex has a topographic meta-organization of acoustic features other than BF (Polley et al. 2007).

Neurons in cortical fields with high spectral resolution also have high-resolution responses to sound localization cues. In most animals, playing sounds from locations to the side of the head generates sound level differences between the two ears (interaural level difference or ILD). Accordingly, ILD varies systematically with horizontal sound position (azimuth) in the rat (Koka et al. 2008). The ILD cues are not sensed at the sensory epithelium of the ear but instead are computed at higher levels of the auditory pathway in rodents (Sanes and Rubel 1988) and other mammals (Tollin 2003). The sensitivities of cortical neurons to ILD cues may be examined by presenting sound to left and right ears (stereo presentation) (Higgins et al. 2010). Using these virtual sound cues, Higgins et al. (2010) have shown that VAF and cSRAF are more sensitive to stereo sound presentation than A1 or rSRAF.

On average, A1 neurons are indifferent to sound presentation from one ear versus two ears. In contrast, VAF and cSRAF neurons respond minimally or not at all when sounds are presented to one ear only (Wu et al. 2006; Higgins et al. 2010). In addition, A1 neurons respond at near maximal spike rates for a broad range of ILDs that cue sound positions on the contralateral side of the body (e.g., Fig. 7.4B). In contrast, VAF and cSRAF respond maximally to a narrow range of ILDs that cue sound positions near the body midline (e.g., Fig. 7.4D). Analogous cortical field differences are evident in the cat where both spectral (Read et al. 2001; Imaizumi and Schreiner 2007; Atencio and Schreiner 2012) and ILD resolution are higher in central versus dorsal subfields of A1 (Semple and Kitzes 1993), when mapped with binaural sound presentation. These findings indicate that neuronal abilities to resolve spatial and spectral cues in sound covary across cortical fields and subfields.

7.5 Segregated Thalamocortical Pathways to Different Cochleotopic Cortices

There is now considerable evidence supporting a new *connectome* or anatomical framework for how sound sensitivity differences emerge in the different cochleotopic cortical fields and subfields (Fig. 7.6). Historically, pathway diagrams depict

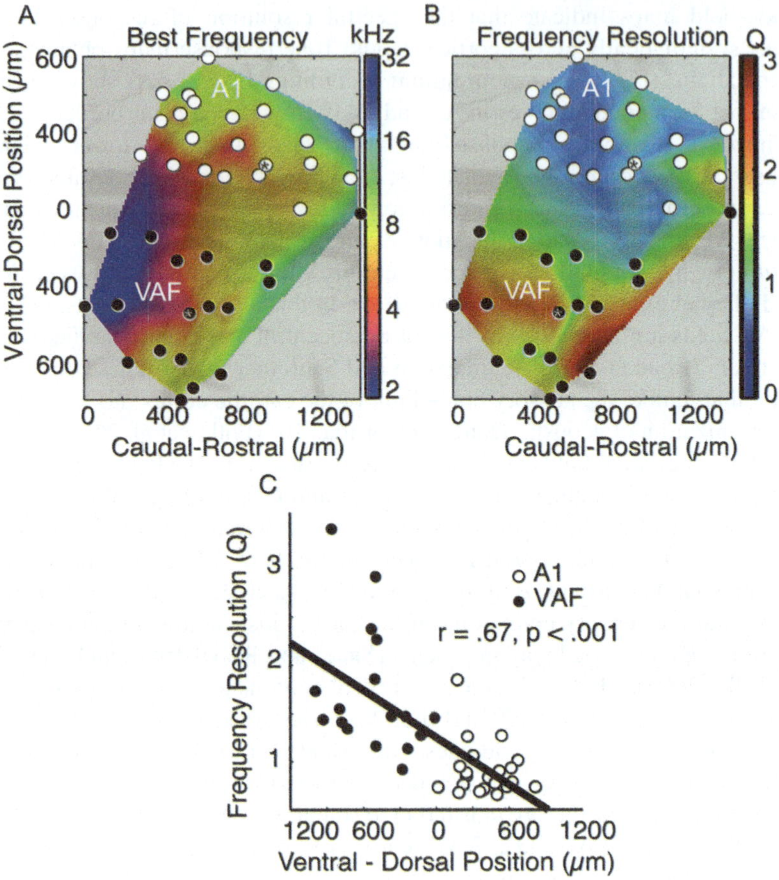

Fig. 7.5 Tone-response best frequency (BF) and resolution (Q) organization in rat A1 (*white circles*) and VAF (*black circles*). *Circles with asterisks* indicate positions for example multi-unit frequency response areas shown in Fig. 7.4A, C. **(A)** BF was recorded from layers 4 and 3B across all positions indicated with the filled circles in one hemisphere of rat temporal cortex. BF is determined from the frequency response areas, as shown in Fig. 7.4A, C. BF is plotted for each anatomical position and resulting maps are tessellated and smoothed. BF's change from 2 to 32 kHz (*see color bar*) along the caudal–rostral (*x-axis*) anatomical axis. **(B)** Q factors estimated at the best sound intensity level (*Q*) change from about 0.5 to 3, indicating a shift from coarse (*blue*) to fine (*red*) spectral resolution. **(C)** Data in panel **B** are collapsed along the caudal–rostral anatomical axis to generate a scatter plot of Q as a function of ventral-to-dorsal position. Note how the spectral resolution (Q) shifts from high to low values moving from ventral to dorsal cortex, or *VAF* to *A1*, respectively. (*r*, correlation coefficient) [**A–C** adapted from Storace et al. (2011)]

the same division of the thalamus with divergent output to all cochleotopic cortical fields (Fig. 7.6A) (Weinberger and Diamond 1987; Kaas and Hackett 2000). To elaborate, sensory input to all of the auditory cortex arises from the medial geniculate body (MGB). The MGB is divided into ventral, dorsal, and medial areas, among others, based on measures such as myelinated pathway density, cytochrome oxidase

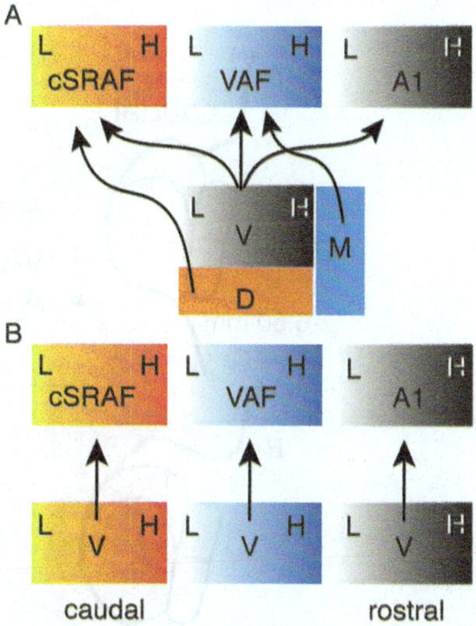

Fig. 7.6 Parallel caudal-to-rostral organization of ascending thalamocortical pathways mirror physiological differences in target cochleotopic cortical fields. **(A)** *Single inherited cochleotopy model*—the historical view of how physiological differences are created through feedforward thalamic pathways from auditory thalamus to cochleotopic cortical fields in rat. According to this schema, all cochleotopic cortical fields inherit their cochleotopic organization from the presumed same population of neurons located in the ventral (*V*) division of the auditory thalamus. Each cortical field then receives a different combination of converging inputs from thalamic divisions outside of the ventral division, such as the medial (*M*) and dorsal (*D*) divisions. **(B)** *Multiple parallel thalamocortical pathway model*—a new model for how physiological differences are created for layer 4 neurons in cochleotopic cortical fields. This model accounts for the reduced label (~20–30%) in *M* and *D* divisions with deposits restricted to layers 4 and 3b (e.g., Fig. 7.8). This model also accounts for the observed lack of double-labeled neurons with injections into frequency-matched regions of different cochleotopic cortical fields as well as the caudal-to-rostral organization of those connections. (*A1*, primary auditory field; cSRAF, caudal suprarhinal auditory field; *L*, low frequency location; *H*, high frequency location)

reaction density, cytoskeletol protein (SMI-32) density, as well as morphology, size, and packing densities of neurons in cat (Morest 1964), tree shrew (*Tupaia glis*) (Oliver and Hall 1975), rat (Paxinos et al. 1980; Clerici and Coleman 1990; Winer et al. 1999), guinea pig (Anderson et al. 2007), gerbil (Cant and Benson 2007), and mouse (Horie et al. 2013). For example, estimated divisional boundaries for a caudal-to-rostral series of coronal sections of rat thalamus are shown in Fig. 7.7.

In all mammals examined, neurons in the ventral division of the thalamus respond selectively to sound frequency (Imig and Morel 1985; Hackett et al. 2011) and are spatially organized in a cochleotopic manner (Cant and Benson 2007; Storace et al. 2011; Horie et al. 2013; Cant and Oliver, Chap. 2). In mammals, the ventral division neurons that project to cochleotopic cortical fields are located within different caudal-to-rostral segments of the thalamus in several species (e.g., rat, Fig. 7.8A). In rat, the majority (70–75%) of thalamic neurons that project to cochleotopic cortex are located in the ventral division of the thalamus (Fig. 7.8B). The remaining

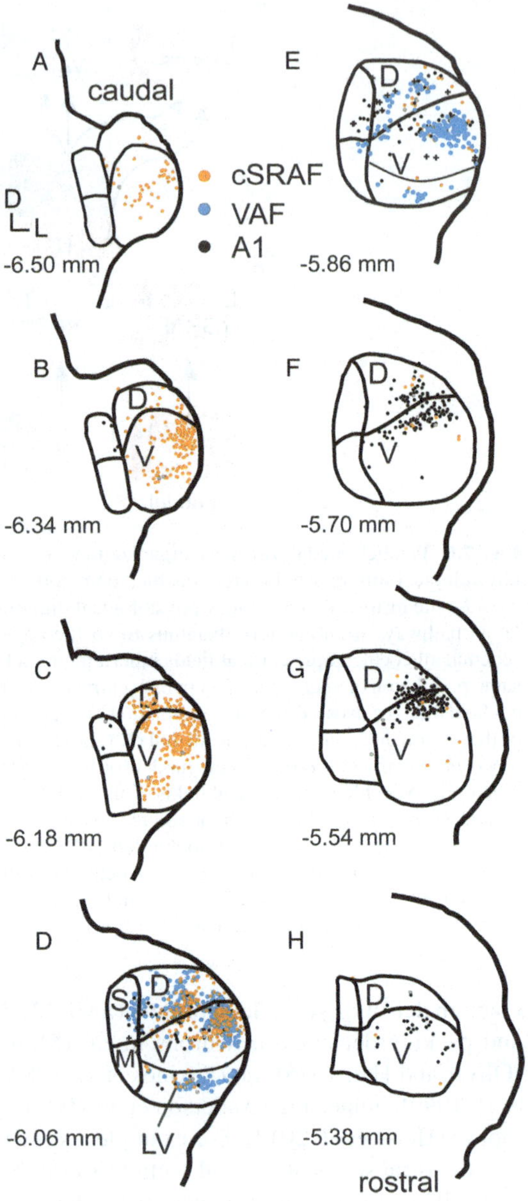

Fig. 7.7 Series of coronal thalamic section outlines illustrating caudal-to-rostral organization of thalamic neurons projecting to three physiologically mapped cortical fields in rat. **(A–H)** Series of 50 μm coronal section outlines depicting the position of retrogradely labeled neurons projecting to *cSRAF (orange circles)*, *VAF (blue circles)* and *A1 (black circles)* along the caudal-to-rostral anatomical axis in the rat. Filled-circles (orange, blue and black) indicate position of somata with retrograde label projecting to each respective cortical field. *Gray-filled circles* indicate position of neurons that are double labeled and project to both A1 and cSRAF. *Black + symbols* indicate position of neurons that are double labeled and project to both A1 and VAF. Data are combined for two animals. (Divisions of the medial geniculate body: *D*, dorsal; *LV* lateral ventral; *M*, medial; *S*, suprageniculate; *V*, ventral) [Figures adapted from Storace et al. (2011, 2012)]

(~25%) thalamic neurons projecting to layers 4 and 3b are located in other thalamic divisions (Fig. 7.8B). Similarly, in the cat the majority of thalamic neurons that project to central and dorsal cochleotopic subfields of A1 (Read et al. 2011) and to PAF (Butler et al. 2016) are located in the ventral division of the thalamus.

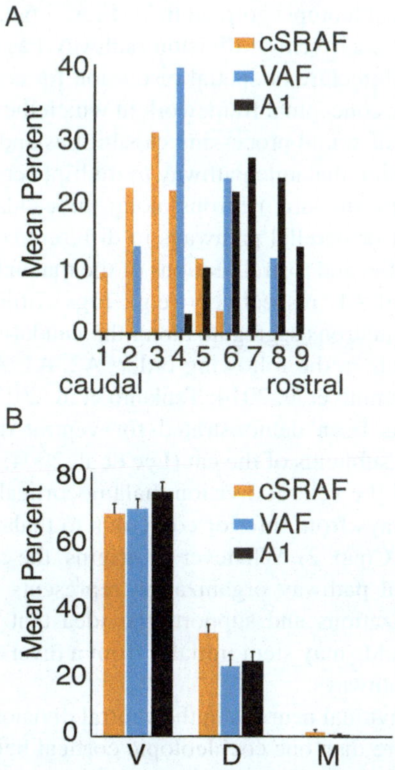

Fig. 7.8 Caudal–to–rostral organization and percentage of thalamic neurons that project to three physiologically mapped cortical fields in rat. (**A**) Ventral division thalamic neurons that project to the physiologically mapped 10–12 kHz domain of *cSRAF (orange bars)*, *VAF (blue bars)*, and *A1 (black bars)* auditory fields are located in *caudal*, *mid-caudal*, and *rostral thalamus*, respectively (*x-axis*). Note that the cSRAF and A1 data come from dual injections in a single animal and the VAF data are from another animal. (**B**) Bar graph indicating the mean percentage of retrogradely labeled neurons located in ventral (*V*), dorsal (*D*), and medial (*M*) thalamic divisions following injections into *cSRAF*, *VAF*, and *A1* (*color code as in* **A**). Note that the largest percentage of neurons that are retrogradely labeled from deposits centered in layers 4 and 3b are located in the ventral division. The three cochleotopic cortical fields receive 70–75% of their input from the ventral division and 20–30% from the dorsal division. Less than 2% of neurons are double labeled following dual retrograde tracer deposits into layers 4 and 3b of frequency-matched domains of different cochleotopic cortical fields. [**A** data are combined from Storace et al. (2011, 2012); **B** is adapted from Storace et al. (2011, 2012)]

A tacit, but often unstated, assumption of prior anatomical models is that a common population of ventral division neurons projects to each of the cortical fields (Fig. 7.6A). However, there is mounting evidence to the contrary—there are multiple cochleotopic subdivisions within the ventral division (Fig. 7.6B). Each thalamic subdivision differs in its caudal–rostral position within the thalamus (Fig. 7.8A) and projects to a different cochleotopic cortical field (Figs. 7.6A, 7.7, 7.8). Moreover, as elaborated in Sect. 7.6, each ventral division pathway has a distinct cochleotopic and thalamocortical architecture or spatial resolution for connections with the cortex. This supports a new conceptual framework in which the thalamus sets the foundation for the diversity of sound processing capabilities and dispels the notion of a monolithic ventral division thalamic pathway to multiple cochleotopic cortices.

Segregation of ventral division neurons along the caudal-to-rostral anatomical axis forms independent or parallel pathways to different cochleotopic cortices. In the rat, the caudal, middle, and rostral sections of the thalamic ventral division project to cSRAF, VAF and A1, respectively (e.g., Figs. 7.6B and 7.8A). Likewise, mouse ventral division neurons segregate along the caudal-to-rostral axis based on their target cortical fields in the following order: A2, A1, AAF, DM/DP, and IAF (Horie et al. 2013; Takemoto et al. 2014; Tsukano et al. 2015). A similar caudal-to-rostral organization has been demonstrated for ventral division connections to cochleotopic fields and subfields of the cat (Lee et al. 2004; Read et al. 2011). This parallel organization of the ventral division thalamocortical pathway likely begins in the ascending pathways from inferior colliculus to thalamus (Cant and Benson 2007; Cant and Oliver, Chap. 2). Whatever its origins, the caudal-to-rostral ventral division thalamocortical pathway organization represents a significant departure from prior conceptualizations and supports the idea that functional diversity of cochleotopic cortical fields may stem initially from a diverse and parallel architecture in the ascending pathways.

Less than 2% of individual neurons in the ventral division in a given segment of thalamus project to more than one cochleotopic cortical field. Anterograde tracing studies indicate that neurons within, or axons passing through, a single rostral–caudal section of ventral division thalamus project primarily to layers 4 and 3b of target cortex (Smith et al. 2012) and can project to two different cortical fields or subfields (Romanski and LeDoux 1993). Thalamocortical connection architecture was further delineated with retrograde label studies that showed deposits restricted to layers 4 and 3b (Storace et al. 2010, 2012). However, the latter studies used retrograde tracers that are not transported by fibers of passage, and they found that single ventral division neurons rarely project to two different cochleotopic auditory fields even when the BF is the same. This lack of double-labeled neurons exists even though 5–10% of single-labeled neurons projecting to different cochleotopic fields can be found within the same 50 μm coronal thalamic section. Indeed, in all studies injecting two retrograde tracers into frequency-matched domains of two different cochleotopic cortical fields or subfields, less than 2% of neurons in the ventral division are double labeled, as demonstrated in cat (Lee et al. 2004; Read et al. 2011), rat (Storace et al. 2010, 2012), and mouse (Takemoto et al. 2014). Together these studies indicate that unique and anatomically segregated populations of neurons in the ventral division of the thalamus are the sources of ascending input to each of the cochleotopically distinct cortical fields (summary diagram in Fig. 7.6B).

7.6 Thalamocortical Pathway Architectures for Different Spectral Resolutions

Combined anatomical and physiological mapping studies reveal distinct cochleotopic organization of thalamic neurons that project to auditory cortical fields with low versus high spectral resolution (Fig. 7.9). In rat VAF, when different retrograde tracers are injected into two BF contours (e.g., 4.5 and 30 kHz) that are separated by 2.5 octaves or 450 μm, two clusters of retrogradely labeled ventral division neurons are labeled (Fig. 7.9B) that have centers separated by 400 μm (Storace et al. 2011). The two clusters of VAF-projecting thalamic cells that project to low and high BF contours are separated by a large gap of unlabeled neurons (Fig. 7.9B). Labeling with nuclear stain confirms the presence of neurons in the gap (Fig. 7.9A), and additional studies indicate that these neurons project to intervening BF contours in VAF. This anatomical architecture indicates that the cochleotopic organization of ventral division input to VAF is of high spatial resolution.

In contrast, dual retrograde tracer deposits into BF contours separated by 2.5 octaves and 450 μm in A1 reveal two overlapping clusters of ventral division neurons with a "salt-and-pepper" organization at the border (Fig. 7.9E). Such A1-projecting labeled clusters have centers that are separated by 200 μm, which is half the distance of comparable VAF-projecting clusters (Storace et al. 2011). Such results indicate a coarse cochleotopy for A1-projecting neurons that cluster close together with no intervening unlabeled neurons. That is, there is a 2-fold higher cochleotopic spatial resolution for thalamocortical input to VAF as compared with A1, and this corresponds to a 2-fold increase in the number of neurons per octave of sound in VAF. As elaborated in Sect. 7.8, simulations indicate that a higher spatial resolution for thalamic projections to cortex can result in a higher spectral resolution of sound by the target cortex. An analogous organization is evident in the cat: thalamic pathways where neurons with low versus high cochleotopic spatial resolution project to dorsal versus central subfields of A1, which in turn resolve sound with low versus high spectral resolution, respectively (Read et al. 2011). Together these findings support a general framework in which sound frequency resolution in each auditory cortical field is determined by the architecture of the thalamocortical pathway.

Though single-axon anterograde studies have not been performed, retrograde tracer studies indicate that individual neurons in the ventral division of the thalamus have terminal fields that are spatially restricted to a circular area with a diameter of 300 μm or less. Less than 2% of thalamic neurons are double labeled when two retrograde tracer deposits are separated by 300 μm in cortex. This is strong evidence that ventral division thalamic neurons have terminal fields that span less than 300 μm in the cochleotopic dimension. The intervening gap of unlabeled VAF-projecting thalamic neurons in experiments using dual tracer injections in VAF (Fig. 7.9B) is not observed following analogous injections in A1 (Fig. 7.9E). Those results suggest that individual terminal fields of VAF-projecting thalamic neurons spread out over an area as much as 2-fold smaller than those of A1.

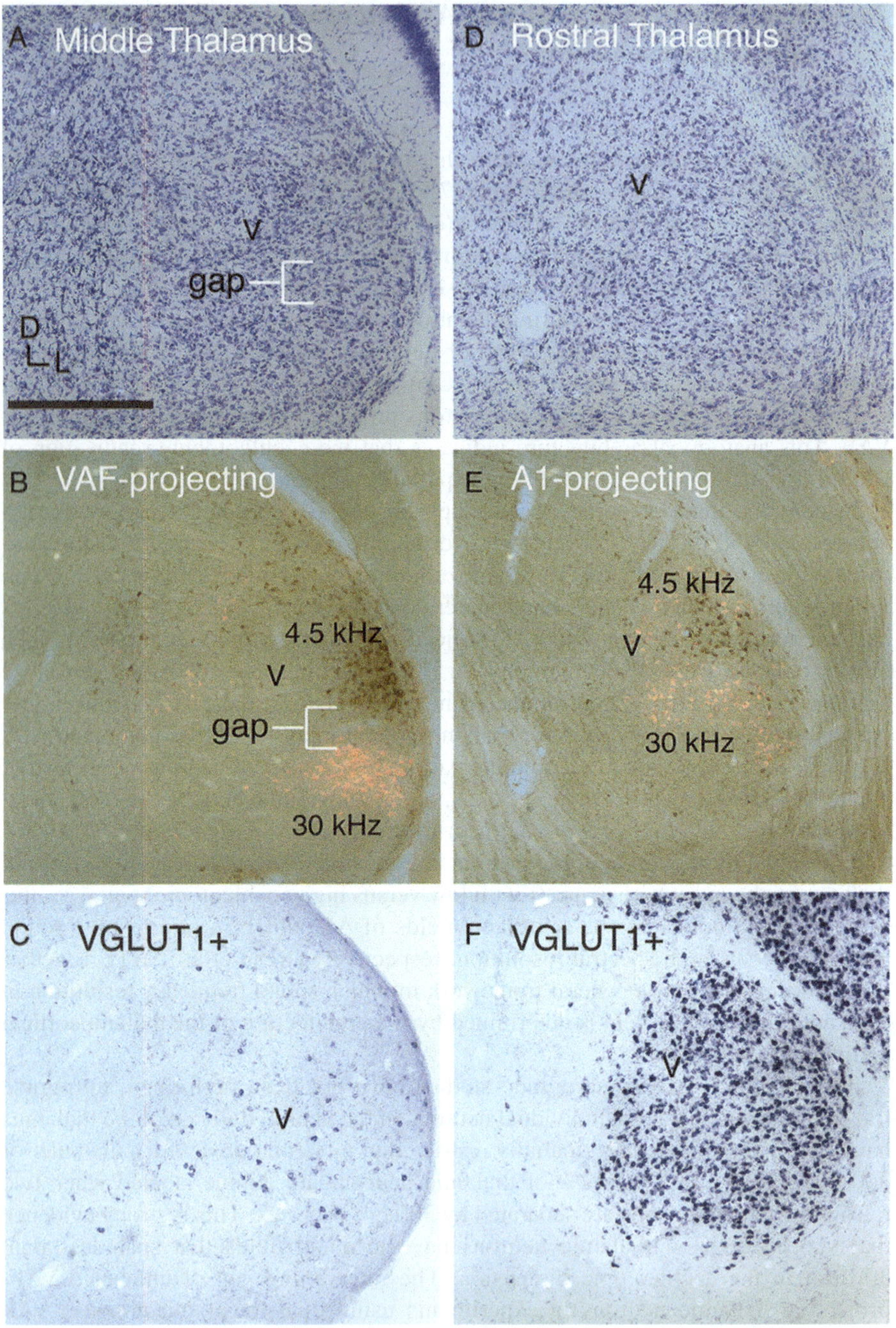

Fig. 7.9 Differences in organization of frequency lamina and gene expression patterns of thalamocortical projection neurons located in caudal versus rostral thalamus. Coronal sections (50 μm) of rat brain illustrate typical label patterns. **(A–C)** Sections taken from the midcaudal region of thalamus.

7.7 Gene Expression Differences Across Thalamocortical Pathways

The segregation of ascending ventral division thalamic pathways is underscored by a robust difference in gene expression patterns in rodents. In all mammals examined thus far, thalamic neurons express type 2 vesicular glutamate transporter (VGLUT2), indicating that glutamate is packaged and is the neurotransmitter released at their synapses with cortical neurons. In the rat, the majority (87%) of rostral ventral division neurons express genes encoding for type 1 transporter (VGLUT1) in addition to VGLUT2 (Ito et al. 2011). VGLUT1 gene expression decreases in the rostral-to-caudal direction of the ventral division in the thalamus (Storace et al. 2012).

Combined physiology, retrograde label, and in situ hybridization studies have found that the majority of ventral division neurons located in the rostral part of the thalamus project to A1 (Fig. 7.9E) and express the gene for the VGLUT1 transporter (Fig. 7.9F). In contrast, virtually none of the neurons from caudal thalamus that project to cSRAF express the VGLUT1 gene in the rat (Fig. 7.9C) as confirmed with combined VGLUT1 gene expression and retrograde label from cSRAF (Storace et al. 2012). A similar rostral-to-caudal gradient for gene expression of the VGLUT1 transporter is evident in the mouse auditory thalamus (Ito et al. 2011). Since two glutamate transporters are needed to accommodate high spike rates in vitro (Wojcik et al. 2004; Siksou et al. 2013), expression of VGLUT1 in A1-projecting neurons may be necessary for neurons to spike reliably to the broad range of sound features. Conversely, two transporters may not be necessary for thalamic pathways to VAF and cSRAF where neurons are sensitive to a narrow range of spectral and spatial sound features. This dual glutamate transporter gene expression pattern does not appear to exist for ventral division thalamic neurons in primates (Ito et al. 2015). It is conceivable that this is due to the characteristically higher spectral resolution of neurons in primate A1 (Cheung et al. 2001). Though the underlying drivers remain elusive, VGLUT1 gene expression is a powerful tool for marking functionally distinct pathways in rodent (Hackett et al. 2016).

Fig. 7.9 (continued) **(D–F)** Sections taken from the rostral region of thalamus. **(A, D)** Nissl nuclear staining reveals density and distribution of neurons in this part of thalamus. **(B, E)** Retrogradely labeled neurons projecting to *VAF* and *A1*, respectively. Neurons that appear brown and white project to 4.5 kHz and 30 kHz domains, respectively. VAF-projecting neurons are located in midcaudal thalamus as shown in **B**. Note that labeled neurons in **B** are densely distributed into dorsal and ventral clusters that are separated by a *gap* that lacks label. Nissl stains of adjacent 50 μm sections in **A** indicate that there are many neurons within the gap that were not labeled, presumably projecting to intervening BF domains in VAF. This indicates a fine spatial resolution for VAF-projecting neurons. **(E)** Retrogradely labeled A1-projecting neurons are located in rostral thalamus. **(C, F)** Typical gene expression patterns for vesicular glutamate transporter type 1 (*VGLUT1+*) in midcaudal versus rostral sections of thalamus from a second animal. (*Scale bar* in **A** is 500 μm) (*D*, dorsal; *L*, lateral; *V*, ventral division of medical geniculate body) [Figure adapted from Storace et al. 2011, 2012]

7.8 Coupling Thalamocortical Afferent Architecture and Cortical Microcircuits

Though the anatomy of the thalamocortical pathways connecting with auditory cortex is considered essentially fixed in adult animals, cochleotopy and sound processing in general can vary under different experimental and behavioral states. For example, the frequency selectivity of neurons and the steepness of the BF gradient vary with sound intensity (Phillips et al. 1994; Tan et al. 2007), the level of anesthesia or states of wakefulness (Guo et al. 2012; Pachitariu et al. 2015), and with behavioral state (Schneider et al. 2014; Zhou et al. 2014). Such flexible sound processing could emerge from a dynamic coupling between thalamic pathway architecture and local cortical circuitry that has been observed physiologically (de la Rocha et al. 2008; Levy and Reyes 2011) and could potentially account for differences in BF gradients and frequency tuning across cochleotopic fields.

The temporal dynamics of synaptic events in cortex is determined by a combination of the thalamic inputs and the recurrent excitatory and inhibitory synaptic inputs from local cortical neurons. Consider the spatial organization of two-dimensional networks of sound-activated excitatory and inhibitory cells in cortex (Fig. 7.10). Excitatory and inhibitory cells make extensive connections with each other that are nonrandom (Oswald et al. 2009). Typically, the connection probability between layer 4 neurons follows a Gaussian profile such that adjacent neurons are more likely to be connected than distal neurons (Fig. 7.11). Consequently, a brief tone will activate thalamic inputs to a subset of cortical excitatory and inhibitory neurons (Fig. 7.10A). A short time later, recruited neurons will provide additional recurrent inhibitory and excitatory inputs to neighboring neurons (Fig. 7.10A). Hence, the balance between excitation and inhibition is determined by the combination of thalamocortical and recruited corticocortical inputs, and the balance changes dynamically following sound activation.

Because of the topography of thalamic inputs and nonrandom local circuitry, the size of the cortical area activated during a tone becomes an important determinant of neuronal firing. The excitation-to-inhibition balance depends on the terminal field area or spatial footprint of the thalamic input in the following way. Thalamic inputs do not innervate cortex uniformly but instead provide excitatory synaptic current (g_{syn}) inputs to excitatory and inhibitory neurons within a relatively confined

---

Fig. 7.10 (continued) In the lateral inhibitory network (***B1***), the excitatory input (*green curve*) covers a narrow cortical distance that is smaller than the distance covered by inhibitory input (*red curve*). In the co-tuned network (***B2***), the excitatory thalamic input covers a large cortical distance (*green curve*) that is the same as the distance for inhibitory input (*red curve*). **(C1, 2)** Rate level functions emerging from simulations of lateral inhibitory configuration (***C1***) and co-tuned configuration (***C2***). **(D1, 2)** Frequency response areas for simulated lateral inhibitory (***D1***) and co-tuned (***D2***) networks. *Color* represents spike rate frequency (code on right). **(E1, 2)** Underlying synaptic currents reveal prominent sideband inhibition for lateral inhibitory (***E1***) but not for co-tuned (***E2***) network configurations. *Color* represents synaptic current (code on right). [See de la Rocha et al. (2008) and Levy and Reyes (2011) for a different rendition of these results and further details]

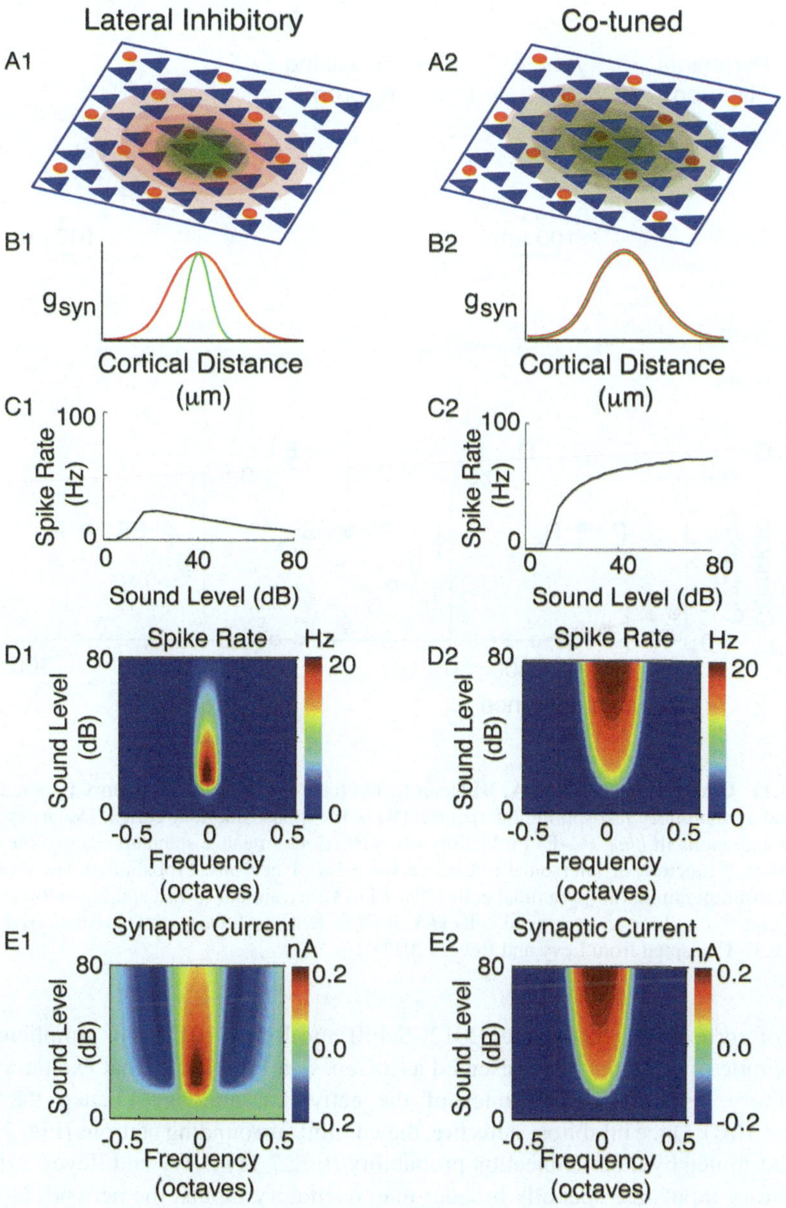

Fig. 7.10 Simulating excitatory-inhibitory configurations for sound activation of the cortex. **(A1, 2)** Schematics of two-dimensional networks consisting of excitatory (*blue triangles*) and inhibitory (*red circles*) cortical neurons for the lateral inhibitory (*A1*) and co-tuned (*A2*) configurations. During a pure tone stimulus, excitatory and inhibitory neurons within a confined area of cortex receive excitatory input from the thalamus (*green area*). When the inhibitory cells become active, they in turn send inhibitory inputs to the surrounding area (*red*). **(B1, 2)** Profiles of the excitatory synaptic conductance (g_{syn}) from thalamus (*green*) and from cortical inhibitory interneurons (*red*).

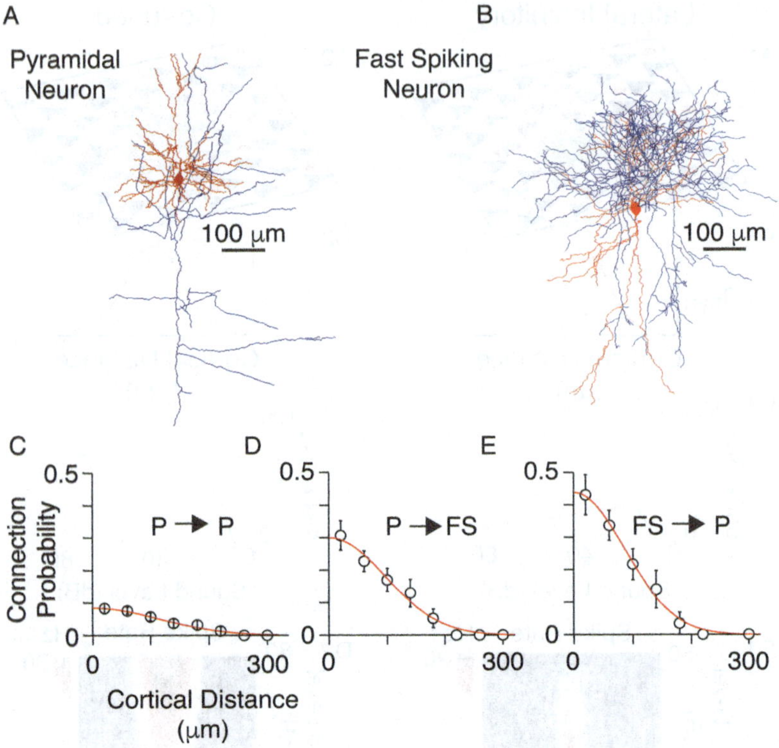

Fig. 7.11 Cortical architecture. **(A, B)** Neurolucida traces of a layer 4 excitatory pyramidal cell (**A**) and an inhibitory fast spiking interneuron (**B**) in the mouse auditory cortex. Dendrites shown in *red* and axons in *blue*. **(C–E)** Probability of connections (mean ± standard error of the mean) plotted as a function of intersomatic distance for pairs of neurons. Probabilities are shown for inputs from: pyramidal-to-pyramidal cells (*P to P*) in **C**; pyramidal-to-fast spiking cells (*P to FS*) in **D**; and fast spiking-to-pyramidal cells (*FS to P*) in **E**. Data fitted with Gaussians (*red lines*). [Panels **B–E** adapted from Levy and Reyes (2012)]

area of cortex (Levy and Reyes 2012; Schiff and Reyes 2012). For simplicity, the magnitude of the input is represented as a Gaussian function so that excitatory and inhibitory neurons in the center of the activation area are excited the most (Fig. 7.10B). Once inhibitory cells fire, they inhibit surrounding neurons (Fig. 7.10B) as determined by their connection probability (Fig. 7.11) (Levy and Reyes 2011). If inhibitory inputs are spatially broader than excitatory inputs, the network is in the lateral inhibitory configuration (Fig. 7.10B1). If both excitatory and inhibitory inputs have equal spatial widths, then the network is in the co-tuned configuration (Fig. 7.10B2). Electrophysiological recordings suggest the presence of both in rodent auditory cortices in presumed layer 4 (Wehr and Zador 2003; Tan and Wehr 2009) and layer 3 neurons (Li et al. 2014). Theoretical analyses and computer simulations suggest that the distribution of thalamic inputs to cortex bias the network toward either the lateral inhibitory or co-tuned configuration (Levy and Reyes 2011).

Accordingly, thalamic inputs with a spatially restricted or narrow terminal field will tend to produce the lateral inhibitory configuration (Fig. 7.10B1), while thalamic inputs with a broad terminal field will tend to produce the responses with a cotuned configuration (Fig. 7.10B2).

Simulations indicate that the coupling of thalamic pathway architecture and local cortical circuitry can determine how a given cortical field resolves sound frequency and a host of other sound properties. Networks that simulate the lateral inhibitory and co-tuned configurations produce unique sets of tonal sound responses commonly found in vivo. For example, the lateral inhibitory configuration can result in sustained firing, nonmonotonic rate level functions (Fig. 7.10C1), narrow frequency response areas (Fig. 7.10D1), and sideband inhibition (Fig. 7.10E1) that leads to lateral frequency response suppression (not shown) (de la Rocha et al. 2008). In contrast, the co-tuned configuration can result in transient firing, monotonic rate level functions (Fig. 7.10C2), V-shaped frequency response areas (Fig. 7.10D1), and minimal sideband inhibition (Fig. 7.10E2). Thus, the relative balance of excitation and inhibition accounts for several sound response properties that are known to covary in cortical neurons and across cortical fields.

The dependence of sound response properties on the effective spatial distribution of thalamocortical inputs has important implications for sound processing. Recall that different populations of neurons in the ventral division of the thalamus project via independent pathways to regions of VAF and A1 that respond to the same cochlear position and sound frequencies (Fig. 7.6B). Based on retrograde tracings, the projection field of ventral division thalamic input to VAF is spatially more restricted than that to A1 (Fig. 7.9B versus 7.9E). Accordingly, the aforementioned computer simulations would predict that VAF would tend to be in the lateral inhibitory configuration and A1 in the co-tuned configuration. The physiologically observed response properties are consistent with these and other model predictions. For example, VAF neurons receive high spatial resolution thalamic input and respond to a narrow range of sound frequencies consistent with lateral inhibition (Fig. 7.4C) as predicted by simulations (Fig. 7.10D1). In contrast, A1 neurons receive low spatial resolution thalamic input and respond to a broad range of sound frequencies consistent with co-tuning (Fig. 7.4A) as found in simulations (Fig. 7.10D2). Physiologically, A1 frequency bandwidth increases with intensity and response areas tend to be V-shaped (Fig. 7.4A); whereas VAF frequency bandwidth tends to be narrow across most sound intensities with a nonmonotonic sound-level dependence (Fig. 7.4C). These intensity dependencies are predicted by corresponding rate versus sound level functions in computer simulations (Fig. 7.10C1, C2). Recent studies find more sustained single unit responses to sound in VAF and cSRAF than in A1 (Lee et al. 2016). The latter results are consistent with simulations that find more sustained temporal responses when the spatial footprint of thalamocortal input is smaller (de la Rocha et al. 2008). These results indicate that cortical field differences in spatial, spectral, and temporal sound processing all can be predicted by the spatial resolution of thalamocortical connections and known local cortical circuitries.

An important implication of these simulations is that the spatial footprint for activation could vary not only with input pathway but also with the acoustic or behavioral conditions. Cortical neurons in layers 4 and 3b receive input from thalamocortical as well as corticocortical pathways that presumably have different terminal field and cortical activation areas. Moreover, the relative contribution of a given sensory pathway is known to vary with sensory engagement (von Trapp et al. 2016), stimulus context, anesthesia, and cortical feedback (Lamme et al. 1998; Haider et al. 2010). This diversity and the flexibility of converging sensory inputs in combination with local cortical circuits provide a mechanism for highly flexible sound processing. Accordingly, the spatial footprints for activating cortex could change dynamically with ongoing changes in sound and sound location as well as the attention state of the listener.

7.9 Summary and Future Directions

Mammals have multiple cochleotopic auditory cortical fields and each has a unique ability to process and sense sound. Segregated populations of ventral division auditory thalamic neurons have different spatial distributions and project to different cortical fields. The spatial distribution of thalamocortical connections combined with local cortical circuitry determine the balance of synaptic excitation and inhibition and could account for sound sensitivity differences observed across cortical fields. The parallel and spatially distinct thalamocortical input to different cortical fields provides a mechanism by which spectral, temporal, and spatial cues could be sensed and perceived simultaneously on multiple spectral, temporal, and spatial scales. Such organization could allow for pathway-specific processing of distinct sound features while reducing correlated synaptic noise across pathways that later converge. Moreover, this organization predicts that populations of neurons activated within and across cortical fields will change with variations in effective thalamic input, sensory context, and behavioral context.

Long-term goals of this line of research include determining underlying mechanisms and health intervention strategies for amelioration of sensory deficits associated with thalamocortical or cortical circuit dysfunction. For example, cortical field-specific deficits emerge after developmental alteration of thalamocortical input to auditory cortex (Escabi et al. 2007). A better understanding of structure-function relationships could promote development of treatment strategies. These goals require a combination of new approaches.

Though wide-field intrinsic and calcium-imaging techniques are refined enough to determine response difference and boundaries between cortical field areas, the spatial (100 μm) and temporal (>100 ms) scales are not refined enough to determine fine scale differences in a spatial footprint for cortical activation. Ultimately, wide-field imaging approaches need to be combined with approaches that allow for a finer spatial and temporal resolution of neuronal activity. Two-photon calcium-imaging techniques provide fine, single neuron spatial scales (<50 mm) and have been used to successfully demonstrate a coarse topographic organization of single neuron responses to simple pure-tone sounds in A1 (Kanold et al. 2014; Budinger and Kanold, Chap. 8). Such coarse organization could emerge from the coarse spatial organization of thalamocortical input to A1 as described in Sect. 7.6. Future studies are needed to determine if cortical fields, such as VAF, have smaller spatial footprints for activation than A1, as predicted. Combined wide-field and fine-scale calcium imaging also might be useful for determining whether sensory feature selectivity (e.g., BF, ILD) is more variable between neighboring neurons in A1 versus VAF (e.g., VAF and cSRAF). For example, higher BF variance across neighboring neurons would be expected for A1 as compared to VAF based on thalamocortical projections.

Another future direction is to develop approaches to determine how the combined parallel and multiscaled organization of auditory cortices impacts sound perception and discrimination abilities. The perceptual advantages for multiple parallel cortical pathways in mammals remain unknown. Optogenetic approaches could be adapted to alter select pathways (Aizenberg et al. 2015; Seybold et al. 2015) and to examine the effects of a modified spatial footprint for cortical activation. Since two-photon calcium-imaging signals are slow, new approaches may be needed to unravel the more dynamic properties of sound processing simultaneously across many neurons. Finally, these approaches would need to be carried out in actively listening animals to determine how the effective spatial footprint of cortical activation may change with behavior and sensory context.

Acknowledgments We acknowledge funding support from the National Science Foundation (NSF) (Award 1355065 to H. L. Read and M. A. Escabí), the National Institutes of Health (NIH) (1R01DC015138-01 to M. A. Escabí, H. L. Read, and I. Stevenson; DC005787-06 to A. D. Reyes), and the University of Connecticut research foundation. This work was supported in part by the Kavli Institute for Theoretical Physics (KITP) at Santa Barbara and the National Science Foundation under Grant No. NSF PHY-1125915.

Compliance with Ethics Requirements Heather Read declares she has no conflict of interest.

Alex Reyes declares that he has no conflict of interest.

References

Aizenberg, M., Mwilambwe-Tshilobo, L., Briguglio, J. J., Natan, R. G., & Geffen, M. N. (2015). Bidirectional regulation of innate and learned behaviors that rely on frequency discrimination by cortical inhibitory neurons. *PLoS Biology, 13*(12), e1002308.

Anderson, L. A., Wallace, M. N., & Palmer, A. R. (2007). Identification of subdivisions in the medial geniculate body of the guinea pig. *Hearing Research, 228*(1-2), 156–167.

Atencio, C. A., & Schreiner, C. E. (2012). Spectrotemporal processing in spectral tuning modules of cat primary auditory cortex. *PLoS One, 7*(2), e31537.

Baumann, S., Joly, O., Rees, A., Petkov, C. I., Sun, L., Thiele, A., et al. (2015). The topography of frequency and time representation in primate auditory cortices. *eLife, 4*. https://doi.org/10.7554/eLife.03256.001.

Bizley, J. K., & King, A. J. (2009). Visual influences on ferret auditory cortex. *Hearing Research, 258*(1-2), 55–63.

Brett-Green, B., Fifkova, E., Larue, D. T., Winer, J. A., & Barth, D. S. (2003). A multisensory zone in rat parietotemporal cortex: Intra- and extra-cellular physiology and thalamocortical connections. *Journal of Comparative Neurology, 460*(2), 223–237.

Butler, B. E., Chabot, N., & Lomber, S. G. (2016). Quantifying and comparing the pattern of thalamic and cortical projections to the posterior auditory field in hearing and deaf cats. *The Journal of Comparative Neurology, 524*(15), 3042–3063.

Cant, N. B., & Benson, C. G. (2007). Multiple topographically organized projections connect the central nucleus of the inferior colliculus to the ventral division of the medial geniculate nucleus in the gerbil, *Meriones unguiculatus. The Journal of Comparative Neurology, 503*(3), 432–453.

Carlyon, R. P., Macherey, O., Frijns, J. H., Axon, P. R., Kalkman, R. K., Boyle, P., et al. (2010). Pitch comparisons between electrical stimulation of a cochlear implant and acoustic stimuli presented to a normal-hearing contralateral ear. *Journal of the Association for Research in Otolaryngology, 11*(4), 625–640.

Cheung, S. W., Bedenbaugh, P. H., Nagarajan, S. S., & Schreiner, C. E. (2001). Functional organization of squirrel monkey primary auditory cortex: Responses to pure tones. *Journal of Neurophysiology, 85*(4), 1732–1749.

Clerici, W. J., & Coleman, J. R. (1990). Anatomy of the rat medial geniculate body: I. Cytoarchitecture, myeloarchitecture, and neocortical connectivity. *The Journal of Comparative Neurology, 297*(1), 14–31.

de la Rocha, J., Marchetti, C., Schiff, M., & Reyes, A. D. (2008). Linking the response properties of cells in auditory cortex with network architecture: Cotuning versus lateral inhibition. *The Journal of Neuroscience, 28*(37), 9151–9163.

Doron, N. N., Ledoux, J. E., & Semple, M. N. (2002). Redefining the tonotopic core of rat auditory cortex: Physiological evidence for a posterior field. *The Journal of Comparative Neurology, 453*(4), 345–360.

Escabi, M. A., Higgins, N. C., Galaburda, A. M., Rosen, G. D., & Read, H. L. (2007). Early cortical damage in rat somatosensory cortex alters acoustic feature representation in primary auditory cortex. *Neuroscience, 150*(4), 970–983.

Escabi, M. A., Read, H. L., Viventi, J., Kim, D. H., Higgins, N. C., Storace, D. A., et al. (2014). A high-density, high-channel count, multiplexed muECoG array for auditory-cortex recordings. *Journal of Neurophysiology, 112*(6), 1566–1583.

Formisano, E., Kim, D. S., Di Salle, F., van de Moortele, P. F., Ugurbil, K., & Goebel, R. (2003). Mirror-symmetric tonotopic maps in human primary auditory cortex. *Neuron, 40*(4), 859–869.

Greenwood, D. D. (1990). A cochlear frequency-position function for several species—29 years later. *The Journal of the Acoustical Society of America, 87*(6), 2592–2605.

Guo, W., Chambers, A. R., Darrow, K. N., Hancock, K. E., Shinn-Cunningham, B. G., & Polley, D. B. (2012). Robustness of cortical topography across fields, laminae, anesthetic states, and neurophysiological signal types. *The Journal of Neuroscience, 32*(27), 9159–9172.

Hackett, T. A., Barkat, T. R., O'Brien, B. M., Hensch, T. K., & Polley, D. B. (2011). Linking topography to tonotopy in the mouse auditory thalamocortical circuit. *The Journal of Neuroscience, 31*(8), 2983–2995.

Hackett, T. A., Clause, A. R., Takahata, T., Hackett, N. J., & Polley, D. B. (2016). Differential maturation of vesicular glutamate and GABA transporter expression in the mouse auditory forebrain during the first weeks of hearing. *Brain Structure and Function, 221*(5), 2619–2673.

Haider, B., Krause, M. R., Duque, A., Yu, Y., Touryan, J., Mazer, J. A., et al. (2010). Synaptic and network mechanisms of sparse and reliable visual cortical activity during nonclassical receptive field stimulation. *Neuron, 65*(1), 107–121.

Hall, A. J., & Lomber, S. G. (2015). High-field fMRI reveals tonotopically organized core auditory cortex in the cat. *Hearing Research, 325*, 1–11.

Harrington, I. A., Stecker, G. C., Macpherson, E. A., & Middlebrooks, J. C. (2008). Spatial sensitivity of neurons in the anterior, posterior, and primary fields of cat auditory cortex. *Hearing Research, 240*(1-2), 22–41.

Heffner, R. S. (2004). Primate hearing from a mammalian perspective. *The Anatomical Record Part A: Discoveries in Molecular, Cellular, and Evolutionary Biology, 281*(1), 1111–1122.
Higgins, N. C., Storace, D. A., Escabi, M. A., & Read, H. L. (2010). Specialization of binaural responses in ventral auditory cortices. *The Journal of Neuroscience, 30*(43), 14522–14532.
Horie, M., Tsukano, H., Hishida, R., Takebayashi, H., & Shibuki, K. (2013). Dual compartments of the ventral division of the medial geniculate body projecting to the core region of the auditory cortex in C57BL/6 mice. *Neuroscience Research, 76*(4), 207–212.
Imaizumi, K., & Schreiner, C. E. (2007). Spatial interaction between spectral integration and frequency gradient in primary auditory cortex. *Journal of Neurophysiology, 98*(5), 2933–2942.
Imaizumi, K., Priebe, N. J., Sharpee, T. O., Cheung, S. W., & Schreiner, C. E. (2010). Encoding of temporal information by timing, rate, and place in cat auditory cortex. *PLoS One, 5*(7), e11531.
Imig, T. J., & Reale, R. A. (1980). Patterns of cortico-cortical connections related to tonotopic maps in cat auditory cortex. *The Journal of Comparative Neurology, 192*(2), 293–332.
Imig, T. J., & Morel, A. (1985). Tonotopic organization in ventral nucleus of medial geniculate body in the cat. *Journal of Neurophysiology, 53*(1), 309–340.
Issa, J. B., Haeffele, B. D., Agarwal, A., Bergles, D. E., Young, E. D., & Yue, D. T. (2014). Multiscale optical Ca^{2+} imaging of tonal organization in mouse auditory cortex. *Neuron, 83*(4), 944–959.
Ito, T., & Oliver, D. L. (2012). The basic circuit of the IC: Tectothalamic neurons with different patterns of synaptic organization send different messages to the thalamus. *Frontiers in Neural Circuits, 6*, 48. https://doi.org/10.3389/fncir.2012.00048.
Ito, T., Bishop, D. C., & Oliver, D. L. (2011). Expression of glutamate and inhibitory amino acid vesicular transporters in the rodent auditory brainstem. *The Journal of Comparative Neurology, 519*(2), 316–340.
Ito, T., Inoue, K., & Takada, M. (2015). Distribution of glutamatergic, GABAergic, and glycinergic neurons in the auditory pathways of macaque monkeys. *Neuroscience, 310*, 128–151.
Jenkins, W. M., & Merzenich, M. M. (1984). Role of cat primary auditory cortex for sound-localization behavior. *The Journal of Neuroscience, 52*(5), 819–847.
Kaas, J. H., & Hackett, T. A. (2000). Subdivisions of auditory cortex and processing streams in primates. *Proceedings of the National Academy of Sciences of the United States of America, 97*(22), 11793–11799.
Kalatsky, V. A., Polley, D. B., Merzenich, M. M., Schreiner, C. E., & Stryker, M. P. (2005). Fine functional organization of auditory cortex revealed by Fourier optical imaging. *Proceedings of the National Academy of Sciences of the United States of America, 102*(37), 13325–13330.
Kanold, P. O., Nelken, I., & Polley, D. B. (2014). Local versus global scales of organization in auditory cortex. *Trends in Neurosciences, 37*(9), 502–510.
Kim, H., & Bao, S. (2013). Experience-dependent overrepresentation of ultrasonic vocalization frequencies in the rat primary auditory cortex. *Journal of Neurophysiology, 110*(5), 1087–1096.
Koka, K., Read, H. L., & Tollin, D. J. (2008). The acoustical cues to sound location in the rat: Measurements of directional transfer functions. *The Journal of the Acoustical Society of America, 123*(6), 4297–4309.
Lamme, V. A., Zipser, K., & Spekreijse, H. (1998). Figure-ground activity in primary visual cortex is suppressed by anesthesia. *Proceedings of the National Academy of Sciences of the United States of America, 95*(6), 3263–3268.
Lee, C. C., Imaizumi, K., Schreiner, C. E., & Winer, J. A. (2004). Concurrent tonotopic processing streams in auditory cortex. *Cerebral Cortex, 14*(4), 441–451.
Lee, C. M., Osman, A. F., Volgushev, M., Escabi, M. A., & Read, H. L. (2016). Neural spike-timing patterns vary with sound shape and periodicity in three auditory cortical fields. *Journal of Neurophysiology, 115*(4), 1886–1904.
LePage, E. L. (2003). The mammalian cochlear map is optimally warped. *The Journal of the Acoustical Society of America, 114*(2), 896–906.
Levy, R. B., & Reyes, A. D. (2011). Coexistence of lateral and co-tuned inhibitory configurations in cortical networks. *PLoS Computational Biology, 7*(10), e1002161.

Levy, R. B., & Reyes, A. D. (2012). Spatial profile of excitatory and inhibitory synaptic connectivity in mouse primary auditory cortex. *The Journal of Neuroscience, 32*(16), 5609–5619.
Li, L. Y., Ji, X. Y., Liang, F., Li, Y. T., Xiao, Z., Tao, H. W., et al. (2014). A feedforward inhibitory circuit mediates lateral refinement of sensory representation in upper layer 2/3 of mouse primary auditory cortex. *The Journal of Neuroscience, 34*(41), 13670–13683.
Lomber, S. G., & Malhotra, S. (2008). Double dissociation of 'what' and 'where' processing in auditory cortex. *Nature Neuroscience, 11*(5), 609–616.
Merzenich, M. M., Knight, P. L., & Roth, G. L. (1973). Cochleotopic organization of primary auditory cortex in the cat. *Brain Research, 63*, 343–346.
Merzenich, M. M., Knight, P. L., & Roth, G. L. (1975). Representation of cochlea within primary auditory cortex in the cat. *Journal of Neurophysiology, 38*(2), 231–249.
Miller, L. M., & Recanzone, G. H. (2009). Populations of auditory cortical neurons can accurately encode acoustic space across stimulus intensity. *Proceedings of the National Academy of Sciences of the United States of America, 106*(14), 5931–5935.
Miller, L. M., Escabi, M. A., Read, H. L., & Schreiner, C. E. (2002). Spectrotemporal receptive fields in the lemniscal auditory thalamus and cortex. *Journal of Neurophysiology, 87*(1), 516–527.
Moerel, M., De Martino, F., & Formisano, E. (2014). An anatomical and functional topography of human auditory cortical areas. *Frontiers in Neuroscience, 8*, 225. https://doi.org/10.3389/fnins.2014.00225.
Moore, B. C. (2003). Coding of sounds in the auditory system and its relevance to signal processing and coding in cochlear implants. *Otology and Neurotology, 24*(2), 243–254.
Morest, D. K. (1964). The neuronal architecture of the medial geniculate body of the cat. *Journal of Anatomy, 98*, 611–630.
Mrsic-Flogel, T. D., Versnel, H., & King, A. J. (2006). Development of contralateral and ipsilateral frequency representations in ferret primary auditory cortex. *European Journal of Neuroscience, 23*(3), 780–792.
Nelken, I., Bizley, J. K., Nodal, F. R., Ahmed, B., King, A. J., & Schnupp, J. W. (2008). Responses of auditory cortex to complex stimuli: Functional organization revealed using intrinsic optical signals. *Journal of Neurophysiology, 99*(4), 1928–1941.
Nishimura, M., & Song, W. J. (2014). Greenwood frequency-position relationship in the primary auditory cortex in guinea pigs. *Neuroimage, 89*, 181–191.
Oliver, D. L., & Hall, W. C. (1975). Subdivisions of the medial geniculate body in the tree shrew (*Tupaia glis*). *Brain Research, 86*(2), 217–227.
Oswald, A. M., Doiron, B., Rinzel, J., & Reyes, A. D. (2009). Spatial profile and differential recruitment of $GABA_B$ modulate oscillatory activity in auditory cortex. *Journal of Neuroscience, 29*(33), 10321–10334.
Pachitariu, M., Lyamzin, D. R., Sahani, M., & Lesica, N. A. (2015). State-dependent population coding in primary auditory cortex. *The Journal of Neuroscience, 35*(5), 2058–2073.
Pandya, P. K., Rathbun, D. L., Moucha, R., Engineer, N. D., & Kilgard, M. P. (2008). Spectral and temporal processing in rat posterior auditory cortex. *Cerebral Cortex, 18*(2), 301–314.
Paxinos, G., Watson, C. R., & Emson, P. C. (1980). AChE-stained horizontal sections of the rat brain in stereotaxic coordinates. *Journal of Neuroscience Methods, 3*(2), 129–149.
Petkov, C. I., Kayser, C., Augath, M., & Logothetis, N. K. (2006). Functional imaging reveals numerous fields in the monkey auditory cortex. *PLoS Biology, 4*(7), e215.
Phillips, D. P., Semple, M. N., Calford, M. B., & Kitzes, L. M. (1994). Level-dependent representation of stimulus frequency in cat primary auditory cortex. *Experimental Brain Research, 102*(2), 210–226.
Polley, D. B., Steinberg, E. E., & Merzenich, M. M. (2006). Perceptual learning directs auditory cortical map reorganization through top-down influences. *The Journal of Neuroscience, 26*(18), 4970–4982.
Polley, D. B., Read, H. L., Storace, D. A., & Merzenich, M. M. (2007). Multiparametric auditory receptive field organization across five cortical fields in the albino rat. *Journal of Neurophysiology, 97*(5), 3621–3638.

Read, H. L., Winer, J. A., & Schreiner, C. E. (2001). Modular organization of intrinsic connections associated with spectral tuning in cat auditory cortex. *Proceedings of the National Academy of Sciences of the United States of America, 98*(14), 8042–8047.

Read, H. L., Nauen, D. W., Escabi, M. A., Miller, L. M., Schreiner, C. E., & Winer, J. A. (2011). Distinct core thalamocortical pathways to central and dorsal primary auditory cortex. *Hearing Research, 274*(1-2), 95–104.

Reale, R. A., & Imig, T. J. (1980). Tonotopic organization in auditory cortex of the cat. *The Journal of Comparative Neurology, 192*(2), 265–291.

Romanski, L. M., & LeDoux, J. E. (1993). Organization of rodent auditory cortex: Anterograde transport of PHA-L from MGv to temporal neocortex. *Cerebral Cortex, 3*(6), 499–514.

Sanes, D. H., & Rubel, E. W. (1988). The ontogeny of inhibition and excitation in the gerbil lateral superior olive. *The Journal of Neuroscience, 8*(2), 682–700.

Schiff, M. L., & Reyes, A. D. (2012). Characterization of thalamocortical responses of regular-spiking and fast-spiking neurons of the mouse auditory cortex in vitro and in silico. *Journal of Neurophysiology, 107*(5), 1476–1488.

Schneider, D. M., Nelson, A., & Mooney, R. (2014). A synaptic and circuit basis for corollary discharge in the auditory cortex. *Nature, 513*(7517), 189–194.

Semple, M. N., & Kitzes, L. M. (1993). Focal selectivity for binaural sound pressure level in cat primary auditory cortex: Two-way intensity network tuning. *Journal of Neurophysiology, 69*(2), 462–473.

Seybold, B. A., Phillips, E. A., Schreiner, C. E., & Hasenstaub, A. R. (2015). Inhibitory actions unified by network integration. *Neuron, 87*(6), 1181–1192.

Siksou, L., Silm, K., Biesemann, C., Nehring, R. B., Wojcik, S. M., Triller, A., et al. (2013). A role for vesicular glutamate transporter 1 in synaptic vesicle clustering and mobility. *European Journal of Neuroscience, 37*(10), 1631–1642.

Smith, P. H., Uhlrich, D. J., Manning, K. A., & Banks, M. I. (2012). Thalamocortical projections to rat auditory cortex from the ventral and dorsal divisions of the medial geniculate nucleus. *The Journal of Comparative Neurology, 520*(1), 34–51.

Storace, D. A., Higgins, N. C., & Read, H. L. (2010). Thalamic label patterns suggest primary and ventral auditory fields are distinct core regions. *The Journal of Comparative Neurology, 518*(10), 1630–1646.

Storace, D. A., Higgins, N. C., & Read, H. L. (2011). Thalamocortical pathway specialization for sound frequency resolution. *The Journal of Comparative Neurology, 519*(2), 177–193.

Storace, D. A., Higgins, N. C., Chikar, J. A., Oliver, D. L., & Read, H. L. (2012). Gene expression identifies distinct ascending glutamatergic pathways to frequency-organized auditory cortex in the rat brain. *Journal of Neuroscience, 32*(45), 15759–15768.

Takemoto, M., Hasegawa, K., Nishimura, M., & Song, W. J. (2014). The insular auditory field receives input from the lemniscal subdivision of the auditory thalamus in mice. *The Journal of Comparative Neurology, 522*(6), 1373–1389.

Tan, A. Y., & Wehr, M. (2009). Balanced tone-evoked synaptic excitation and inhibition in mouse auditory cortex. *Neuroscience, 163*(4), 1302–1315.

Tan, A. Y., Atencio, C. A., Polley, D. B., Merzenich, M. M., & Schreiner, C. E. (2007). Unbalanced synaptic inhibition can create intensity-tuned auditory cortex neurons. *Neuroscience, 146*(1), 449–462.

Tollin, D. J. (2003). The lateral superior olive: A functional role in sound source localization. *Neuroscientist, 9*(2), 127–143.

Tsukano, H., Horie, M., Bo, T., Uchimura, A., Hishida, R., Kudoh, M., et al. (2015). Delineation of a frequency-organized region isolated from the mouse primary auditory cortex. *Journal of Neurophysiology, 113*(7), 2900–2920.

Updyke, B. V. (1986). Retinotopic organization within the cat's posterior suprasylvian sulcus and gyrus. *Journal of Comparative Neurology, 246*(2), 265–280.

von Trapp, G., Buran, B. N., Sen, K., Semple, M. N., & Sanes, D. H. (2016). A decline in response variability improves neural signal detection during auditory task performance. *The Journal of Neuroscience, 36*(43), 11097–11106.

Wehr, M., & Zador, A. M. (2003). Balanced inhibition underlies tuning and sharpens spike timing in auditory cortex. *Nature, 426*(6965), 442–446.
Weinberger, N. M., & Diamond, D. M. (1987). Physiological plasticity in auditory cortex: Rapid induction by learning. *Progress in Neurobiology, 29*(1), 1–55.
Winer, J. A., Kelly, J. B., & Larue, D. T. (1999). Neural architecture of the rat medial geniculate body. *Hearing Research, 130*(1-2), 19–41.
Wojcik, S. M., Rhee, J. S., Herzog, E., Sigler, A., Jahn, R., Takamori, S., et al. (2004). An essential role for vesicular glutamate transporter 1 (VGLUT1) in postnatal development and control of quantal size. *Proceedings of the National Academy of Sciences of the United States of America, 101*(18), 7158–7163.
Wu, G. K., Li, P., Tao, H. W., & Zhang, L. I. (2006). Nonmonotonic synaptic excitation and imbalanced inhibition underlying cortical intensity tuning. *Neuron, 52*(4), 705–715.
Zhou, M., Liang, F., Xiong, X. R., Li, L., Li, H., Xiao, Z., et al. (2014). Scaling down of balanced excitation and inhibition by active behavioral states in auditory cortex. *Nature Neuroscience, 17*(6), 841–850.

Chapter 8
Auditory Cortex Circuits

Eike Budinger and Patrick O. Kanold

Abstract The auditory cortex (ACX) is the site of transformation from an acoustic analysis of the auditory scene to its perceptual representation. The circuits intrinsic to the ACX are crucial for creating auditory objects and the auditory processing related to higher cognitive information. Considerable progress has been made in the last decades to unravel the complexity of ACX. This chapter includes a description of the various cell types within the layers of ACX and a discussion of the anatomical and functional dissection of auditory cortical microcircuits in vitro and in vivo. Finally, those aspects of the cortex are considered with respect to the functional maps in ACX and auditory processing.

Keywords Connection probability · Cortical cell types · Cortical development · Feedback · Feedforward · Multisensory integration · Object formation · Subplate neurons · Synaptic integration · Synaptic strength · Tonotopic organization

8.1 Introduction

Located at the nexus between the ascending and descending auditory pathways and the higher order cortical system, the auditory cortex (ACX) represents the most central processing stage of auditory information along the auditory pathway. The anatomical position of the ACX and its multiple connections with other sensory and nonsensory brain structures reflect the key role of the ACX in bottom-up (ascending, feedforward, stimulus-driven) and top-down (descending, feedback,

E. Budinger
Department of Systems Physiology of Learning, Leibniz Institute for Neurobiology, Magdeburg, Germany
e-mail: budinger@lin-magdeburg.de

P. O. Kanold (✉)
Department of Biology, Institute for Systems Research, University of Maryland, College Park, MD, USA
e-mail: pkanold@umd.edu

© Springer International Publishing AG 2018

D. L. Oliver et al. (eds.), *The Mammalian Auditory Pathways*, Springer Handbook of Auditory Research 65,
https://doi.org/10.1007/978-3-319-71798-2_8

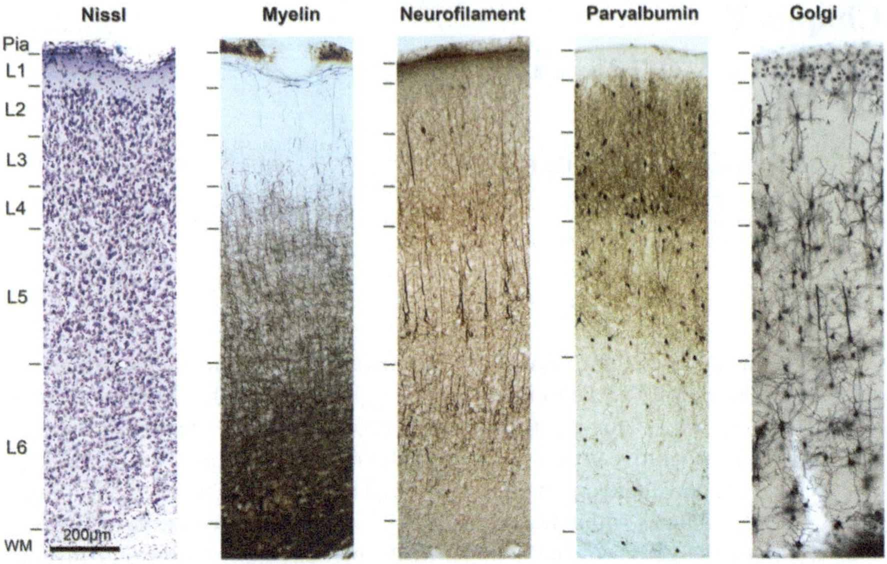

Fig. 8.1 Basic laminar organization of primary ACX: six layers (L1–6), bordered by the pia and white matter (*WM*), can be distinguished by their specific cyto-, fiber- and chemoarchitectonic features. Horizontal sections through rodent (Mongolian gerbil) primary ACX were stained for (from left to right): *Nissl* substance (labels cell bodies; note the cell-dense L4), *myelin*, myelinated fibers (Gallyas stain, note horizontal association fibers in L1), *neurofilament* protein (SMI-32 antibody, labels mainly principal cells), calcium-binding protein *parvalbumin* (often colocalized with GABA in interneurons), and *Golgi* (visualizes cell morphology, see Fig. 8.2)

task-dependent) processing of relevant auditory and nonauditory information (Fritz et al. 2007; Scheich et al. 2007). Accordingly, the ACX most likely represents the site of transformation from an acoustic scene analysis to a perceptual representation of the acoustic environment. The circuits intrinsic to the ACX are expected to be involved in creating "auditory objects" in the brain (Griffiths and Warren 2004; Bizley and Cohen 2013).

The ACX is part of the temporal neocortex. Classically, six cortical layers are distinguished from layer 1 (L1), bordering the pia mater, to layer 6 (L6), bordering the white matter (WM) (Fig. 8.1). Each layer can be further subdivided into sublayers. Layers 1–3 (L1–3) are called supragranular, layer 4 (L4) is granular (in ACX, including sublayer 3b; Winer 2011a; see Sect. 8.2.4), and layers 5–6 (L5–6) are infragranular layers. A thin layer 7 (or 6b) of persisting subplate cells has been identified, but its prominence is both species and region dependent (Reep 2000; Kanold and Luhmann 2010).

Primary auditory cortex (field A1) displays a koniocortical architecture (Greek: *konios*, dust; Fig. 8.1), which means that it has a pronounced granular L4 (von Economo and Koskinas 1925) similar to primary somatosensory (S1) and visual (V1) sensory cortex. Primary (also called core) ACX is surrounded by higher-order auditory areas (belt, parabelt), which have a less well-developed L4 and different cyto-, fiber-, and chemoarchitectonic characteristics (Hackett et al. 2001; Budinger and Heil 2005). Another fundamental principle of neocortical architecture is its columnar arrangement perpendicular to the cortical layers (Mountcastle 1997; Rockland and Ichinohe 2004), which is also reflected in functional modules like ocular dominance columns (Hubel and Wiesel 1977) and columns of auditory tuning preferences (Linden and Schreiner 2003).

Neurons within and across these cortical columns are connected by a complex system of intrinsic (i.e., local and mesoscale) and extrinsic (i.e., long ranging and larger scale) (Nieuwenhuys 1994; Bannister 2005), excitatory and inhibitory (Douglas and Martin 2004; Thomson and Lamy 2007), feedforward and feedback (Felleman and Van Essen 1991; Rouiller et al. 1991) connections (also see Read and Reyes, Chap. 7). In order to understand the connectional networks of the ACX, an overview of the various cell types within the six layers is provided. Then, the anatomical and functional dissection of auditory cortical microcircuits is discussed. All abbreviations are defined in Table 8.1.

Table 8.1 Abbreviations

5-HT3a	Ionotropic serotonin receptor 3a
A1	Primary auditory cortex
ACX	Auditory cortex
AMPA	Alpha-amino-3-hydroxy-5-methyl-4-isoxazolepropionic acid
BF	Best frequency
CB	Calbindin
CBP	Calcium binding protein
CCK	Cholecystokinin
CR	Calretinin
GABA	Gamma-aminobutyric acid
GAD2	Glutamate decarboxylase 2
L1-6	Cortical layers 1 through 6
LSPS	Laser-scanning photostimulation
M2	Motor cortex
NMDA	N-methyl-D-aspartate
nNOS	Neuronal nitric oxide synthase
NPY	Neuropetide Y
PPD	Preprodynorphin
PV	Parvalbumin
S1	Primary somatosensory cortex
SST	Somatostatin
V1	Primary visual cortex
VIP	Vasointestinal peptide
WM	White matter

8.2 Laminar Distribution, Neurotransmitters, and Basic Physiological Properties of Auditory Cell Types

In the 1992 edition of this book, Winer distinguished 47 different neuronal cell types within the ACX based largely on his experimental results in cats (mainly A1) and by comparisons with earlier studies in other mammalian species (Fig. 8.2) (Winer 1992). This extraordinary, detailed morphological description cemented the foundation for the basic anatomy of the ACX. Winer's first morphological classification of cell types was close to the numbers obtained by recent big data analyses of the neocortex and its microcircuitry that identified 55 morphological (and 207 morpho-electrical) neuronal types in the juvenile rat somatosensory cortex (Markram et al. 2015). Because the amount of structural detail was overwhelming, the earlier classification was streamlined to approximately 16 different ACX cell types in 2011 (Winer 2011b). Unfortunately, mainly due to the unavailability of sufficient axon staining, Winer's classification did not really match the developing nomenclature of cortical neurons (DeFelipe et al. 2013; Markram et al. 2015); thus, it led to various difficulties in comparing cell types of A1 with those of other cortical areas and of species other than cats. This chapter attempts to synthesize these two classifications.

There are two major classes of neurons in the neocortex: principal cells and interneurons (DeFelipe et al. 2013). Principal cells have a long axon (projecting neurons), bear many spines on their dendrites, and are excitatory, usually using the neurotransmitter glutamate. The typical example is the pyramidal cell; however, the spiny stellate cell with a rather short axon is the principal neuron of the granular layers.

Interneurons have short axons (intrinsic neurons), usually a nonpyramidal soma, are aspiny or sparsely spinous, and use the neurotransmitter GABA (gamma-aminobutyric acid); thus, they are mainly inhibitory. There are many exceptions, such as spiny, long-range projecting, pyramidal-shaped, and/or glutamatergic interneurons; hence, a common nomenclature has been difficult to establish. Currently, interneurons are classified according to five features centered around the nature of the axon: (1) intra- or trans-laminar projections; (2) intra- or trans-columnar projections; (3) centered (overlapping) or displaced (nonoverlapping) axonal and dendritic branches; (4) ascending or descending axons, or both; and (5) other morphological features described as "arcade" (= "willow"), "basket", "Cajal-Retzius", "chandelier", "horsetail" (= "double bouquet"), "Martinotti", or "neurogliaform" (see illustrations in Fig. 8.2B) (DeFelipe et al. 2013). These cells are discussed in more detail in subsequent sections. Cells with an unusual shape and without oriented dendritic or axonal arbors are called "common type." Additional classifications of GABAergic interneurons (Table 8.2) have been established recently that are based on the expression of specific proteins (e.g., calcium-binding proteins, receptor proteins, other neurotransmitters) and on electrical properties (e.g., onset responses, firing patterns, accommodation, adaptation, threshold) (Markram et al. 2015; Tremblay et al. 2016).

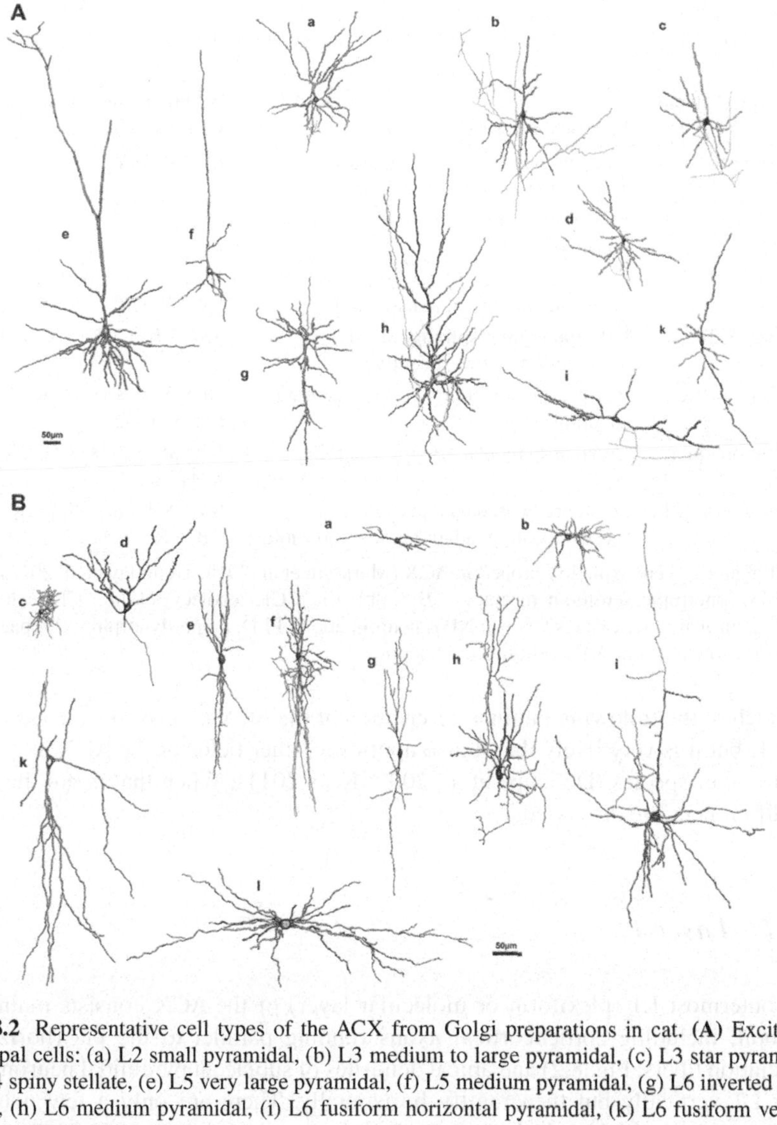

Fig. 8.2 Representative cell types of the ACX from Golgi preparations in cat. **(A)** Excitatory principal cells: (a) L2 small pyramidal, (b) L3 medium to large pyramidal, (c) L3 star pyramidal, (d) L4 spiny stellate, (e) L5 very large pyramidal, (f) L5 medium pyramidal, (g) L6 inverted pyramidal, (h) L6 medium pyramidal, (i) L6 fusiform horizontal pyramidal, (k) L6 fusiform vertical pyramidal cell. **(B)** Inhibitory interneurons: (a) L1 horizontal, (b) L1 small multipolar, (c) L2 neurogliaform, (d) L2 extraverted multipolar, (e) L3 bitufted, (f) L3 double-bouquet, (g) L4 bipolar, (h) L4 chandelier, (i) L4 large basket, (k) L6 Martinotti, (l) L6 bitufted horizontal cell. *Scale bars* are 50 μm for all drawings; there is no "j" for (A) or (B). [Reproduced with permission from Winer (1984a, 1984b, 1984c, 1985), Prieto et al. (1994b), Prieto and Winer (1999) and Winer and Prieto (2001)]

Not all of the same chemical and electrophysiological features were explicitly probed in the ACX but most likely also apply to it (Rockel et al. 1980; Rudy et al. 2011).

Table 8.2 Major GABAergic interneurons of the cortex: Electrical properties to sustained and step current injections, additional neurotransmitters, calcium-binding proteins (CBPs), and receptors

Cell type (Chapter Fig.)	Electrical properties	Additional neurotransmitters, CBPs, receptors[a]
Neurogliaform (8.2Bc)	Late and nonfast spiking, largely nonaccommodating, nonadapting	nNOS, NPY
Chandelier (8.2Bh)	Fast spiking, nonaccommodating, nonadapting	PV, CB
Basket (8.2Bi)	Fast spiking, sometimes stuttering, nonaccommodating, nonadapting	Many neuropeptides, CCK, PV, CB, CR
Bipolar (8.2Bg)	Late and regular spiking, mostly nonaccommodating, adapting	SST, VIP, CCK, CR, 5-HT3a
Bitufted (8.2Be)	Regular, burst spiking, accommodating, adapting	NPY, VIP, SST, CCK, CB, CR, 5-HT3a
Double-bouquet (8.2Bf)	Irregular to regular spiking, adapting	VIP, SST, CCK, CB, CR, 5-HT3a
Martinotti (8.2Bk)	Mostly regular, burst spiking, accommodating, adapting, low threshold	SST, NPY, nNOS, CCK, PPD, CB, CR

Not all features were explicitly probed in ACX (Markram et al. 2015; Tremblay et al. 2016)
[a]5-HT3a, ionotropic serotonin receptor; CB, calbindin; CCK, cholecystokinin; CR, calretinin; nNOS, neuronal nitric oxide synthase; NPY, neuropetide Y; PPD, preprodynorphin; PV, parvalbumin; SST, somatostatin; VIP, vasointestinal peptide

Much of the following laminar description of the ACX is derived from studies of cat A1, but it is very likely that it also applies to other fields of the ACX in general and to other species (DeFelipe et al. 2002; Kaas 2011). When that is not the case, the differences are discussed.

8.2.1 Layer 1

The outermost L1 (plexiform or molecular layer) of the ACX consists mainly of neuropil, including corticocortical axons running parallel to the pia (horizontal association fibers; Fig. 8.2) and apical dendrites of subjacent pyramidal neurons that enter L1 vertically but then ramify horizontally. There are only a few neurons, always nonpyramidal, with short axons (i.e., interneurons); more than 90% of them are inhibitory and GABAergic (Winer and Larue 1989). The density of GABA receptors and GABAergic axonal terminals in L1 is also the highest of all the cortical layers (Prieto et al. 1994a). Among the GABAergic interneurons is a conspicuous horizontal cell (Fig. 8.2Ba) whose dendrites and axon are restricted to L1 (i.e., intralaminar but transcolumnar) as well as diverse GABAergic multipolar cells (cat: Mitani et al. 1985; Prieto et al. 1994b). The dendrites and axons of the smaller multipolar cells are intralaminar (Fig. 8.2Bb). In contrast, the dendrites and axons of the larger cells can reach up to L3. In many neocortical areas, and thus probably also in ACX, the GABAergic interneurons of L1 form a gap-junction-coupled network (specifically between similar cell types) (Hestrin and Galarreta 2005) and provide inhibitory inputs to the apical pyramidal cell dendrites.

Cajal-Rezius cells might constitute the remaining 10% of excitatory neurons (Marin-Padilla 1984; Baloyannis et al. 1993), although their existence in the adult brain is controversial (Meyer et al. 1999; Kirischuk et al. 2014). Cajal-Rezius cells secrete the extracellular matrix glycoprotein *reelin* and are involved in laminar development. They also express the calcium-binding protein *parvalbumin* (PV; otherwise typical for GABAergic neurons) (Celio 1986; Markram et al. 2015). The dendrites of Cajal-Rezius cells are horizontally oriented, receiving inputs from various ascending fibers and from local interneurons. The axons of Cajal-Rezius cells are restricted to L1 and synapse on the apical dendrites of pyramidal neurons (Marin-Padilla 1984).

8.2.2 Layer 2

L2 (outer granular layer) harbors small and medium pyramidal neurons (mainly in its deeper half, sublayer 2b) as well as a wide range of nonpyramidal neurons. Pyramidal cells have many spines on all but the proximal parts of the apical and basal dendrites. Their axons branch out extensively locally (mainly L1–3) to serve intracolumnar connectivities and send out modest projections to adjacent auditory areas (Fig. 8.2Aa) (Mitani et al. 1985; Winer 1985). Pyramidal cells of L2 (and L3) are regular spiking or intrinsically bursting (Metherate and Aramakis 1999; Huggenberger et al. 2009). Their firing rate either remains constant (nonaccommodating; RS1 type) or continuously decreases (accommodating; RS2) to a sustained current injection (Agmon and Connors 1992). Likewise, with increasing current intensity, the duration of the spike/burst train either remains relatively constant (*nonadapting*) or decreases (*adapting*). So far, there is no information to support a correlation of the different physiological features and particular anatomical features of pyramidal cells in L2 (and L3).

8.2.2.1 Major Types of Interneurons

Approximately 24% of the neurons in L2 are GABAergic interneurons (Winer 2011b), although other neurotransmitters are also used (Table 8.2). The following interneurons usually occur in L2, but most of them also occur in other layers.

Neurogliaform Cells

Neurogliaform ("spiderweb") cells (Fig. 8.2Bc; Table 8.2) predominate in L2–4 of ACX, but they also appear in all other layers (L5–6) (Prieto et al. 1994b). They form largely overlapping and symmetrical dendritic and axonal fields. Their dendrites receive excitation from L2–5 afferents and local inhibition, and their axons inhibit nearby dendrites of other cells.

Chandelier Cells

Chandelier cells (Fig. 8.2Bh; Table 8.2) are rare but can be found in L2–6 (Woodruff et al. 2010; Inan and Anderson 2014); in the ACX, they are primarily located in L2–3 (Prieto et al. 1994b). Their multipolar to bipolar dendrites can extend throughout L1–4; their axons form candle-shaped synaptic plexuses with initial axonal segments of supragranular pyramidal neurons (axoaxonic contacts).

Basket Cells

Basket cells constitute about 50% of all cortical inhibitory interneurons and are common in L2–6. There are different types of basket cells, including large (Fig. 8.2Bi), small, and nest basket cells. Large and small basket cells have bipolar to multipolar, aspiny to sparsely spinous dendrites, and (sometimes "curvy") axons with long horizontal collaterals. Most likely a subtype of the small basket cell is the arcade (or "willow") cell (Peters and Saint Marie 1984; Wang et al. 2002). It has an axon that gives rise to arcades with predominantly vertical arbors and relatively long descending collaterals. Nest basket cells have an appearance like a bird's nest with a more local axonal cluster. All basket cell axons target somata and proximal dendrites of nearby (but also laterally distant) pyramidal cells and interneurons, thereby forming the typical perisomatic "baskets".

Bipolar and Bitufted Cells

Bipolar and bitufted (including double-bouquet) cells are translaminar but intracolumnar neurons with two main dendrites running in opposite directions with a relatively short trajectory (DeFelipe et al. 2013). Since those features relate only to dendritic (and not axonal) morphology, the classification of bipolar and bitufted cells is inconsistent throughout the literature except for the double-bouquet cells with their distinctive horsetail axon.

Bipolar cells (Fig. 8.2Bg, Table 8.2) occur in L2–6 and appear to be evenly distributed in the cortex, each cell having its own columnar domain that is 30 μm across on average (rat visual cortex: Morrison et al. 1984). The two principal dendrites extend from the opposite poles of the soma and branch in narrow vertical fields that span up to five layers. Bipolar cells receive thalamocortical inputs and inputs from basket cells. They terminate on both pyramidal cells and interneurons, thereby acting on them in an inhibitory or excitatory way. Comparative studies suggest that rodents have more bipolar cells throughout all layers (L2–6) and that bipolar cells are less common in primates, where they are mainly located in supragranular layers (DeFelipe et al. 2002).

Bitufted cells (Fig. 8.2Be, Table 8.2) can be distinguished from bipolar neurons by their already branched proximal dendritic portions. These neurons occur in L2–4, receive corticocortical and thalamocortical synapses, and send axons toward the dendrites of surrounding cells (even across columns) and the somata of infragranular interneurons.

Double-bouquet (horsetail) cells (Fig. 8.2Bf, Table 8.2) also have a bitufted dendritic morphology and occur in L2–5, preferentially in granular and supragranular layers. Their axons display a distinctive pattern of long, tightly interwoven vertical bundles that resemble a horsetail. They heavily contact the dendrites of surrounding pyramidal cells within a very narrow column. There are interspecies differences in the morphology of double-bouquet cells (Somogyi and Cowey 1984), and they seem to be most numerous in primates, rarer in carnivores, and probably absent in rodents, lagomorphs, and artiodactyls (DeFelipe 2002; Ballesteros-Yanez et al. 2005).

Of the above mentioned interneuron cell types, neurogliaform, bipolar, and bitufted cells were explicitly described in L2 of cat ACX (Winer 1985; Prieto et al. 1994b). In addition, the small, smooth multipolar cell may be analogous to the chandelier cell; the large, sparsely spinous multipolar cell may be analogous to the large basket cell; and one of the bitufted forms may be analogous to the double-bouquet cell (Winer 1985). The other multipolar (Winer 1985; Prieto et al. 1994b) and stellate (Mitani et al. 1985) types probably correspond to several types of basket cells.

Extraverted Multipolar Cells

A type of neuron apparently unique to L2 is the extraverted multipolar cell (Fig. 8.2Bd) (Winer 1985; Prieto et al. 1994b). Its dendrites ramify into L1 and form candelabra-like arbors (which should not be confused with the candelabra-like axonal arborization patterns of chandelier or arcade cells). Because this morphological cell type occurs more frequently in archicortical and allocortical areas than in neocortical areas and occurs more frequently in marsupials and insectivores than in carnivores and primates, it was hypothesized that the extraverted multipolar cell represents a developmental precursor of pyramidal cells (Sanides and Sanides 1972; Nieuwenhuys 1994). However, the extraverted multipolar cell described in the cat ACX is GABAergic (Prieto et al. 1994b). This argues against the developmental scenario because pyramidal cells are glutamatergic.

8.2.3 *Layer 3*

Within L3 (outer pyramidal layer), pyramidal neurons occur throughout the layer but are especially common in its deeper half (sublayer 3b). Generally, L3 pyramidal cells have more complex dendritic arbors than L2 pyramidal cells. The apical dendrites of small L3 pyramidal cells, which are more frequent in L3a, barely reach L2, while those of larger pyramidal cells (Fig. 8.2Ab), which are more frequent in L3b, extend up into L1 (Winer 1984a; Ojima et al. 1991). The larger pyramidal neurons provide and receive the majority of intrinsic, ipsilateral, and contralateral (interhemispheric, commissural) cortical and thalamic connections (Code and Winer 1985; Winguth and Winer 1986). In cortical L5, L3 pyramidal neurons primarily target the large (intrinsically bursting) pyramidal cells (Sect. 8.2.5.1) and only rarely the smaller (regular spiking) ones (Thomson and Bannister 1998). In cat ACX, there are also two types of star pyramidal cells (Fig. 8.2Ac) with more radiate dendrites: one with spiny and one with rather smooth dendrites (Winer 1984a). Together with the sparsely spinous stellate cells (Winer 1984b), they could form a group of neurons having more or less equivalent functions like the classical L4 spiny stellate cells of the visual cortex (Lund 1984) (see also Sect. 8.2.4).

GABAergic neurons of L3, which make up to approximately 24% of all neurons within this layer, include neurogliaform, bipolar, bitufted (Fig. 8.2Be), double bouquet (Fig. 8.2Bf; termed sparsely spinous, small multipolar cell by Prieto et al. 1994b), chandelier (small, smooth stellate cell by Winer 1984b), and basket cells (medium-sized, smooth stellate cell by Winer 1984b).

8.2.4 Layer 4

In contrast to the visual and somatosensory cortex, L4 (inner granular layer) of the ACX is almost devoid of pyramidal cells (Winer 1984c; Prieto et al. 1994b). It could be expected that the principal (excitatory) neurons are the densely packed spiny stellate cells (Lund 1984), since their concentration in L4 led to early descriptions of primary cortical areas as "granulous" or "koniocortical" (von Economo and Koskinas 1925). However, compared to the visual (da Costa and Martin 2011) and somatosensory cortex (Staiger et al. 2004), spiny stellate cells are relatively rare in the ACX of bats (Fitzpatrick and Henson 1994), rabbits (McMullen and Glaser 1982), cats (Winer 1984c; Mitani et al. 1985; one study found none: Smith and Populin 2001), and humans (Meyer et al. 1989). In these species, spiny stellate cells primarily occur in sublayers 4a and 3b (Fig. 8.2Ad). In cat ACX, spiny stellate cells, which have a small spherical dendritic field and an axon targeting L3–5, are driven by thalamic, corticocortical, and commissural stimulation; other physiological properties are unknown (Mitani et al. 1985). In visual and somatosensory cortex, spiny stellate cells are mostly regular spiking but also intrinsically bursting, and they accommodate to sustained current injections (Smith and Populin 2001; Andjelic et al. 2009).

A range of GABAergic cells, much like those in L3, is found in L4, including neurogliaform, bipolar (Fig. 8.2Bg), bitufted, double-bouquet, chandelier (Fig. 8.2Bh; small or medium tufted cell by Winer 1984c), and basket cells (Fig. 8.2Bi; e.g., large multipolar cell by Winer 2011b). These cells constitute about 26% of all neurons within L4.

Together, there is considerable evidence for species- and modality-specific differences in L4 organization. In addition to the near absence of pyramidal and spiny stellate cells in ACX, the weak sublamination (particularly compared to the visual cortex) hints at a reduced intralaminar circuitry. A related question is that of the putative target cells of the massive, mainly lemniscal, thalamic inputs to auditory L4 and L3b (Huang and Winer 2000). Certainly, the few principal cells are located among the targets, thereby serving thalamocortical feedforward excitation, but these inputs most likely also terminate on the GABAergic interneurons, providing strong feedforward inhibition (Verbny et al. 2006; Lee and Sherman 2008).

8.2.5 Layer 5

8.2.5.1 Pyramidal Cells

L5 (inner pyramidal or ganglionic layer) can be subdivided into a cell-sparse upper sublayer 5a and a cell-rich lower sublayer 5b. The somata of particularly large pyramidal cells (Fig. 8.2Ae), whose apical dendrites extend to L1 and form curved tufts parallel to the pia, are found close to the border of the two sublayers (Ojima et al. 1992; Winer and Prieto 2001). Other large pyramidal neurons are mainly located in sublayer 5a; their apical dendrites extend also to L1. The axons of the large and very large pyramidal cells project subcortically and also arborize locally in the infragranular layers. The small and medium pyramidal neurons are common throughout L5 (Fig. 8.2Af). Their apical dendrites barely reach L1; their axons project subcortically and branch extensively in the supragranular layers. Physiologically, the large pyramidal cells correspond to intrinsically bursting, mostly nonaccommodating, nonadapting cells as described in rat and mouse A1 (Hefti and Smith 2000; Sun et al. 2013). The smaller pyramidal cells are regular spiking and nonaccommodating but sometimes adapting (Hefti and Smith 2000). Compared to L2/3, there are relatively more intrinsically bursting pyramidal cells in L5 (and L6), and regular-spiking pyramidal cells have lower firing thresholds (Atzori et al. 2004; Huggenberger et al. 2009). Star pyramidal cells have a more radiate dendritic domain and many local axonal branches (Winer and Prieto 2001). They also are either intrinsically bursting or regular spiking (rat S1: Cowan and Stricker 2004; Staiger et al. 2004). Generally, the local axonal branches of L5 pyramidal cells are simpler than those of their counterparts in L2, L3, and L6.

Together, the different L5 pyramidal cells are a main source of corticofugal projections to the striatum, thalamus, midbrain, and brainstem (Winer 2006). Notably, corticothalamic axons of L5 pyramidal neurons in nonlemniscal MGD (medial geniculate body, dorsal division) bear giant (2–10 μm) boutons (presynapses), probably serving transthalamic feedforward activation of other cortical areas, whereas axons of L6 pyramidal neurons (Sect. 8.2.6) in lemniscal MGV (medial geniculate body, ventral division) bear small (<1 μm) boutons, probably resulting in thalamic feedback control of the cortical area of origin (Ojima 1994; Rouiller and Welker 2000).

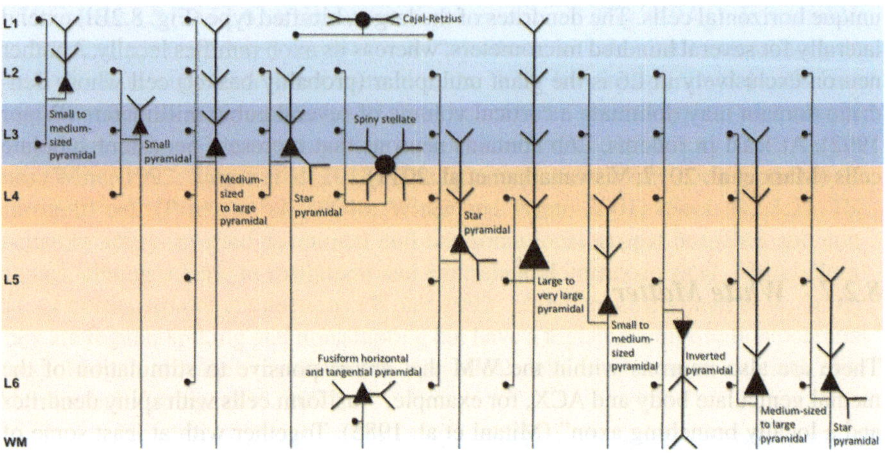

Fig. 8.3 Wiring diagram of major excitatory cell types (principal cells) within the ACX. Depicted are the laminar locations (L1–6, *WM*, white matter) of their cell bodies (*triangles* for pyramidal cells, *circles* for stellate and Cajal-Retzius cells), apical (or other typical) dendrites, and axonal terminations (*dots*). Axonal projections may end in additional layers as described for other sensory cortices

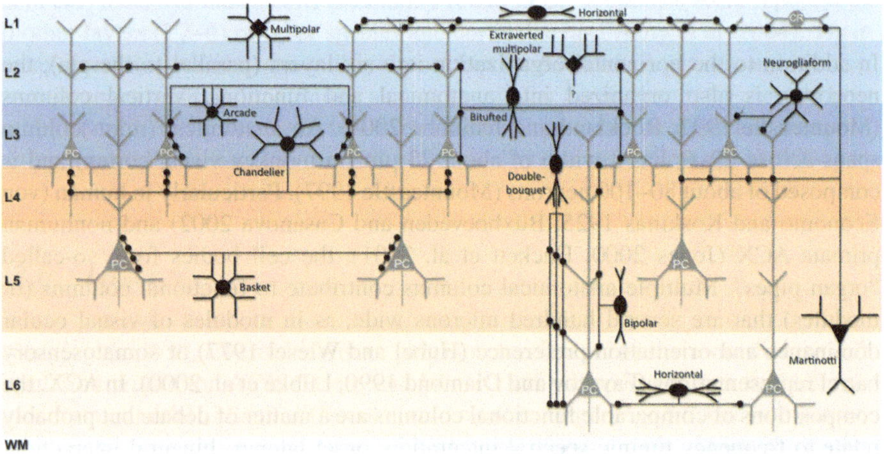

Fig. 8.4 Wiring diagram of major inhibitory cell types (interneurons) within the ACX. Depicted are the laminar locations (L1–6) of their cell bodies (individual cell types indicated), the gross morphology of their dendrites, and their major axonal targets on principal cells. For clarity, only some types of target principal cells are shown (*PC*, pyramidal cells; *CR*, Cajal-Retzius cells; *WM*, white matter)

2011) (also see Read and Reyes, Chap. 7). Also, L2 and L3 are the main targets of short local corticocortical connections, although deeper layers get substantial local inputs as well. Long-range corticocortical afferents terminate in the granular layer (and usually originate in supragranular layers) when they are feedforward, terminate in upper supragranular and deep infragranular layers (and originate in infragranular layers) when they are feedback, or terminate across all layers (and originate equally in infragranular and supragranular layers) when they are intrinsic-like (often termed lateral, e.g., Cappe and Barone 2005; Budinger and Scheich 2009).

Even primary ACX processes nonauditory sensory (e.g., visual, somatosensory) and nonsensory (e.g., motor, higher cognitive, neurotransmitter-specific modulatory) information as shown, for example, in rats and mice (Wallace et al. 2004; Nelson et al. 2013), ferrets (Bizley et al. 2007; Meredith and Allman 2015), monkeys (Fu et al. 2003; Brosch et al. 2005), and humans (Murray et al. 2005; Noesselt et al. 2010). The anatomical pathways underlying multimodal processing at this early cortical level include several possibilities that are not mutually exclusive. One possibility is the local integration of different information in ACX via direct subcortical (in particular thalamic) and intracortical inputs, both of which have been demonstrated by anatomical tracing studies in various rodent (Campi et al. 2010; Henschke et al. 2015), carnivore (Bizley et al. 2007; Chabot et al. 2015), and nonhuman primate species (de la Mothe et al. 2006a; Falchier et al. 2010).

Nonauditory thalamic afferents to the ACX arise, for example, from the visual suprageniculate nucleus and pulvinar complex as well as from the somatosensory posterior nucleus (Budinger et al. 2006; de la Mothe et al. 2006b). They terminate mainly in superficial and deep layers; thereby, single axons often innervate multiple layers of the ACX and form asymmetric (excitatory) synapses on spines and dendrites of mainly nonGABAergic cells (Smith et al. 2010).

Nonauditory cortical inputs to the ACX come from areas of virtually all sensory and nonsensory modalities in rodents (Budinger and Scheich 2009; Campi et al. 2010) but from a more restricted set of areas in primates (Smiley and Falchier 2009; Meredith and Lomber 2017). These inputs might terminate in specific layers of the ACX and on specific cell types. In mice, for example, axons from motor cortex (M2) project across all auditory cortical laminae and make excitatory synapses on pyramidal cells and PV+ interneurons, thereby exerting a primarily global suppressive effect on ACX (Nelson et al. 2013).

Collectively, the various excitatory principal cells and inhibitory interneurons differentially contribute to the intracortical microcircuitry of the ACX (Figs. 8.3 and 8.4). Pyramidal cells usually have a dense intracolumnar axonal domain, crossing several layers, and an extrinsically projecting axonal part (ipsilateral, contralateral) that also mostly terminates in a columnar fashion. This was demonstrated, for example, in cat (Ojima et al. 1991; Wallace et al. 1991), ferret (Wallace and Harper 1997), and gerbil (Budinger et al. 2000). In addition, there are always local axon collaterals projecting transcolumnarly and thus across the tonotopic gradient of the ACX, for example, of cats (Matsubara and Phillips 1988), ferrets (Wallace and Bajwa 1991), and gerbils (Kurt et al. 2008). Cajal-Retzius cells mainly serve transcolumnar and intralaminar connections, whereas spiny stellate cells mainly serve transcolumnar and translaminar connections. Interneurons have, by definition, short axonal branches, but many of them show clear orientation preferences. Thus, for example, horizontal cells of L1 and L6, multipolar cells (L1), large basket, arcade, chandelier, bitufted, and Martinotti cells connect multiple columnar domains, usually within a given layer, whereas the axons of bipolar and double-bouquet cells are restricted to one or a few columns but cover many layers (DeFelipe et al. 2013). In addition to chemical synapses, interneurons also form electrical contacts with each other via gap junctions (Hestrin and Galarreta 2005; Wallace and He 2011).

8.4 Physiological Studies of Intracortical Circuits

Recent physiological analyses of ACX circuits have relied on two complementary approaches. The first is dissection of circuits in vitro using electrophysiological and optical techniques, and the other is in vivo recordings and the inference of circuits from sound-evoked synaptic currents.

8.4.1 In Vitro Studies

Because of their accessibility, the functional microcircuitry of the ACX has been primarily studied in rodent brain slices. Importantly, because slices can be made in various orientations, these studies are performed either in coronal or in thalamocortical (tilted horizontal) planes (Cruikshank et al. 2002). While the former slice plane is oriented roughly perpendicular to the tonotopic axis in rodents (Stiebler et al. 1997), the latter is oriented parallel to it, and thus allows the study of the synaptic basis of frequency integration.

8.4.2 Results of Paired Recordings

Paired whole-cell patch recordings can reveal the properties of synaptic connections between neurons. Typically, these recordings involve patching two or more neurons within a local area (typically <100–200 μm apart). Such recordings revealed that within supragranular L2/3 pairs of neurons could have a variety of synaptic dynamics (Atzori et al. 2001; Oswald and Reyes 2008). Synaptic connections between cells showed either high release probability and adaptation or low release probability and little adaptation (Atzori et al. 2001). Such circuit differences were speculated to be advantageous for separating static versus dynamic aspects of sensory inputs (Atzori et al. 2001). Spatial analysis of synaptic connectivity between pyramidal cells revealed that cells with somata located within 200 μm showed appreciable connection probability, and neighboring cells showed a connection probability of 0.1–0.2 (Oswald and Reyes 2008; Levy and Reyes 2012). Inputs from inhibitory neurons tend to arise from a slightly more constrained area spatially as compared to close-by inputs from inhibitory neurons that have a larger connection probability of 0.3–0.5 (Levy and Reyes 2012). Overall, paired recording studies in the supragranular layers showed high overlap between anatomically predicted connectivity and observed connectivity.

8.4.3 Results of Functional Optical Circuit Mapping

Optical techniques combined with electrophysiological techniques have been used to study mesoscale translaminar and intralaminar connectivity. The first such technique was laser-scanning photostimulation (LSPS) using the focal photolysis of caged glutamate combined with patch clamp recordings (Callaway and Katz 1993; Shepherd et al. 2003). This technique utilizes local photoactivation of the caged glutamate via a UV laser beam that is scanned over the tissue using a set of galvanometer mirrors such that hundreds to thousands of putative presynaptic locations over a large area (>1 mm^2 depending on magnification of the objective) can be activated rapidly. Thus, a functional connectivity map over large regions of the ACX can be assembled on the single cell level. With LSPS one can clamp cells at the reversal potential of GABA or glutamate receptors (e.g., 0 mV, −70 mV) and sample excitatory or inhibitory inputs to each neuron to assess the spatial balance of excitatory and inhibitory inputs (Watkins et al. 2014; Meng et al. 2015, 2017b) and evaluate "silent" synapses that contain NMDA but not AMPA receptors (Meng et al. 2014) (see Table 8.1 for abbreviations).

Another recent approach that can be used to study mesoscale circuits is the application of optogenetics coupled with slice recordings. By utilizing mice that express opsins in molecularly identified cell types, specific connections can be probed (Lee et al. 2012; Petrus et al. 2015).

8.4.4 Layer 2/3 Neurons

Consistent with the anatomy of the ACX, LSPS studies in coronal (Barbour and Callaway 2008), thalamocortical (Oviedo et al. 2010; Meng et al. 2015), or horizontal (Watkins et al. 2014) slices revealed that supragranular L2/3 pyramidal neurons received most of their inputs from within L2/3 and from L4. In some studies, inputs from L5b and L6 were detected (Oviedo et al. 2010; Meng et al. 2015, 2017b). The difference in the existence of L6 inputs is likely due to differences in slice orientation as the projection from deep layers to L3 pyramidal cells originates somewhat rostral to the columnar location of the soma (home column) of the L3 pyramidal cell (Oviedo et al. 2010); this would be lost in coronal slices (Barbour and Callaway 2008). Pyramidal cells in L2/3 could be further subdivided based on the amount of input from L4 (Oviedo et al. 2010). Based on the laminar functional connections, L2/3 cells can be subdivided in separate sublaminae (Meng et al. 2017b). While L2 pyramidal cells received larger inputs from L6, L3 cells received smaller inputs. Moreover, L3 cells received intralaminar input that was biased toward the rostral direction (Oviedo et al. 2010). L2 and L3 cells also differ in the amount of input from L4, with deeper L3 cells receiving strong L4 input while superficial cells received less (Meng et al. 2017b). This difference in L4 input is also reflected in sublaminar differences in frequency integration (Meng et al. 2017b).

Laser-scanning photostimulation studies also showed that L2/3 pyramidal cells integrate information from other excitatory L2/3 cells that can be located more than 600 μm away (Oviedo et al. 2010; Meng et al. 2015, 2017b). In mouse A1, the tonotopic frequency gradient is about 3 octaves/mm (Stiebler et al. 1997; Bandyopadhyay et al. 2010; Guo et al. 2012); thus, L2/3 cells integrate information over more than two octaves. The intralaminar integration in L2/3 has been studied further using tangential sections of A1, which preserve the dendritic arbors of L2/3 cells in both the rostrocaudal axis and the dorsoventral/mediolateral direction (since slices are tilted by about 50°) (Watkins et al. 2014). These studies showed that the pyramidal cells received excitatory inputs from somewhat further away in the rostrocaudal (tonotopic) direction than in the mediolateral (isofrequency) direction. Moreover, while in general the connection probability decreased with distance from the soma, a spatially patchy distribution of connection probability was observed, indicating that certain locations have higher connection probabilities than others. This patchy connectivity was present in both the mediolateral (isofrequency) direction and also in the rostrocaudal (tonotopic) direction (Watkins et al. 2014). This suggests that intermingled, spatially precise circuits are present between pyramidal cells of L2/3 of A1. The patchy connectivity was not present in slices of visual cortex (Watkins et al. 2014), indicating that these patchy connections represent a specialization of A1, perhaps related to spectral tuning properties (Read et al. 2001).

Recordings from L2/3 pyramidal neurons in tangential slices revealed that, in contrast to the patchy nature of excitatory connections, inhibitory connections were more uniform with the highest probability for connections close to the soma and with radially decreasing probability for connections, spanning about 800 μm (Watkins et al. 2014; Meng et al. 2015). Connection probability extended further into the rostrocaudal than mediolateral direction (half widths: 288 μm versus 510 μm), indicating integration across frequency bands (Watkins et al. 2014). Recordings in thalamocortical slices showed that most inhibitory input to L2/3 cells originated from within L2/3 with some inputs that arose in L4 and L5 (Meng et al. 2015). Intralaminar inputs could originate from an area of around 800 μm surrounding the soma; L4 inputs arose from a slightly more focal area (about 400 μm range).

8.4.5 Layer 4 Neurons

Studies of excitatory L4 neurons in coronal slices using LSPS showed that the dominant excitatory input originated from L4 and L5a (Barbour and Callaway 2008), while studies in thalamocortical slices (Kratz and Manis 2015; Meng et al. 2017a) showed that L4 cells receive excitatory input from L2/3 as well as additional input from L5b and L6, which is slightly offset from the columnar soma position toward the rostral direction. The magnitude of the L6 input to L4 neurons in those studies tended to be smaller than that originating in other layers. In contrast, others reported that the excitatory input to L4 neurons from L6 exceeded that of L4 (Lee and Imaizumi 2013). Optogenetic studies revealed that specific subpopulations of L6 intracortical and corticothalamic projection neurons (neurotensin receptor 1 positive neurons) project to L4 (Lee et al. 2012) and suggested that convergent input from many L6 neurons is sufficient to cause spiking in L4 neurons. During development, in particular before thalamocortical inputs are mature, L4 receives input from subplate neurons (Zhao et al. 2009).

On average, intralaminar inputs from L4 originated over a range of about 1.2 mm in the rostrocaudal direction (Kratz and Manis 2015; Meng et al. 2017a). In contrast, inputs from superficial L2/3 originated from a narrower region and inputs from L6 originated from a broader region. Detailed spatial analysis of the input patterns showed that inputs to individual L4 cells from L4, L5, and L6 arose from within the home column and could originate up to 1 mm away from the soma (in either the rostral or caudal direction). However, in addition to the home column, a dominant location for inputs was observed from an area approximately 450–500 μm away from the soma. Displacement of the L6 input from the home column to L4 cells also was observed by Lee and Imaizumi (2013). This patchy input pattern is reminiscent of the input pattern observed in L2/3 (Watkins et al. 2014) and suggests that L4 cells integrate information from different spectral regions. However, the spatial scale of the displaced inputs seems to vary between layers. Inputs to L4 neurons seem to originate from further away than those to L2/3 cells (Watkins et al. 2014; Kratz and Manis 2015).

8.4.6 Layer 5/6

Few studies have focused on L5/6. The excitatory neurons in deeper L5 and L6 differ in their local connections. Putative excitatory L5 cells receive a large fraction of their inputs from outside their layer of origin (Llano and Sherman 2009; Joshi et al. 2015); in contrast, inputs to L6 neurons were predominantly local (Llano and Sherman 2009; Lee and Imaizumi 2013). Moreover, the amount of integration over the tonotopic axis was larger in L5 neurons than for L6 neurons, suggesting a higher degree of frequency integration by L5 neurons.

8.4.7 The Subplate

During prenatal and neonatal development, additional neurons, the so-called "subplate neurons", are present in the cerebral cortex (Kanold and Luhmann 2010). These excitatory neurons are among the earliest born cortical neurons, and they play a key role in cortical development (also see Sects. 8.2.6 and 8.2.7). They reside in areas that will become the adult WM. Depending on species, some of the subplate neurons are retained as L6b as well (Tervo et al. 2016; Marx et al. 2017). Subplate neurons receive excitatory input from the thalamus (Zhao et al. 2009) and also excitatory and inhibitory input from the developing cortical layers, such as L5/6 and L4 (Viswanathan et al. 2012; Meng et al. 2014). Similar to intralaminar circuits within L4 and L2/3, intra-subplate connections can be patchy (Meng et al. 2014). Importantly, L4 inputs to subplate neurons are present at very early ages and are mediated via silent synapses (see Sect. 8.4.3) (Meng et al. 2014). Subplate neurons, and not L4 neurons, are the first ACX neurons in development to show sound evoked responses and a nascent topographic organization (Wess et al. 2017). Importantly in altricial animals, such as ferrets, these responses can be elicited via close ear canals indicating that the auditory system is functioning before ear opening (Wess et al. 2017). Subplate neurons project to excitatory and inhibitory neurons in L4 (Zhao et al. 2009; Deng et al. 2017) and thus form an early sensory processing layer and a relay of sensory information into the future thalamorecipient layer L4. Ablation studies in the somatosensory and visual cortices have shown that subplate neurons are required for the thalamocortical refinement (Ghosh et al. 1990; Ghosh and Shatz 1992), normal functional maturation of thalamocortical (Kanold et al. 2003; Tolner et al. 2012) and inhibitory connections (Kanold and Shatz 2006), and for plasticity during the critical period (Kanold and Shatz 2006; Kanold and Luhmann 2010). However, a direct role for subplate neurons in ACX has not been demonstrated. A fraction of subplate neurons survive in the adult brain and form L6b (Marx et al. 2017; Viswanathan et al. 2017).

8.4.8 Inhibitory Neurons

Few electrophysiological studies have been performed on the connectivity of inhibitory neurons in ACX. Paired whole-cell patch recordings from L4 interneurons showed that excitatory inputs to inhibitory neurons arose from within 250 μm of the soma and had a connection probability of about 0.3 for neighboring cells (Levy and Reyes 2012). Studies of inhibitory neurons in L2/3 and L4 in coronal slices using LSPS showed that these neurons received most input from within L2/3 and L4 and little input from deeper layers (Barbour and Callaway 2008). The GAD2+ (glutamate decarboxylase 2) inhibitory neurons in L4 and L5 formed two populations: one received more restricted inputs while the other received extensive excitatory input from essentially all cortical layers (Deng et al. 2017). The PV+ interneurons across L2/3 to L6 received thalamic inputs while thalamic input to other interneuron types was restricted to L4 (Ji et al. 2016).

8.5 Comparisons of Physiological Techniques

While both paired recordings and LSPS have shown that on average there is a higher connection probability for neighboring cells and a lower connection probability for cells further apart, there are important differences. Paired recordings tend to show a smooth decay of connection probability (Oswald and Reyes 2008; Levy and Reyes 2012) while LSPS studies tend to show patchy connections (Watkins et al. 2014; Kratz and Manis 2015). Moreover, LSPS reveals connections between neurons that can be almost 1 mm apart, which is not seen in paired recordings. The differences are likely due to selective sampling using paired recordings versus more unbiased optical activation of neurons with LSPS. Moreover, LSPS enables the probing of hundreds of presynaptic sites (frequently up to 900) (Meng et al. 2014) for each probed postsynaptic neuron. On the other hand, LSPS, using single photon excitation, does not have single cell resolution; thus, it is likely that more than one neuron is activated at each stimulation site. The spatial resolution depends on the optical characteristics of the microscope, the size of the laser beam, and the objective used. Moreover, LSPS does not allow sampling of presynaptic cells close (<100 μm) to the recorded neurons due to direct activation of glutamate receptors on the recorded neuron.

8.6 In Vivo Studies of Cortical Circuits

Anatomical and brain slice studies can reveal circuits, but to understand the function of these circuits it is necessary to know how they relate to sound processing. In vivo patch clamp studies in rodents have been the first step in bridging this gap. These studies are typically performed in anesthetized rodents and allow the targeted recording of genetically identified neurons by combining in vivo imaging with electrophysiology (Wu et al. 2011). An assessment of the spiking properties of targeted neurons is possible through the use of loose patch recording. Whole-cell recordings can identify synaptic inputs to the cell that are evoked by different sound stimuli and thus can be used to measure the auditory receptive fields of synaptic inputs. When cells are voltage-clamped at the reversal potential of GABA or glutamate receptors, separate receptive fields for excitatory and inhibitory inputs, respectively, can be derived. By measuring the latencies of the synaptic inputs or silencing cortical inputs, the synaptic inputs that arose from direct thalamocortical activation can be distinguished from those that arose from intracortical sources. Thus, this approach can aid in assembling a functional circuit diagram under in vivo conditions; however, a general drawback of those studies is that when the soma is patched, distal inputs are not sampled due to electrotonic attenuation, and therefore, the extent of visible inputs is limited. This is especially an issue for cells with extensive dendritic trees.

8.6.1 Excitatory Neurons

L4 cells receive excitatory and inhibitory inputs that are approximately balanced. Excitation and inhibition have similar best frequencies (BFs) and somewhat similar frequency ranges with inhibition being slightly broader near the peak of the tuning curve (Wehr and Zador 2003; Zhang et al. 2003). Thalamocortical inputs provided broad excitation while intracortical excitatory inputs displayed much sharper tuning, suggesting that the frequency selectivity of the spiking output is determined by the intracortical input (Liu et al. 2007). The laminar source of this input cannot be determined in vivo, but in vitro studies have shown that L4 cells receive excitation from within L4 as well as from L2/3 and L6 (Barbour and Callaway 2008; Meng et al. 2017a).

L2/3 pyramidal cells receive excitatory inputs with longer latency but similar bandwidth as L4, presumably reflecting direct input from L4 (Li et al. 2014). In contrast, inhibitory input to L4 is significantly broader (Li et al. 2014) and is thought to be from L2/3 PV+ interneurons. Indeed, L2/3 neurons receive inhibitory inputs from different tonotopic locations (Oviedo et al. 2010; Meng et al. 2015, 2017b); thus, the broad tuning of inhibition to L2/3 cells could be due to the widespread integration from inhibitory neurons at different positions in the A1 frequency map. Moreover, in vivo imaging has revealed that single spines on L2/3 pyramidal cells can have BFs that are multiple octaves apart (Chen et al. 2011), which was not appreciated in patch clamp studies. With respect to the temporal response properties, intracellular recordings reveal a sequence of early excitation and delayed inhibition in subsets of L2/3 neurons (Ojima and Murakami 2002; Tan et al. 2007), which is consistent with the conclusion that these neurons receive excitation from L4 and disynaptic inhibition. Paired recordings in rodent slices across developmental periods showed that the connection probability of L2/3 neurons stayed constant between P10 (postnatal day for onset of low threshold hearing) and P29 (Oswald and Reyes 2008).

L5 cells can be divided into narrow tuned, regular-spiking cells and more widely tuned, intrinsic-bursting cells (Hefti and Smith 2000; Sun et al. 2013; also see Sect. 8.2.5). Like L4 cells, intrinsic-bursting cells receive short-latency excitatory input and feedforward inhibitory input, probably from fast-spiking neurons. In contrast, regular-spiking cells receive long-latency excitatory input from intracortical sources as well as intracortical inhibitory input, probably from nonfast spiking cells (Sun et al. 2013).

Excitatory L6 cells also seem to fall into two categories (Zhou et al. 2010). One group of L6 neurons is similar to L4 cells, showing sound responses and approximately matched receptive fields for excitation and inhibition, probably due to direct thalamocortical inputs and feedforward inhibitory input. Another group did not show responses to sound, due to a large and strong inhibitory receptive field, and showed only weak long-latency excitatory input, probably from intracortical sources.

8.6.2 *Inhibitory Neurons*

Various classes of inhibitory neurons can be identified using cell type-specific promoters (e.g., PV; somatostatin, SST), spiking properties (e.g., fast-spiking), or posthoc histology (Table 8.2). The receptive field structure of PV cells is under debate. Two studies reported that PV cells in L2/3 show broader tonal receptive fields compared to pyramidal cells (Cohen and Mizrahi 2015; Li et al. 2015), which is consistent with PV cells representing fast-spiking cells (Wu et al. 2008; Sun et al. 2013) and is consistent with the extensive dendritic tree of PV cells (Wu et al. 2011; Sun et al. 2013). Fast-spiking cells in cat also have short latencies and large receptive fields (Atencio and Schreiner 2008). In contrast, another study reported that PV cells had narrower receptive fields (Moore and Wehr 2013). The excitatory input to PV neurons is broader and shorter latency than the excitatory input to SST cells (Atencio and Schreiner 2008; Li et al. 2015). Since synaptic latencies of PV neurons and excitatory neurons were similar, both probably receive ascending input from L4, and excitatory input to SST cells might originate within L2/3. Thus, the broad tuning of L2/3 PV cells could also contribute to the broad inhibition seen in L2/3 pyramidal cells (Li et al. 2014).

8.7 Relation of Microcircuits to Functional Maps

Anatomical and physiological studies have started to unravel the bewildering complexity of auditory cortical circuits. How are these circuits related to the functional properties of A1? The hallmark of A1 organization is the tonotopic organization of frequency preference (Kanold et al. 2014). While smooth on a large scale, this map is fractured on the fine scale with L2/3 (Bandyopadhyay et al. 2010; Rothschild et al. 2010) displaying a much higher heterogeneity than L4 (Winkowski and Kanold 2013; Kanold et al. 2014), such that neighboring neurons in L2/3 can have very different frequency preferences, while neighboring L4 neurons are more similar in their frequency preference. The local homogeneity of frequency preference in L4 is probably due to shared thalamocortical inputs between neighboring neurons, which is consistent with the broad tuning of thalamocortical inputs (Liu et al. 2007) and suggests convergent thalamic input onto single neurons.

The heterogeneity in frequency preferences within L2/3 is higher for excitatory pyramidal than for PV inhibitory neurons (Maor et al. 2016). The heterogeneity of frequency tuning for excitatory L2/3 neurons can be generated by distinct circuits that show sublaminar differences in frequency integration (Meng et al. 2017b). Ascending circuits from L4 to L2/3 could arise from different locations within the tonotopic map. Indeed, LSPS studies show that, on average, excitatory input from L4 to L2/3 is broad (Meng et al. 2015) and diverse (Meng et al. 2017b); thus, L2/3 neurons have access to multiple locations on the tonotopic map. Moreover, L2/3 neurons receive intralaminar inputs; therefore, excitatory input originating from other tonotopic locations within L2/3 has the potential to alter the frequency preference of L2/3 neurons. Alternatively, since L2/3 neurons receive broad inhibition (Li et al. 2014; Meng et al. 2015, 2017b), the tuning of this inhibitory input could alter the frequency preference of L2/3 neurons; thus, neighboring L2/3 neurons might receive varying inhibitory input. However, the latter scenario is less likely because inhibitory neurons within L2/3 show similar tuning (Maor et al. 2016). The lower heterogeneity of frequency preferences of PV neurons is likely due to PV neurons receiving broad excitatory input (Li et al. 2015) and being coupled via gap junctions, which can regulate inhibitory network strength (Deans et al. 2001; Postma et al. 2011).

8.8 Patchy Intracortical Connections as a Substrate for Auditory Object Formation

One presumed function of A1 is to extract behaviorally meaningful signals, for example, conspecific vocalizations. Thus, it might be predicted that underlying neural circuits are structured to facilitate such function and, in essence, start to assemble auditory objects (Brosch et al. 2013; Bizley et al. 2016). Natural vocalizations frequently contain energy in harmonically related frequency areas and many neurons in the ACX of primates (Kadia and Wang 2003), cats (Sutter and Schreiner 1991; Norena et al. 2008), and rodents (Winkowski and Kanold 2013) can have multipeaked receptive fields. To extract the presence of such an object, single cells must integrate information from different spectral areas that correspond to different locations in the tonotopic map. In mice, the represented frequency doubles about every 330 μm (Stiebler et al. 1997; Guo et al. 2012). Thus, if a single neuron at a particular tonotopic location also receives input from an area tuned to an octave lower (or higher), this would be visible as inputs from about 350 μm away (Fig. 8.5). These distances are within the range of what is seen with LSPS in L4 and L2/3 (Watkins et al. 2014; Kratz and Manis 2015), suggesting that cells in these layers might perform such an integrative function. There is the possibility that L2/3 contains different circuits in L2 and L3 in rodents (Oviedo et al. 2010), and it is conceivable that even more circuit diversity is present than has been unmasked so far.

8.9 Summary and the Road Ahead

Given the complexity of the ACX, most of what needs to be known to understand how ACX processes sound remains unknown. The enormous, even puzzling, anatomical diversity of neuronal morphologies is not yet reflected in the physiological studies, likely due to under-sampling of neurons and averaging of connectivity profiles. Moreover, most studies of ACX circuits paint a rather static picture of cortical circuits. Cortical circuits assemble and change during development and can further

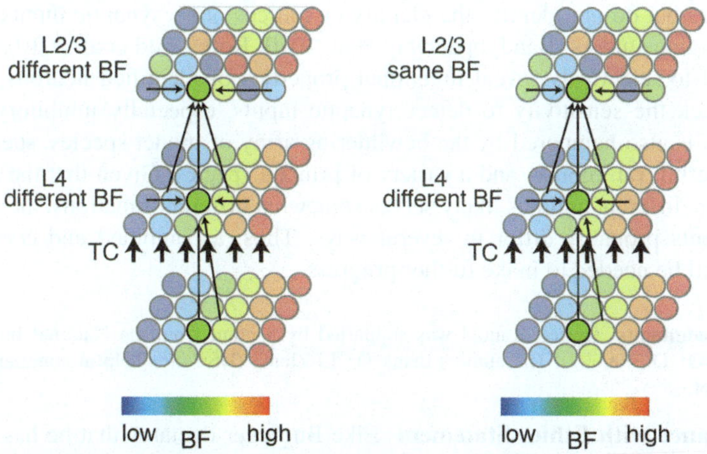

Fig. 8.5 Potential functional relationships of patchy connections within the ACX. Colorbar indicates relative best frequency (*BF*). L4 (*middle*) and L2/3 (*upper*) neurons are depicted with the BF (*color*) represented along the two-dimensional extent of A1. The tonotopic organization in L2/3 is more heterogeneous than in L4. Arrows indicate intralaminar and interlaminar connections. Since L4 has a more homogeneous frequency organization with respect to the tonotopic gradient, intralaminar connections within L4 likely connect cells with different BFs. In contrast, due to the spatial heterogeneity of BFs in L2/3, intralaminar connections in L2/3 can either connect cells with different BFs (*left*) or similar BFs (*right*). Given the multipeaked nature of some L2/3 receptive fields, intralaminar as well as interlaminar connections could originate from harmonically related locations in L4 or L2/3

be modified due to, for example, attention, memory, and learning. Given that sensory systems can adapt to their environment, developmental studies have the potential of informing us about the function of particular circuits (Henschke et al. 2017). Anatomical studies in development thus far have not taken advantage of modern viral tracers because the survival time required is too long to assess early developmental stages in rodents. Functional studies in development in vivo as well as in vitro will have to take into account that excitatory transmission can be mediated via silent synapses (i.e., those containing only NMDA receptors; Meng et al. 2014; Deng et al. 2017).

The ACX can adapt rapidly to changes in the auditory (Froemke et al. 2007; Froemke and Schreiner 2015) or visual (Petrus et al. 2014) environment, sound statistics (Taaseh et al. 2011; Khouri and Nelken 2015), or behavioral demands, as seen in ferrets (Fritz et al. 2003; David et al. 2012) and rodents (Ohl and Scheich 1997; Otazu et al. 2009); however, only a few studies have shown how ACX circuits adjust (Meng et al. 2015; Petrus et al. 2015).

Although anatomical studies can show the totality of all possible connections, they cannot assess synaptic connection strength and functional properties, especially the timing, dynamics, and coactivation of inputs. In contrast, brain slice studies can reveal connection profiles and connection strength but lack information about the functional properties of connected neurons and the greater resolution of anatomical studies (Fig. 8.5). In vivo patch clamp studies can identify tuning properties but do not identify the identity or sources of presynaptic inputs. In vivo Ca^{2+} imaging studies (Bandyopadhyay et al. 2010; Rothschild et al. 2010) have the potential to selectively reveal functional properties of identified neurons but currently lack the sensitivity to detect synaptic inputs, especially inhibitory inputs. Progress is also hampered by the bewildering array of model species, such as cat, ferret, gerbil, rat, mouse, and a variety of primate species. Given that the auditory system in different animals likely serves somewhat different functions, the underlying circuits probably differ in several ways. Thus, a combined and coordinated effort will be needed to make further progress.

Acknowledgments Patrick Kanold was supported by a grant from the National Institutes of Health (RO1 DC009607). The authors thank Dr. Li Zhang (USC) for helpful comments on the manuscript.

Compliance with Ethics Statement Eike Budinger declares that he has no conflict of interest.

Patrick Kanold declares that he has no conflict of interest.

References

Agmon, A., & Connors, B. W. (1992). Correlation between intrinsic firing patterns and thalamocortical synaptic responses of neurons in mouse barrel cortex. *The Journal of Neuroscience, 12*(1), 319–329.

Andjelic, S., Gallopin, T., Cauli, B., Hill, E. L., Roux, L., Badr, S., et al. (2009). Glutamatergic nonpyramidal neurons from neocortical layer VI and their comparison with pyramidal and spiny stellate neurons. *Journal of Neurophysiology, 101*(2), 641–654.

Atencio, C. A., & Schreiner, C. E. (2008). Spectrotemporal processing differences between auditory cortical fast-spiking and regular-spiking neurons. *The Journal of Neuroscience, 28*(15), 3897–3910.

Atzori, M., Flores Hernandez, J., & Pineda, J. C. (2004). Interlaminar differences of spike activation threshold in the auditory cortex of the rat. *Hearing Research, 1899*(1-2), 101–106.

Atzori, M., Lei, S., Evans, D. I., Kanold, P. O., Phillips-Tansey, E., McIntyre, O., & McBain, C. J. (2001). Differential synaptic processing separates stationary from transient inputs to the auditory cortex. *Nature Neuroscience, 4*(12), 1230–1237.

Ballesteros-Yanez, I., Munoz, A., Contreras, J., Gonzalez, J., Rodriguez-Veiga, E., & DeFelipe, J. (2005). Double bouquet cell in the human cerebral cortex and a comparison with other mammals. *The Journal of Comparative Neurology, 486*(4), 344–360.

Baloyannis, S. J., Manolides, S., Arzoglou, L., Costa, V., & Manolides, L. (1993). The structural organization of layer I of the adult human acoustic cortex. A Golgi and electron microscopy study. *Acta Oto-Laryngologica, 113*(4), 502–506.

Bandyopadhyay, S., Shamma, S. A., & Kanold, P. O. (2010). Dichotomy of functional organization in the mouse auditory cortex. *Nature Neuroscience, 13*(3), 361–368.

Bannister, A. P. (2005). Inter- and intra-laminar connections of pyramidal cells in the neocortex. *Neuroscience Research, 53*(2), 95–103.

Barbour, D. L., & Callaway, E. M. (2008). Excitatory local connections of superficial neurons in rat auditory cortex. *The Journal of Neuroscience, 28*(44), 11174–11185.

Bizley, J. K., & Cohen, Y. E. (2013). The what, where and how of auditory-object perception. *Nature Reviews Neuroscience, 14*(10), 693–707.

Bizley, J. K., Maddox, R. K., & Lee, A. K. (2016). Defining auditory-visual objects: behavioral tests and physiological mechanisms. *Trends in Neurosciences, 399*(2), 74–85.
Bizley, J. K., Nodal, F. R., Bajo, V. M., Nelken, I., & King, A. J. (2007). Physiological and anatomical evidence for multisensory interactions in auditory cortex. *Cerebral Cortex, 17*(9), 2172–2189.
Brosch, M., Budinger, E., & Scheich, H. (2013). Different synchronization rules in primary and nonprimary auditory cortex of monkeys. *Journal of Cognitive Neuroscience, 25*(9), 1517–1526.
Brosch, M., Selezneva, E., & Scheich, H. (2005). Nonauditory events of a behavioral procedure activate auditory cortex of highly trained monkeys. *The Journal of Neuroscience, 25*(29), 6797–6806.
Budinger, E., & Heil, P. (2005). Anatomy of the auditory cortex. In S. Greenberg & W. A. Ainsworth (Eds.), *Listening to speech* (pp. 91–113). Mahwah, NJ: Lawrence Erlbaum Associates.
Budinger, E., Heil, P., Hess, A., & Scheich, H. (2006). Multisensory processing via early cortical stages: Connections of the primary auditory cortical field with other sensory systems. *Neuroscience, 143*(4), 1065–1083.
Budinger, E., Heil, P., & Scheich, H. (2000). Functional organization of auditory cortex in the Mongolian gerbil (*Meriones unguiculatus*). III. Anatomical subdivisions and corticocortical connections. *European Journal of Neuroscience, 12*(7), 2425–2451.
Budinger, E., & Scheich, H. (2009). Anatomical connections suitable for the direct processing of neuronal information of different modalities via the rodent primary auditory cortex. *Hearing Research, 258*(1-2), 16–27.
Buxhoeveden, D. P., & Casanova, M. F. (2002). The minicolumn hypothesis in neuroscience. *Brain, 125*(5), 935–951.
Callaway, E. M., & Katz, L. C. (1993). Photostimulation using caged glutamate reveals functional circuitry in living brain slices. *Proceedings of the National Academy of Sciences of the United States of America, 990*(16), 7661–7665.
Campi, K. L., Bales, K. L., Grunewald, R., & Krubitzer, L. (2010). Connections of auditory and visual cortex in the prairie vole (*Microtus ochrogaster*): Evidence for multisensory processing in primary sensory areas. *Cerebral Cortex, 20*(1), 89–108.
Cappe, C., & Barone, P. (2005). Heteromodal connections supporting multisensory integration at low levels of cortical processing in the monkey. *European Journal of Neuroscience, 22*(11), 2886–2902.
Celio, M. R. (1986). Parvalbumin in most gamma-aminobutyric acid-containing neurons of the rat cerebral cortex. *Science, 231*, 995–997.
Chabot, N., Butler, B. E., & Lomber, S. G. (2015). Differential modification of cortical and thalamic projections to cat primary auditory cortex following early- and late-onset deafness. *The Journal of Comparative Neurology, 523*(15), 2297–2320.
Chen, X., Leischner, U., Rochefort, N. L., Nelken, I., & Konnerth, A. (2011). Functional mapping of single spines in cortical neurons in vivo. *Nature, 475*(7357), 501–505.
Code, R. A., & Winer, J. A. (1985). Commissural neurons in layer III of cat primary auditory cortex (AI): Pyramidal and non-pyramidal cell input. *The Journal of Comparative Neurology, 242*(4), 485–510.
Cohen, L., & Mizrahi, A. (2015). Plasticity during motherhood: Changes in excitatory and inhibitory layer 2/3 neurons in auditory cortex. *The Journal of Neuroscience, 35*(4), 1806–1815.
da Costa, N. M., & Martin, K. A. (2011). How thalamus connects to spiny stellate cells in the cat's visual cortex. *The Journal of Neuroscience, 31*(8), 2925–2937.
Cowan, A. I., & Stricker, C. (2004). Functional connectivity in layer IV local excitatory circuits of rat somatosensory cortex. *Journal of Neurophysiology, 992*(4), 2137–2150.
Cruikshank, S. J., Rose, H. J., & Metherate, R. (2002). Auditory thalamocortical synaptic transmission in vitro. *Journal of Neurophysiology, 87*(1), 361–384.
David, S. V., Fritz, J. B., & Shamma, S. A. (2012). Task reward structure shapes rapid receptive field plasticity in auditory cortex. *Proceedings of the National Academy of Sciences of the United States of America, 1099*(6), 2144–2149.

Deans, M. R., Gibson, J. R., Sellitto, C., Connors, B. W., & Paul, D. L. (2001). Synchronous activity of inhibitory networks in neocortex requires electrical synapses containing connexin36. *Neuron, 31*(3), 477–485.

DeFelipe, J. (2002). Cortical interneurons: From Cajal to 2001. *Progress in Brain Research, 136*, 215–238.

DeFelipe, J., Alonso-Nanclares, L., & Arellano, J. I. (2002). Microstructure of the neocortex: Comparative aspects. *Journal of Neurocytology, 31*(3-5), 299–316.

DeFelipe, J., Lopez-Cruz, P. L., Benavides-Piccione, R., Bielza, C., Larrañaga, P., Anderson, S., et al. (2013). New insights into the classification and nomenclature of cortical GABAergic interneurons. *Nature Reviews Neuroscience, 14*(3), 202–216.

Deng, R., Kao, J. P.-Y., & Kanold, P. O. (2017). Distinct translaminar glutamatergic circuits to GABAergic interneurons in the neonatal auditory cortex. *Cell Reports, 199*(6), 1141–1150.

Douglas, R. J., & Martin, K. A. (2004). Neuronal circuits of the neocortex. *Annual Review of Neuroscience, 27*, 419–451.

von Economo, C., & Koskinas, G. N. (1925). *Die Cytoarchitektonik des erwachsenen Menschen*. Berlin: Springer.

Falchier, A., Schroeder, C. E., Hackett, T. A., Lakatos, P., Nascimento-Silva, S., Ulbert, I., et al. (2010). Projection from visual areas V2 and prostriata to caudal auditory cortex in the monkey. *Cerebral Cortex, 20*(7), 1529–1538.

Favorov, O. V., & Diamond, M. E. (1990). Demonstration of discrete place-defined columns—segregates—in the cat SI. *The Journal of Comparative Neurology, 2998*(1), 97–112.

Felleman, D. J., & Van Essen, D. C. (1991). Distributed hierarchical processing in the primate cerebral cortex. *Cerebral Cortex, 1*(1), 1–47.

Fitzpatrick, D. C., & Henson, O. W., Jr. (1994). Cell types in the mustached bat auditory cortex. *Brain, Behavior and Evolution*, 43(2), 79–91.

Fritz, J. B., Elhilali, M., David, S. V., & Shamma, S. A. (2007). Auditory attention—focusing the searchlight on sound. *Current Opinion in Neurobiology, 17*(4), 437–455.

Fritz, J., Shamma, S., Elhilali, M., & Klein, D. (2003). Rapid task-related plasticity of spectrotemporal receptive fields in primary auditory cortex. *Nature Neuroscience, 6*(11), 1216–1223.

Froemke, R. C., Merzenich, M. M., & Schreiner, C. E. (2007). A synaptic memory trace for cortical receptive field plasticity. *Nature, 450*(7168), 425–429.

Froemke, R. C., & Schreiner, C. E. (2015). Synaptic plasticity as a cortical coding scheme. *Current Opinion in Neurobiology, 35*, 185–199.

Fu, K. M., Johnston, T. A., Shah, A. S., Arnold, L., Smiley, J., Hackett, T. A., et al. (2003). Auditory cortical neurons respond to somatosensory stimulation. *The Journal of Neuroscience, 23*(20), 7510–7515.

Ghosh, A., Antonini, A., McConnell, S. K., & Shatz, C. J. (1990). Requirement for subplate neurons in the formation of thalamocortical connections. *Nature, 347*(6289), 179–181.

Ghosh, A., & Shatz, C. J. (1992). Involvement of subplate neurons in the formation of ocular dominance columns. *Science, 255*(5050), 1441–1443.

Griffiths, T. D., & Warren, J. D. (2004). What is an auditory object? *Nature Reviews Neuroscience, 5*(11), 887–892.

Guo, W., Chambers, A. R., Darrow, K. N., Hancock, K. E., Shinn-Cunningham, B. G., & Polley, D. B. (2012). Robustness of cortical topography across fields, laminae, anesthetic states, and neurophysiological signal types. *The Journal of Neuroscience, 32*(27), 9159–9172.

Hackett, T. A., Barkat, T. R., O'Brien, B. M., Hensch, T. K., & Polley, D. B. (2011). Linking topography to tonotopy in the mouse auditory thalamocortical circuit. *The Journal of Neuroscience, 31*(8), 2983–2995.

Hackett, T. A., Preuss, T. M., & Kaas, J. H. (2001). Architectonic identification of the core region in auditory cortex of macaques, chimpanzees, and humans. *The Journal of Comparative Neurology, 441*(3), 197–222.

Hefti, B. J., & Smith, P. H. (2000). Anatomy, physiology, and synaptic responses of rat layer V auditory cortical cells and effects of intracellular GABA(A) blockade. *Journal of Neurophysiology, 83*(5), 2626–2638.

Henschke, J. U., Noesselt, T., Scheich, H., & Budinger, E. (2015). Possible anatomical pathways for short-latency multisensory integration processes in primary sensory cortices. *Brain Structure and Function, 220*(2), 955–977.

Henschke, J. U., Oelschlegel, A. M., Angenstein, F., Ohl, F. W., Goldschmidt, J., Kanold, P. O., & Budinger, E. (2017). Early sensory experience influences the development of multisensory thalamocortical and intracortical connections, *Brain Structure and Function,* doi: 10.1007/s00429-017-1549-1. [Epub ahead of print].

Hestrin, S., & Galarreta, M. (2005). Electrical synapses define networks of neocortical GABAergic neurons. *Trends in Neurosciences, 28*(6), 304–309.

Huang, C. L., & Winer, J. A. (2000). Auditory thalamocortical projections in the cat: Laminar and areal patterns of input. *The Journal of Comparative Neurology, 427*(2), 302–331.

Hubel, D. H., & Wiesel, T. N. (1977). Ferrier lecture. Functional architecture of macaque monkey visual cortex. *Proceedings of the Royal Society of London B: Biological Sciences, 1998*(1130), 1–59.

Huggenberger, S., Vater, M., & Deisz, R. A. (2009). Interlaminar differences of intrinsic properties of pyramidal neurons in the auditory cortex of mice. *Cerebral Cortex, 199*(5), 1008–1018.

Imig, T. J., & Adrian, H. O. (1977). Binaural columns in the primary field (A1) of cat auditory cortex. *Brain Research, 138*, 241–257.

Inan, M., & Anderson, S. A. (2014). The chandelier cell, form and function. *Current Opinion in Neurobiology, 26*, 142–148.

Ji, X. Y., Zingg, B., Mesik, L., Xiao, Z., Zhang, L. I., & Tao, H. W. (2016). Thalamocortical innervation pattern in mouse auditory and visual cortex: Laminar and cell-type specificity. *Cerebral Cortex, 26*(6), 2612–2625.

Jones, E. G. (2000). Microcolumns in the cerebral cortex. *Proceedings of the National Academy of Sciences of the United States of America, 997*(10), 5019–5021.

Joshi, A., Middleton, J. W., Anderson, C. T., Borges, K., Suter, B. A., Shepherd, G. M., & Tzounopoulos, T. (2015). Cell-specific activity-dependent fractionation of layer 2/3—5B excitatory signaling in mouse auditory cortex. *The Journal of Neuroscience, 35*(7), 3112–3123.

Kaas, J. H. (2011). The evolution of auditory cortex: The core areas. In J. A. Winer & C. E. Schreiner (Eds.), *The auditory cortex* (pp. 407–427). New York: Springer.

Kadia, S. C., & Wang, X. (2003). Spectral integration in A1 of awake primates: Neurons with single- and multi-peaked tuning characteristics. *Journal of Neurophysiology, 899*(3), 1603–1622.

Kanold, P. O., Kara, P., Reid, R. C., & Shatz, C. J. (2003). Role of subplate neurons in functional maturation of visual cortical columns. *Science, 301*(5632), 521–525.

Kanold, P. O., & Luhmann, H. J. (2010). The subplate and early cortical circuits. *Annual Review of Neuroscience, 33*, 23–48.

Kanold, P. O., Nelken, I., & Polley, D. B. (2014). Local versus global scales of organization in auditory cortex. *Trends in Neurosciences, 37*(9), 502–510.

Kanold, P. O., & Shatz, C. J. (2006). Subplate neurons regulate maturation of cortical inhibition and outcome of ocular dominance plasticity. *Neuron, 51*(5), 627–638.

Khouri, L., & Nelken, I. (2015). Detecting the unexpected. *Current Opinion in Neurobiology, 35*, 142–147.

Kirischuk, S., Luhmann, H. J., & Kilb, W. (2014). Cajal-Retzius cells: Update on structural and functional properties of these mystic neurons that bridged the 20th century. *Neuroscience, 275*, 33–46.

Kratz, M. B., & Manis, P. B. (2015). Spatial organization of excitatory synaptic inputs to layer 4 neurons in mouse primary auditory cortex. *Frontiers in Neural Circuits, 99*, 17. https://doi.org/10.3389/fncir.2015.00017.

Kurt, S., Deutscher, A., Crook, J. M., Ohl, F. W., Budinger, E., Moeller, C. K., et al. (2008). Auditory cortical contrast enhancing by global winner-take-all inhibitory interactions. *PLoS One, 3*(3), e1735.

Lee, C. C., & Imaizumi, K. (2013). Functional convergence of thalamic and intrinsic projections to cortical layers 4 and 6. *Neurophysiology, 45*(5-6), 396–406.

Lee, C. C., Lam, Y. W., & Sherman, S. M. (2012). Intracortical convergence of layer 6 neurons. *Neuroreport, 23*(12), 736–740.
Lee, C. C., & Sherman, S. M. (2008). Synaptic properties of thalamic and intracortical inputs to layer 4 of the first- and higher-order cortical areas in the auditory and somatosensory systems. *Journal of Neurophysiology, 100*(1), 317–326.
Levy, R. B., & Reyes, A. D. (2012). Spatial profile of excitatory and inhibitory synaptic connectivity in mouse primary auditory cortex. *Journal of Neuroscience, 32*(16), 5609–5619.
Li, L. Y., Ji, X. Y., Liang, F., Li, Y. T., Xiao, Z., Tao, H. W., & Zhang, L. I. (2014). A feedforward inhibitory circuit mediates lateral refinement of sensory representation in upper layer 2/3 of mouse primary auditory cortex. *The Journal of Neuroscience, 34*(41), 13670–13683.
Li, L. Y., Xiong, X. R., Ibrahim, L. A., Yuan, W., Tao, H. W., & Zhang, L. I. (2015). Differential receptive field properties of parvalbumin and somatostatin inhibitory neurons in mouse auditory cortex. *Cerebral Cortex, 25*(7), 1782–1791.
Linden, J. F., & Schreiner, C. E. (2003). Columnar transformations in auditory cortex? A comparison to visual and somatosensory cortices. *Cerebral Cortex, 13*(1), 83–89.
Liu, B. H., Wu, G. K., Arbuckle, R., Tao, H. W., & Zhang, L. I. (2007). Defining cortical frequency tuning with recurrent excitatory circuitry. *Nature Neuroscience, 10*(12), 1594–1600.
Llano, D. A., & Sherman, S. M. (2009). Differences in intrinsic properties and local network connectivity of identified layer 5 and layer 6 adult mouse auditory corticothalamic neurons support a dual corticothalamic projection hypothesis. *Cerebral Cortex, 199*(12), 2810–2826.
Lubke, J., Egger, V., Sakmann, B., & Feldmeyer, D. (2000). Columnar organization of dendrites and axons of single and synaptically coupled excitatory spiny neurons in layer 4 of the rat barrel cortex. *The Journal of Neuroscience, 20*(14), 5300–5311.
Lund, J. S. (1984). Spiny stellate neurons. In A. Peters & E. G. Jones (Eds.), *Cerebral cortex: Vol. 1. Cellular components of the cerebral cortex* (pp. 255–308). New York: Plenum Press.
Maor, I., Shalev, A., & Mizrahi, A. (2016). Distinct spatiotemporal response properties of excitatory versus inhibitory neurons in the mouse auditory cortex. *Cerebral Cortex, 26*(11), 4242–4252.
Marin-Padilla, M. (1984). Neurons of layer I. In A. Peters & E. G. Jones (Eds.), *Cerebral cortex: Vol. 1. Cellular components of the cerebral cortex* (pp. 447–478). New York: Plenum Press.
Markram, H., Muller, E., Ramaswamy, S., Reimann, M. W., Abdellah, M., Sanchez, C. A., et al. (2015). Reconstruction and simulation of neocortical microcircuitry. *Cell, 163*(2), 456–492.
Marx, M., Qi, G., Hanganu-Opatz, I. L., Kilb, W., Luhmann, H. J., & Feldmeyer, D. (2017). Neocortical layer 6b as a remnant of the subplate—a morphological comparison. *Cerebral Cortex, 27*(2), 1011–1026.
Matsubara, J. A., & Phillips, D. P. (1988). Intracortical connections and their physiological correlates in the primary auditory cortex (AI) of the cat. *The Journal of Comparative Neurology, 268*, 38–48.
McMullen, N. T., & Glaser, E. M. (1982). Morphology and laminar distribution of nonpyramidal neurones in the auditory cortex of the rabbit. *The Journal of Comparative Neurology, 208*, 85–106.
Mendizabal-Zubiaga, J. L., Reblet, C., & Bueno-Lopez, J. L. (2007). The underside of the cerebral cortex: Layer V/VI spiny inverted neurons. *Journal of Anatomy, 211*(2), 223–236.
Meng, X., Kao, J. P., & Kanold, P. O. (2014). Differential signaling to subplate neurons by spatially specific silent synapses in developing auditory cortex. *The Journal of Neuroscience, 34*(26), 8855–8864.
Meng, X., Kao, J. P., Lee, H. K., & Kanold, P. O. (2015). Visual deprivation causes refinement of intracortical circuits in the auditory cortex. *Cell Reports, 12*(6), 955–964. https://doi.org/10.1016/j.celrep.2015.07.018.
Meng, X., Kao, J. P., Lee, H. K., & Kanold, P. O. (2017). Intracortical circuits in thalamorecipient layers of auditory cortex refine after visual deprivation. *eNeuro, 4*(2). doi:https://doi.org/10.1523/ENEURO.0092-17.201.
Meng, X., Winkowski, D. E., Kao, J. P., & Kanold, P. O. (2017). Sublaminar subdivision of mouse auditory cortex layer 2/3 based on functional translaminar connections. *The Journal of Neuroscience, 37*(42), 10200–10214. https://doi.org/10.1523/JNEUROSCI.1361-17.2017.

Meredith, M. A., & Allman, B. L. (2015). Single-unit analysis of somatosensory processing in the core auditory cortex of hearing ferrets. *European Journal of Neuroscience, 41*(5), 686–698.

Meredith, M. A., & Lomber, S. G. (2017). Species-dependent role of crossmodal connectivity among the primary sensory cortices. *Hearing Research, 343*, 83–91.

Metherate, R., & Aramakis, V. B. (1999). Intrinsic electrophysiology of neurons in thalamorecipient layers of developing rat auditory cortex. *Developmental Brain Research, 115*(2), 131–144.

Meyer, G., Goffinet, A. M., & Fairen, A. (1999). What is a Cajal-Retzius cell? A reassessment of a classical cell type based on recent observations in the developing neocortex. *Cerebral Cortex, 99*(8), 765–775.

Meyer, G., Gonzalez-Hernandez, T. H., & Ferres-Torres, R. (1989). The spiny stellate neurons in layer IV of the human auditory cortex. A Golgi study. *Neuroscience, 33*(3), 489–498.

Mitani, A., Shimokouchi, M., Itoh, K., Nomura, S., Kudo, M., & Mizuno, N. (1985). Morphology and laminar organization of electrophysiologically identified neurons in the primary auditory cortex in the cat. *The Journal of Comparative Neurology, 235*, 430–447.

Moore, A. K., & Wehr, M. (2013). Parvalbumin-expressing inhibitory interneurons in auditory cortex are well-tuned for frequency. *The Journal of Neuroscience, 33*(34), 13713–13723.

Morel, A., Garraghty, P. E., & Kaas, J. H. (1993). Tonotopic organization, architectonic fields, and connections of auditory cortex in macaque monkeys. *The Journal of Comparative Neurology, 335*, 437–459.

Morrison, J. H., Magistretti, P. J., Benoit, R., & Bloom, F. E. (1984). The distribution and morphological characteristics of the intracortical VIP-positive cell: An immunohistochemical analysis. *Brain Research, 2992*(2), 269–282.

de la Mothe, L. A., Blumell, S., Kajikawa, Y., & Hackett, T. A. (2006a). Thalamic connections of the auditory cortex in marmoset monkeys: Core and medial belt regions. *The Journal of Comparative Neurology, 4996*(1), 72–96.

de la Mothe, L. A., Blumell, S., Kajikawa, Y., & Hackett, T. A. (2006b). Cortical connections of the auditory cortex in marmoset monkeys: Core and medial belt regions. *The Journal of Comparative Neurology, 4996*(1), 27–71.

Mountcastle, V. B. (1997). The columnar organization of the neocortex. *Brain, 120*(4), 701–722.

Murray, M. M., Molholm, S., Michel, C. M., Heslenfeld, D. J., Ritter, W., Javitt, D. C., et al. (2005). Grabbing your ear: Rapid auditory-somatosensory multisensory interactions in low-level sensory cortices are not constrained by stimulus alignment. *Cerebral Cortex, 15*(7), 963–974.

Nelson, A., Schneider, D. M., Takatoh, J., Sakurai, K., Wang, F., & Mooney, R. (2013). A circuit for motor cortical modulation of auditory cortical activity. *The Journal of Neuroscience, 33*(36), 14342–14353.

Nieuwenhuys, R. (1994). The neocortex. An overview of its evolutionary development, structural organization and synaptology. *Anatomy and Embryology (Berlin), 1990*(4), 307–337.

Noesselt, T., Tyll, S., Boehler, C. N., Budinger, E., Heinze, H. J., & Driver, J. (2010). Sound-induced enhancement of low-intensity vision: Multisensory influences on human sensory-specific cortices and thalamic bodies relate to perceptual enhancement of visual detection sensitivity. *The Journal of Neuroscience, 30*(41), 13609–13623.

Norena, A. J., Gourevitch, B., Pienkowski, M., Shaw, G., & Eggermont, J. J. (2008). Increasing spectrotemporal sound density reveals an octave-based organization in cat primary auditory cortex. *The Journal of Neuroscience, 28*(36), 8885–8896.

Ohl, F. W., & Scheich, H. (1997). Learning-induced dynamic receptive field changes in primary auditory cortex of the unanaesthetized Mongolian gerbil. *Journal of Comparative Physiology A: Neuroethology, Sensory, Neural, and Behavioral Physiology, 181*(6), 685–696.

Ojima, H. (1994). Terminal morphology and distribution of corticothalamic fibers originating from layers 5 and 6 of cat primary auditory cortex. *Cerebral Cortex, 4*(6), 646–663.

Ojima, H., Honda, C. N., & Jones, E. G. (1991). Patterns of axon collateralization of identified supragranular pyramidal neurons in the cat auditory cortex. *Cerebral Cortex, 1*, 80–94.

Ojima, H., Honda, C. N., & Jones, E. G. (1992). Characteristics of intracellularly injected infragranular pyramidal neurons in cat primary auditory cortex. *Cerebral Cortex, 2*(3), 197–216.

Ojima, H., & Murakami, K. (2002). Intracellular characterization of suppressive responses in supragranular pyramidal neurons of cat primary auditory cortex in vivo. *Cerebral Cortex, 12*(10), 1079–1091.

Oswald, A. M., & Reyes, A. D. (2008). Maturation of intrinsic and synaptic properties of layer 2/3 pyramidal neurons in mouse auditory cortex. *Journal of Neurophysiology, 99*(6), 2998–3008.

Otazu, G. H., Tai, L. H., Yang, Y., & Zador, A. M. (2009). Engaging in an auditory task suppresses responses in auditory cortex. *Nature Neuroscience, 12*(5), 646–654.

Oviedo, H. V., Bureau, I., Svoboda, K., & Zador, A. M. (2010). The functional asymmetry of auditory cortex is reflected in the organization of local cortical circuits. *Nature Neuroscience, 13*(11), 1413–1420.

Peters, A., & Saint Marie, R. L. (1984). Smooth and sparsely spinous nonpyramidal cells forming local axonal plexuses. In A. Peters & E. G. Jones (Eds.), *Cerebral cortex: Vol. 1. Cellular components of the cerebral cortex* (pp. 419–445). New York: Plenum Press.

Petrus, E., Isaiah, A., Jones, A. P., Li, D., Wang, H., Lee, H. K., & Kanold, P. O. (2014). Crossmodal induction of thalamocortical potentiation leads to enhanced information processing in the auditory cortex. *Neuron, 81*(3), 664–673.

Petrus, E., Rodriguez, G., Patterson, R., Connor, B., Kanold, P. O., & Lee, H. K. (2015). Vision loss shifts the balance of feedforward and intracortical circuits in opposite directions in mouse primary auditory and visual cortices. *The Journal of Neuroscience, 35*(23), 8790–8801.

Postma, F., Liu, C. H., Dietsche, C., Khan, M., Lee, H. K., Paul, D., & Kanold, P. O. (2011). Electrical synapses formed by connexin36 regulate inhibition- and experience-dependent plasticity. *Proceedings of the National Academy of Sciences of the United States of America, 108*(33), 13770–13775.

Prieto, J. J., Peterson, B. A., & Winer, J. A. (1994a). Laminar distribution and neuronal targets of GABAergic axon terminals in cat primary auditory cortex (AI). *The Journal of Comparative Neurology, 344*, 383–402.

Prieto, J. J., Peterson, B. A., & Winer, J. A. (1994b). Morphology and spatial distribution of GABAergic neurons in cat primary auditory cortex (AI). *The Journal of Comparative Neurology, 344*, 349–382.

Prieto, J. J., & Winer, J. A. (1999). Layer VI in cat primary auditory cortex: Golgi study and sublaminar origins of projection neurons. *The Journal of Comparative Neurology, 404*(3), 332–358.

Read, H. L., Winer, J. A., & Schreiner, C. E. (2001). Modular organization of intrinsic connections associated with spectral tuning in cat auditory cortex. *Proceedings of the National Academy of Sciences of the United States of America, 998*(14), 8042–8047.

Reep, R. L. (2000). Cortical layer VII and persistent subplate cells in mammalian brains. *Brain, Behavior and Evolution, 56*(4), 212–234.

Rockel, A. J., Hiorns, R. W., & Powell, T. P. (1980). The basic uniformity in structure of the neocortex. *Brain, 103*(2), 221–244.

Rockland, K. S., & Ichinohe, N. (2004). Some thoughts on cortical minicolumns. *Experimental Brain Research, 158*(3), 265–277.

Rothschild, G., Nelken, I., & Mizrahi, A. (2010). Functional organization and population dynamics in the mouse primary auditory cortex. *Nature Neuroscience, 13*(3), 353–360.

Rouiller, E. M., Simm, G. M., Villa, A. E. P., de Ribaupierre, Y., & de Ribaupierre, F. (1991). Auditory corticocortical interconnections in the cat: Evidence for parallel and hierarchical arrangement of the auditory cortical areas. *Experimental Brain Research, 86*, 483–503.

Rouiller, E. M., & Welker, E. (2000). A comparative analysis of the morphology of corticothalamic projections in mammals. *Brain Research Bulletin, 53*(6), 727–741.

Rudy, B., Fishell, G., Lee, S., & Hjerling-Leffler, J. (2011). Three groups of interneurons account for nearly 100% of neocortical GABAergic neurons. *Developmental Neurobiology, 71*(1), 45–61.

Saldeitis, K., Happel, M. F., Ohl, F. W., Scheich, H., & Budinger, E. (2014). Anatomy of the auditory thalamocortical system in the Mongolian gerbil: Nuclear origins and cortical field-, layer-, and frequency-specificities. *The Journal of Comparative Neurology, 522*(10), 2397–2430.

Sanides, F., & Sanides, D. (1972). The "extraverted neurons" of the mammalian cerebral cortex. *Zeitschrift für Anatatomie und Entwicklungsgeschichte, 136*(3), 272–293.

Scheich, H., Brechmann, A., Brosch, M., Budinger, E., & Ohl, F. W. (2007). The cognitive auditory cortex: Task-specificity of stimulus representations. *Hearing Research, 2299*(1-2), 213–224.

Shepherd, G. M., Pologruto, T. A., & Svoboda, K. (2003). Circuit analysis of experience-dependent plasticity in the developing rat barrel cortex. *Neuron, 38*(2), 277–289.

Smiley, J. F., & Falchier, A. (2009). Multisensory connections of monkey auditory cerebral cortex. *Hearing Research, 258*(1-2), 37–46.

Smith, P. H., Manning, K. A., & Uhlrich, D. J. (2010). Evaluation of inputs to rat primary auditory cortex from the suprageniculate nucleus and extrastriate visual cortex. *The Journal of Comparative Neurology, 518*(18), 3679–3700.

Smith, P. H., & Populin, L. C. (2001). Fundamental differences between the thalamocortical recipient layers of the cat auditory and visual cortices. *The Journal of Comparative Neurology, 436*(4), 508–519.

Smith, P. H., Uhlrich, D. J., Manning, K. A., & Banks, M. I. (2012). Thalamocortical projections to rat auditory cortex from the ventral and dorsal divisions of the medial geniculate nucleus. *The Journal of Comparative Neurology, 520*(1), 34–51.

Somogyi, P., & Cowey, A. (1984). Double bouquet cells. In A. Peters & E. G. Jones (Eds.), *Cerebral cortex: Vol. 1. Cellular components of the cerebral cortex* (pp. 337–360). New York: Plenum Press.

Staiger, J. F., Flagmeyer, I., Schubert, D., Zilles, K., Kötter, R., & Luhmann, H. J. (2004). Functional diversity of layer IV spiny neurons in rat somatosensory cortex: Quantitative morphology of electrophysiologically characterized and biocytin labeled cells. *Cerebral Cortex, 14*(6), 690–701.

Steger, R. M., Ramos, R. L., Cao, R., Yang, Q., Chen, C. C., Dominici, J., & Brumberg, J. C. (2013). Physiology and morphology of inverted pyramidal neurons in the rodent neocortex. *Neuroscience, 248*, 165–179.

Stiebler, I., Neulist, R., Fichtel, I., & Ehret, G. (1997). The auditory cortex of the house mouse: Left-right differences, tonotopic organization and quantitative analysis of frequency representation. *Journal of Comparative Physiology A: Neuroethology, Sensory, Neural, and Behavioral Physiology, 181*(6), 559–571.

Sun, Y. J., Kim, Y. J., Ibrahim, L. A., Tao, H. W., & Zhang, L. I. (2013). Synaptic mechanisms underlying functional dichotomy between intrinsic-bursting and regular-spiking neurons in auditory cortical layer 5. *The Journal of Neuroscience, 33*(12), 5326–5339.

Sutter, M. L., & Schreiner, C. E. (1991). Physiology and topography of neurons with multipeaked tuning curves in cat primary auditory cortex. *Journal of Neurophysiology, 65*(5), 1207–1226.

Taaseh, N., Yaron, A., & Nelken, I. (2011). Stimulus-specific adaptation and deviance detection in the rat auditory cortex. *PLoS One, 6*(8), e23369.

Tan, A. Y., Atencio, C. A., Polley, D. B., Merzenich, M. M., & Schreiner, C. E. (2007). Unbalanced synaptic inhibition can create intensity-tuned auditory cortex neurons. *Neuroscience, 146*(1), 449–462.

Tervo, D. G., Hwang, B. Y., Viswanathan, S., Gaj, T., Lavzin, M., Ritola, K. D., et al. (2016). A designer AAV variant permits efficient retrograde access to projection neurons. *Neuron, 992*(2), 372–382.

Thomson, A. M., & Bannister, A. P. (1998). Postsynaptic pyramidal target selection by descending layer III pyramidal axons: Dual intracellular recordings and biocytin filling in slices of rat neocortex. *Neuroscience, 84*(3), 669–683.

Thomson, A. M., & Lamy, C. (2007). Functional maps of neocortical local circuitry. *Frontiers in Neuroscience, 1*(1), 19–42. https://doi.org/10.3389/neuro.01.1.1.002.2007.

Tolner, E. A., Sheikh, A., Yukin, A. Y., Kaila, K., & Kanold, P. O. (2012). Subplate neurons promote spindle bursts and thalamocortical patterning in the neonatal rat somatosensory cortex. *The Journal of Neuroscience, 32*(2), 692–702.

Tremblay, R., Lee, S., & Rudy, B. (2016). GABAergic interneurons in the neocortex: From cellular properties to circuits. *Neuron, 991*(2), 260–292.

Verbny, Y. I., Erdelyi, F., Szabo, G., & Banks, M. I. (2006). Properties of a population of GABAergic cells in murine auditory cortex weakly excited by thalamic stimulation. *Journal of Neurophysiology, 996*(6), 3194–3208.
Viswanathan, S., Bandyopadhyay, S., Kao, J. P., & Kanold, P. O. (2012). Changing microcircuits in the subplate of the developing cortex. *The Journal of Neuroscience, 32*(5), 1589–1601.
Viswanathan, S., Sheikh, A., Looger, L. L., & Kanold, P. O. (2017). Molecularly defined subplate neurons project both to thalamocortical recipient layers and thalamus. *Cerebral Cortex, 27*(10), 4759–4768.
Wallace, M. N., & Bajwa, S. (1991). Patchy intrinsic connections of the ferret primary auditory cortex. *Neuroreport, 2*(8), 417–420.
Wallace, M. N., & Harper, M. S. (1997). Callosal connections of the ferret primary auditory cortex. *Experimental Brain Research, 116*(2), 367–374.
Wallace, M. N., & He, J. (2011). Intrinsic connections of the auditory cortex. In J. A. Winer & C. E. Schreiner (Eds.), *The auditory cortex* (pp. 133–145). New York: Springer.
Wallace, M. N., Kitzes, L. M., & Jones, E. G. (1991). Intrinsic inter- and intra-laminar connections and their relationship to the tonotopic map in cat primary auditory cortex. *Experimental Brain Research, 86*, 527–544.
Wallace, M. T., Ramachandran, R., & Stein, B. E. (2004). A revised view of sensory cortical parcellation. *Proceedings of the National Academy of Sciences of the United States of America, 101*(7), 2167–2172.
Wang, Y., Gupta, A., Toledo-Rodriguez, M., Wu, C. Z., & Markram, H. (2002). Anatomical, physiological, molecular and circuit properties of nest basket cells in the developing somatosensory cortex. *Cerebral Cortex, 12*(4), 395–410.
Wang, Y., Toledo-Rodriguez, M., Gupta, A., Wu, C., Silberberg, G., Luo, J., & Markram, H. (2004). Anatomical, physiological and molecular properties of Martinotti cells in the somatosensory cortex of the juvenile rat. *Journal of Physiology, 561*(1), 65–90.
Watkins, P. V., Kao, J. P., & Kanold, P. O. (2014). Spatial pattern of intra-laminar connectivity in supragranular mouse auditory cortex. *Frontiers in Neural Circuits, 8*, 15. https://doi.org/10.3389/fncir.2014.00015.
Wehr, M., & Zador, A. M. (2003). Balanced inhibition underlies tuning and sharpens spike timing in auditory cortex. *Nature, 426*(6965), 442–446.
Wess, J. M., Isaiah, A., Watkins, P. V., & Kanold, P. O. (2017). Subplate neurons are the first cortical neurons to respond to sensory stimuli. *Proceedings of the National Academy of Sciences of the United States of America, 114*(47), 12602–12607. https://doi.org/10.1073/pnas.1710793114. Epub 2017 Nov 7.
Winer, J. A. (1984a). The pyramidal neurons in layer III of cat primary auditory cortex (AI). *The Journal of Comparative Neurology, 2299*, 476–496.
Winer, J. A. (1984b). The non-pyramidal cells in layer III of cat primary auditory cortex (AI). *The Journal of Comparative Neurology, 2299*, 512–530.
Winer, J. A. (1984c). Anatomy of layer IV in cat primary auditory cortex (AI). *The Journal of Comparative Neurology, 224*, 535–567.
Winer, J. A. (1985). Structure of layer II in cat primary auditory cortex (AI). *The Journal of Comparative Neurology, 238*, 10–37.
Winer, J. A. (1992). The functional architecture of the medial geniculate body and the primary auditory cortex. In D. B. Webster, A. N. Popper, & R. R. Fay (Eds.), *The mammalian auditory pathway: Neuroanatomy* (pp. 222–409). New York: Springer.
Winer, J. A. (2006). Decoding the auditory corticofugal systems. *Hearing Research, 212*(1-2), 1–8.
Winer, J. A. (2011a). A profile of auditory forebrain connections and circuits. In J. A. Winer & C. E. Schreiner (Eds.), *The auditory cortex* (pp. 41–74). New York: Springer.
Winer, J. A. (2011b). Neurochemical organization of the medial geniculate body and auditory cortex. In J. A. Winer & C. E. Schreiner (Eds.), *The auditory cortex* (pp. 209–234). New York: Springer.
Winer, J. A., & Larue, D. T. (1989). Populations of GABAergic neurons and axons in layer I of rat auditory cortex. *Neuroscience, 33*(3), 499–515.

Winer, J. A., & Prieto, J. J. (2001). Layer V in cat primary auditory cortex (AI): Cellular architecture and identification of projection neurons. *The Journal of Comparative Neurology, 434*(4), 379–412.

Winguth, S. D., & Winer, J. A. (1986). Corticocortical connections of cat primary auditory cortex (AI): Laminar organization and identification of supragranular neuron projecting to area AII. *The Journal of Comparative Neurology, 248*, 36–56.

Winkowski, D. E., & Kanold, P. O. (2013). Laminar transformation of frequency organization in auditory cortex. *The Journal of Neuroscience, 33*(4), 1498–1508.

Woodruff, A. R., Anderson, S. A., & Yuste, R. (2010). The enigmatic function of chandelier cells. *Frontiers in Neuroscience, 4*, 201. https://doi.org/10.3389/fnins.2010.00201.

Wu, G. K., Arbuckle, R., Liu, B. H., Tao, H. W., & Zhang, L. I. (2008). Lateral sharpening of cortical frequency tuning by approximately balanced inhibition. *Neuron, 58*(1), 132–143.

Wu, G. K., Tao, H. W., & Zhang, L. I. (2011). From elementary synaptic circuits to information processing in primary auditory cortex. *Neuroscience and Biobehavioral Reviews, 35*(10), 2094–2104.

Zhang, L. I., Tan, A. Y., Schreiner, C. E., & Merzenich, M. M. (2003). Topography and synaptic shaping of direction selectivity in primary auditory cortex. *Nature, 424*(6945), 201–205.

Zhao, C., Kao, J. P., & Kanold, P. O. (2009). Functional excitatory microcircuits in neonatal cortex connect thalamus and layer 4. *The Journal of Neuroscience, 299*(49), 15479–15488.

Zhou, Y., Liu, B. H., Wu, G. K., Kim, Y. J., Xiao, Z., Tao, H. W., & Zhang, L. I. (2010). Preceding inhibition silences layer 6 neurons in auditory cortex. *Neuron, 65*(5), 706–717.

Chapter 9
Circuits for Modulation of Auditory Function

Brett R. Schofield and Laura Hurley

Abstract This chapter discusses anatomical, physiological, and functional aspects of circuits associated with four major neuromodulators: acetylcholine, serotonin, noradrenaline, and dopamine. These neuromodulators occur in nearly all auditory structures from the cochlea of the inner ear to the cortex of the brain. A review of the anatomy is focused on the origins of modulatory inputs to auditory structures and the patterns of termination in those areas. Sources of the modulatory inputs include widely recognized cell groups in the basal forebrain and pontomesencephalic tegmentum (for acetylcholine), raphe nuclei (for serotonin), locus coeruleus (for noradrenaline), and ventral tegmental area (for dopamine), as well as smaller cell groups in the brainstem. In addition, there are numerous examples of cells within the auditory system that release one or more of these neuromodulators. Physiology and function are discussed from several perspectives, starting with a brief overview of methods used for assessing modulatory function. Neuromodulators are directly involved in regulating auditory processing according to both internal state and stimulus salience. Many mechanisms are likely involved. Neuromodulators can reconfigure auditory circuitry through multiple receptor types and in multiple auditory regions. Furthermore, multiple neuromodulators may converge at the level of single neuron types. This makes the effects of neuromodulators complex but confers the ability to produce a range of behaviorally appropriate outputs from auditory circuitry. In addition, neuromodulators facilitate long-term plasticity. Such plasticity plays a role in many adaptive responses, including numerous changes that may play a role in the auditory dysfunction that follows hearing loss.

Keywords Acetylcholine · Behavioral context · Basal forebrain · Dopamine · Internal state · Laterodorsal tegmental nucleus · Locus coeruleus · Norepinephrine ·

B. R. Schofield (✉)
Department of Anatomy and Neurobiology, Northeast Ohio Medical University, Rootstown, OH, USA
e-mail: bschofie@neomed.edu

L. Hurley
Department of Biology, Indiana University, Bloomington, IN, USA
e-mail: lhurley@indiana.edu

© Springer International Publishing AG 2018

D. L. Oliver et al. (eds.), *The Mammalian Auditory Pathways*,
Springer Handbook of Auditory Research 65,
https://doi.org/10.1007/978-3-319-71798-2_9

Nucleus basalis · Pedunculopontine tegmental nucleus · Plasticity · Raphe nuclei · Salience · Serotonin · Ventral tegmental area

9.1 Introduction

Neuromodulation of hearing refers to mechanisms that alter the way sounds are processed in auditory circuits. As a consequence, neuromodulators play important and varied roles in hearing, including mediating the effects of behavioral state and external events on auditory perception. Neuromodulation is also important for plasticity during development and learning as well as in response to damage or dysfunction of the nervous system.

There are many ways to define "neuromodulator" (see discussion by Descarries and Mechawar 2008) and recent interpretations include a growing list of interneuronal signaling molecules. Neuromodulators that act on auditory circuits include monoamines and acetylcholine as well as various peptides, including opioids, substance P, vasoactive intestinal polypeptide, and cholecystokinin, and gases such as nitric oxide. Even glutamate and GABA (gamma-aminobutyric acid) can be seen as neuromodulators in some circuits (e.g., Lee and Sherman 2010).

The following discussion focuses on four "conventional" neuromodulators: acetylcholine (ACh), serotonin (5-hydroxytryptamine or 5-HT), noradrenaline (NA or norepinephrine), and dopamine (DA)(all abbreviations in Table 9.1). Numerous neuroscience texts can provide relevant background. For example, Cooper et al. (2003) provide an overview of basic neuropharmacology (including synthesis, receptors, etc.) for each of the neuromodulators. The present and previous volumes in the Springer Handbook of Auditory Research series provide background on auditory circuits (e.g., Webster et al. 1992; Ryugo et al. 2011). In addition, Paxinos (1995) provides a useful introduction with detailed descriptions of many of the nuclei not usually considered in discussions of the traditional auditory pathways.

9.2 Anatomy of Modulatory Circuits

9.2.1 Common Properties of Modulatory Nuclei

Each of the modulatory systems discussed here is associated with one or more nuclei in the brainstem or basal forebrain. These nuclei are often diffusely organized with poorly defined boundaries that can vary between species. Moreover, while a nucleus can be associated with a particular modulator, that nucleus invariably contains neurons with a variety of neurotransmitter phenotypes. For example, the "cholinergic nuclei" of the basal forebrain contain cholinergic as well as glutamatergic and GABAergic cells. In some cases, the different phenotypes have different projections, but multiple phenotypes also can project to a single target. The fact that the different phenotypes are typically intermingled complicates both anatomical and physiological

Table 9.1 Abbreviations

5-HT	5-Hydroxytryptamine or serotonin
5-HT1B	Serotonin receptor type 1B
5-HT2A	Serotonin receptor type 2A
ACh	Acetylcholine
CA	Cerebral aqueduct
CG	Central gray of the midbrain
DA	Dopamine
DRN	Dorsal raphe nucleus
GABA	Gamma-aminobutyric acid
LC	Locus coeruleus
LDT	Laterodorsal tegmental nucleus
LPGi	Lateral paragigantocellular nucleus
MGm	Medial geniculate body, medial subdivision
MGv	Medial geniculate body, ventral subdivision
NA	Noradrenaline
NB	Nucleus basalis
PMT	Pontomesencephalic tegmentum
PPT	Pedunculopontine tegmental nucleus
VTA	Ventral tegmental area

studies because extra steps must be taken to attribute specific features (e.g., response properties, axonal projection patterns) to cells that use a specific neurotransmitter.

Another feature common among the modulatory nuclei is for a relatively small number of cells, from a few thousand up to tens of thousands, to innervate very wide expanses of the nervous system. Those numbers appear to apply across mammals; generally, the data are most complete for rats (Descarries and Mechawar 2008). Branching axonal projections play a role in some pathways, but the extent of branching varies between systems. Finally, there is evidence that some cells within the traditional auditory pathways make and release various modulators (although they are not always portrayed from that perspective). The best known example is the release of acetylcholine by olivocochlear cells. Some of these cells also have projections to the cochlear nucleus. Consequently, cholinergic effects in the cochlear nucleus must represent a combination of effects of cholinergic projections from auditory nuclei (i.e., the superior olivary complex) and cholinergic projections from the nonauditory cholinergic nuclei in the brainstem. Similar examples can be cited for dopaminergic projections. This issue emphasizes the likelihood that a given modulator serves a wide variety of functions.

9.2.2 Acetylcholine

Cholinergic cells are located in numerous brain regions including both brainstem and forebrain areas (Woolf 1991). Cholinergic innervation of the auditory system originates in four regions. Two such regions are associated with widespread cholinergic innervation of the central nervous system: the basal forebrain (Sect. 9.2.2.1) and the pontomesencephalic tegmentum (PMT) (Sect. 9.2.2.2). The third region is within the superior olivary complex and is a source of cholinergic projections to lower auditory centers. The final region is a small nucleus of the reticular formation, the lateral paragigantocellular nucleus (LPGi; also known as the rostral ventrolateral medulla), that has close ties to auditory nuclei and other brainstem regions.

9.2.2.1 Cholinergic Groups in the Basal Forebrain

Cholinergic groups in the basal forebrain include the nucleus basalis, septal nuclei, and the vertical and horizontal limbs of the diagonal band. They have been associated with cognitive function and selective processing of sensory stimuli (Sarter and Bruno 1997). These cell groups provide the main sources of cholinergic input to the neocortex and may provide some cholinergic input to the thalamus (Descarries and Mechawar 2008; Varela 2014). The cholinergic innervation of the auditory cortex originates from this basal forebrain group, but the distribution of the cholinergic cells can vary across species. For example, in ferrets the majority of cells are located in nucleus basalis, but in cats the cells are distributed across the nucleus basalis and laterally into the putamen and globus pallidus (Kamke et al. 2005; Bajo et al. 2014).

Cholinergic axons terminate across auditory cortical areas and in all cortical layers, although the relative density varies with layer, cortical area, and species (Miller et al. 2013; Bajo et al. 2014). The available evidence suggests that cholinergic receptors are located on the cell bodies or dendrites of pyramidal and nonpyramidal cortical cells, on a variety of axon terminals within the cortex including ascending inputs from the thalamus, and on excitatory and inhibitory inputs from other cortical neurons (Metherate 2011; Edeline 2012).

9.2.2.2 Cholinergic Groups in the Pontomesencephalic Tegmentum

The PMT consists of the pedunculopontine tegmental nucleus (PPT) and the laterodorsal tegmental nucleus (LDT), which together are the main sources of cholinergic projections to the thalamus and the ventral tegmental area as well as numerous other brainstem regions. These widespread projections are associated with a variety of functions. As part of the ascending reticular activating system, the PMT has been associated with arousal (Woolf 1991; although see Fuller et al. 2011 for a discussion of arousal and glutamatergic versus cholinergic projections). Arousal and the related control of the sleep–wake cycle are often discussed in concert with PMT projections to the thalamus and the basal forebrain. The PMT is the primary source of cholinergic projections to the thalamus, although little attention has been focused on auditory nuclei (Steriade et al. 1988; Motts and Schofield 2010).

In general (across species and thalamic nuclei), the projections are bilateral with an ipsilateral predominance, and they originate from more cells in the PPT than in the LDT. Studies focused on the medial geniculate nucleus suggest possible species differences, with acetylcholine input arising only in the PPT in rats but in both PPT and LDT in guinea pigs (discussed in Motts and Schofield 2010). Whether the apparent difference is real or results from technical issues is unclear, but it highlights an unresolved issue about differences between the LDT and the PPT. It has been suggested on connectional grounds that the LDT is biased toward limbic circuits (Woolf 1991); the utility of this distinction remains to be determined.

The identity of thalamic cells targeted by cholinergic projections has been studied in more detail in nonauditory than in auditory nuclei. In general, cholinergic inputs target both thalamocortical cells and GABAergic interneurons as well as GABAergic neurons in the thalamic reticular nucleus. Depending on the thalamic nucleus, these inputs are positioned to modulate ascending sensory inputs to the thalamus or descending inputs from the cortex (e.g., Patel and Bickford 1997). Interestingly, there is also evidence for cholinergic inputs to the *axons* of medial geniculate thalamocortical cells (Kawai et al. 2007). These inputs activate nicotinic receptors at nodes on myelinated axons in the thalamic radiations, serving to increase the efficacy of transmission of sensory information to auditory cortex. This unusual mode of action represents a rarely recognized possibility for neuromodulation. It is likely that these cholinergic inputs originate from the PMT or the basal forebrain, but the sources have not been identified directly.

In addition to the projections to the thalamus, the PMT provides cholinergic input to the inferior colliculus and regions of the cochlear nucleus (reviewed in Schofield et al. 2011). The projections to each of these areas originate from more cells in the PPT than in the LDT, but each area receives projections from both cholinergic nuclei. The PMT is the predominant source of cholinergic projections to the inferior colliculus; however, recent studies have revealed a projection from the LPGi as well (Motts and Schofield 2009; Stornetta et al. 2013). Cholinergic fibers terminate throughout the inferior colliculus and a majority of collicular cells are affected by locally applied cholinergic agents. Thus far, GABAergic inferior collicular cells are the only ones identified as receiving direct cholinergic inputs (Yigit et al. 2003). The PMT projections to the cochlear nucleus are known to terminate in the dorsal cochlear nucleus (Mellott et al. 2011), but the specific cell types contacted and whether the PMT projects at all to the ventral cochlear nucleus are unknown.

The PMT is also a source of cholinergic projections to the caudal pontine reticular nucleus. While not a component of the ascending auditory system, the caudal pontine reticular nucleus is a critical premotor component of the startle circuit and is activated during the acoustic startle reflex. The cholinergic projection from the PMT to the caudal pontine reticular nucleus is critical for prepulse inhibition of acoustic startle (Bosch and Schmid 2008). Thus cholinergic projections from the PMT may allow for enhanced (or protected) sensory processing via projections to auditory nuclei while suppressing the motor component of a startle response via projections to premotor nuclei.

9.2.2.3 Cholinergic Cells in the Lateral Paragigantocellular Nucleus

The lateral paragigantocellular nucleus (LPGi) is a small nucleus of the reticular formation located just caudal to the facial nucleus and superior olivary complex and lateral to the pyramids; it is also called the rostral ventrolateral medulla (see discussion in Bellintani-Guardia et al. 1996). This nucleus has been associated with polymodal sensory integration, with the autonomic nervous system, and with control of cardiorespiratory function (Van Bockstaele et al. 1993). Stornetta et al. (2013) used chemically selective tracing techniques to demonstrate that the cholinergic cells in the LPGi project to numerous auditory nuclei but not to the autonomic and cardiorespiratory centers. These cholinergic projections terminate in the dorsal cochlear nucleus as well as parts of the superior olivary complex and inferior colliculus. The target cells in these areas are unknown. The LPGi receives input from several auditory nuclei (cochlear nucleus, inferior colliculus, auditory cortex), but the relationships of these inputs to the cholinergic cells are unknown.

9.2.2.4 Cholinergic Cells in the Superior Olivary Complex

The superior olivary complex is the origin of the most studied cholinergic projections in the auditory brainstem (Ryugo et al. 2011). Targets of these projections are primarily auditory structures (Brown 2011). The best known cholinergic projection from the superior olivary complex is the olivocochlear projection. The olivocochlear system consists of medial and lateral divisions that have different connections and different functions. Most of our knowledge about this system is associated with the medial olivocochlear system, which acts on outer hair cells to modulate cochlear function. Medial olivocochlear axons have collateral branches that terminate in the cochlear nucleus where the cholinergic inputs likely modulate the activity of stellate cells (Benson and Brown 1990; Oertel et al. 2011). Lateral olivocochlear cells terminate on primary afferent fibers associated with inner hair cells in the cochlea and presumably modulate input at the origin of the auditory pathway. These lateral olivocochlear cells may also have collateral projections to the cochlear nucleus but much less is known about them. The superior olivary complex also contains a group of cholinergic cells that innervate the cochlear nucleus, but they do not project to the cochlea (Sherriff and Henderson 1994). The targets of these nonolivocochlear projections appear to include the cochlear root neurons (Gómez-Nieto et al. 2008) and perhaps other parts of the ventral cochlear nucleus. The roles of these various inputs and the possibility that multiple inputs converge on the same cells in the cochlear nucleus have only begun to be explored.

9.2.3 Noradrenaline

Noradrenergic projections terminate in all auditory centers from the cochlear nucleus to the cortex. The details of termination patterns have been described for some areas, including auditory cortex (Levitt and Moore 1978; Campbell et al.

1987), cochlear nucleus and inferior colliculus (Klepper and Herbert 1991), and superior olivary complex (Mulders and Robertson 2001). In the auditory cortex, noradrenergic fibers terminate in all cortical layers with the densest termination in layer I (Levitt and Moore 1978; Campbell et al. 1987). There is a high degree of collateralization, suggesting that single fibers terminate on many target cells in multiple layers. Little is known about the specific cells targeted by noradrenergic inputs to auditory cortex.

Noradrenergic fibers terminate in multiple areas of the cochlear nucleus with only the granule cell area and the molecular layer of the dorsal cochlear nucleus singled out as receiving minimal noradrenergic innervation. Thus, there is ample opportunity for a majority of the cell types in the cochlear nucleus to receive noradrenergic input but, to date, only cochlear root neurons have been identified specifically as likely targets of noradrenergic axons (Gómez-Nieto et al. 2008). Noradrenaline also broadly innervates the superior olivary complex (Mulders and Robertson 2005a). The innervation density varies across nuclei and in some cases within nuclei, suggesting varying levels of noradrenergic effects on different olivary circuits. Thus far, noradrenergic inputs have been associated with olivocochlear cells (Mulders and Robertson 2000) and with olivary cells that project to the cochlear nucleus (Behrens et al. 2002). Finally, noradrenergic fibers terminate throughout the inferior colliculus, where the density of fibers varies both across and within subdivisions (Klepper and Herbert 1991). The same authors described noradrenergic fibers terminating throughout the nuclei of the lateral lemniscus, but they did not describe termination patterns in detail. In none of these areas (lemniscal nuclei or inferior colliculus) have the targets of the noradrenergic fibers been identified.

The major source of noradrenergic innervation is the locus coeruleus (Berridge and Waterhouse 2003). For some areas (e.g., auditory cortex), the locus coeruleus is the sole source of noradrenergic innervation, but other areas (e.g., cochlear nucleus) receive smaller contributions (depending on species) that originate in other nuclei of the reticular formation (Klepper and Herbert 1991).

9.2.4 Dopamine

Dopaminergic fibers or dopamine receptors have been described in the cochlea, cochlear nucleus, nuclei of the lateral lemniscus, inferior colliculus, and auditory cortex (Tong et al. 2005; Descarries and Mechawar 2008). Dopaminergic fibers are reportedly absent from the superior olivary complex (Mulders and Robertson 2005a).

Dopaminergic input to the cochlea is associated with lateral olivocochlear fibers that terminate on primary afferent fibers receiving input from inner hair cells (Mulders and Robertson 2004; Darrow et al. 2006). Studies suggest that the dopaminergic efferents inhibit responses in auditory nerve fibers and may provide some protection against acoustic trauma (Le Prell et al. 2005; Niu et al. 2007).

The sources of dopaminergic innervation for the rest of the auditory system are less clear. Dopaminergic projections to much of the central nervous system originate in the ventral midbrain, including the substantia nigra and ventral tegmental area along with several adjacent areas (Yetnikoff et al. 2014). It is likely that these nuclei innervate auditory cortex and perhaps some subcortical auditory regions. In addition, there is evidence for dopaminergic cells within several auditory regions, including the inferior colliculus, nuclei of the lateral lemniscus, and as described previously, the superior olivary complex (Altschuler and Shore 2010). Other than the olivocochlear projections, the projections of dopaminergic cells located within auditory nuclei are unknown.

9.2.5 Serotonin

Serotonin neurons are located in a series of raphe nuclei that are distributed on or near the midline from the medulla to the midbrain (Descarries and Mechawar 2008). Nine nuclei are usually distinguished. Together, these nuclei project throughout much of the central nervous system, from spinal cord to neocortex. The nuclei have been divided into superior and inferior groups (Jacobs and Azmitia 1992). The *superior group* consists of the dorsal raphe, median raphe, and caudal linear nuclei, as well as group B9. The *inferior group* consists of nuclei raphe obscurus, raphe pallidus, and raphe magnus, as well as the LPGi and the area postrema. There is a rough topography such that the inferior group nuclei project to the medulla and spinal cord, whereas the superior nuclei project to the forebrain. More refined distinctions that could relate to functional differences might apply to projections from different cell groups within individual nuclei (Commons 2015).

As a group, the serotonergic nuclei appear to project to all auditory nuclei. Details of the origins of projections to the auditory cortex and auditory thalamus are limited; most of the information is available within broader studies not focused on the auditory system (Descarries and Mechawar 2008). The data suggest that the dorsal and median raphe nuclei, which are two of the largest serotonergic nuclei, provide the main innervation of auditory forebrain.

Origins of serotonin innervation of brainstem auditory nuclei have been studied in more detail. Serotonergic fibers terminate throughout the cochlear nucleus with the densest terminations in the molecular layer of the dorsal cochlear nucleus and the granule cell area. The inferior colliculus also receives serotonin inputs that terminate across all subdivisions but terminate most heavily in the dorsal and external cortex. The cochlear nucleus and the inferior colliculus receive predominant input from the dorsal raphe with small contributions from other raphe nuclei (Klepper and Herbert 1991). The smaller contributions originate mostly from the superior group but include some contributions from inferior group nuclei. One such nucleus is the LPGi, described previously for its contingent of cholinergic cells. Serotonergic cells in the LPGi project to the cochlear nuclei or the inferior colliculus and appear to receive direct inputs from the cochlear nucleus (Bellintani-Guardia et al. 1996). These inputs arise in part from cochlear root neurons, which could provide for rapid activation of the serotonergic cells by acoustic stimuli.

Serotonin fibers also innervate the nuclei of the lateral lemniscus and superior olivary complex (Klepper and Herbert 1991; Thompson and Hurley 2004). The sources of this innervation are assumed to be among the raphe nuclei, but they have yet to be identified directly. The terminations vary in density between different nuclei, and the patterns may also differ between species (Woods and Azeredo 1999; Hurley and Thompson 2001). The target cells in lemniscal nuclei are unknown. Serotonin-targeted cells in the superior olivary complex are likely to include cells that project to the cochlea or to the cochlear nucleus (Brown 2011).

9.2.6 Some Remaining Issues Regarding Modulatory Anatomy

Many questions remain to be addressed about the anatomical organization of modulatory inputs to auditory circuits. The previous discussions included relatively little detail on modulatory circuits in the auditory cortex. To a degree this reflects a common perspective that many circuits and actions are similar across cortical areas and thalamic nuclei (acknowledging, for example, a distinction between first- and higher-order nuclei in the thalamus) (Sherman and Guillery 2011; Varela 2014). Of the neuromodulators discussed in the present chapter, acetylcholine has been studied most extensively in auditory thalamus and cortex. Many additional insights for all four modulators might be gained by considering work in other systems. Studies in multiple cortical areas have emphasized modulatory effects on different subclasses of GABAergic interneurons (e.g., reviewed by Bacci et al. 2005).

Recent work supports the distinction of interneuron types in auditory cortex and suggests that the different types have distinct physiological characteristics (Li et al. 2015; Mesik et al. 2015). In several cortical areas a given modulator, such as noradrenaline, can excite or inhibit different types of interneurons. Because different interneuron types have different projection patterns within the cortex, a simple (i.e., relatively nonspecific) modulatory input can have dramatic effects on information flow within the cortex. Such effects are proposed to switch cortical processing between an intracolumnar versus a horizontal (i.e., transcolumnar) mode (Bacci et al. 2005). If such a process occurs in auditory cortex, one could predict modulation that, for example, could promote cross-frequency (horizontal) integration versus columnar processing that might promote frequency discrimination. An interesting possibility is that the different modulators take advantage of the same GABAergic circuitry to dynamically shift cortical processing strategies. Differences between the modulators, then, would depend primarily upon the different circumstances under which each modulatory system is active.

Highly collateralized projections have long been associated with modulatory systems whereby individual axons branch many times to innervate many different areas. To some extent such collateralization is implied by the widespread innervation of the central nervous system by a relatively small number of neurons. Numerous studies have identified widespread collaterals in serotonergic and noradrenergic systems (see discussions in Berridge and Waterhouse 2003; Descarries and

Mechawar 2008). The extent of collateral projections to auditory nuclei has not been studied extensively, but collateral projections are common in cholinergic innervation of the auditory brainstem (Schofield et al. 2011). However, there is evidence that broad collateralization is not universal among modulatory projections, such as in the quite limited collateral branching among dopaminergic projections (Descarries and Mechawar 2008). This means that the dopaminergic projections to different targets originate from separate groups of cells (that may or may not be intermingled). The key point is that branched axons can allow for broad actions but with limited opportunity for differential effects on targets, whereas innervation by separate sets of source neurons can facilitate distinct effects in different targets. Many questions remain about the degree of branching of modulatory projections to auditory targets.

Another issue related to breadth versus specificity of action is that of synaptic release versus volume transmission (discussed by Descarries and Mechawar 2008). Volume transmission implies slower onset and longer duration of action compared to synaptic transmission and often has been associated with modulatory circuits. The frequency with which axonal swellings form traditional synapses varies according to target area, modulator, and perhaps species. Furthermore, the assumption of volume transmission has often followed from an inability to identify synapses that include a traditional synaptic junction (with visible postsynaptic density). Recent work suggests that cholinergic synapses may include those with traditional densities as well as some without such densities (e.g., Takács et al. 2013). These nontraditional synapses can be associated with typical clusters of postsynaptic receptors and otherwise allow all the specificity associated with traditional synapses. Indeed, physiological studies argue that ACh can exert highly specific effects in neocortex (Muñoz and Rudy 2014).

9.3 Physiology and Function of Modulatory Circuits

9.3.1 Neuromodulatory Anatomy Provides a Blueprint for Function

The anatomical pathways connecting modulatory nuclei with their inputs provide a blueprint for understanding their function. Most centralized neuromodulatory systems receive projections from an impressive variety of brain regions. These range from primary sensory areas to more integrative neural centers that respond to sensory information as it is filtered by factors such as motivational state or top-down cognitive processing (e.g., Sarter et al. 2005; Yetnikoff et al. 2014). These inputs converge directly on neuromodulatory neurons or onto local interneurons, providing a substrate for the multifactorial control of spiking activity (Challis et al. 2013; Yetnikoff et al. 2014). These patterns of anatomical connection give rise to a model in which neuromodulatory centers receive information from many sources, sort and

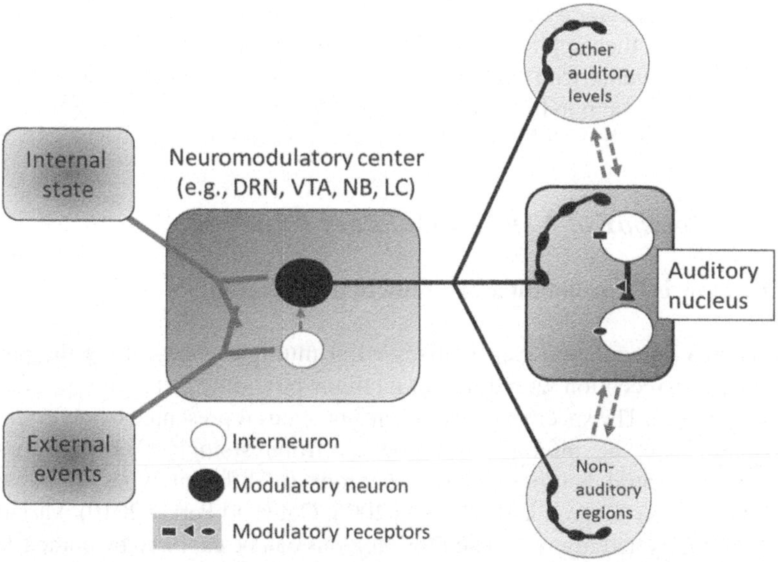

Fig. 9.1 Depiction of a model of neuromodulatory function, emphasizing the integration of information from diverse sources by neuromodulatory centers and its subsequent projection to multiple auditory and other brain regions. *DRN*, dorsal raphe nucleus; *LC*, locus coeruleus; *NB*, nucleus basalis; *VTA*, ventral tegmental area [Taken from Velho et al. (2012) with permission]

prioritize it, and then send it to multiple auditory destinations in the form of specific neurochemicals like dopamine and noradrenaline (both catecholamines), acetylcholine, and serotonin (Fig. 9.1). As a result of this process, neuromodulatory neurons are in a prime position to signal salient aspects of behavioral context to the auditory system.

The conveyance of salient information by neuromodulatory pathways is only half of the equation; auditory neurons must also interpret this information. This is accomplished through the expression of neuromodulatory receptors by auditory neurons themselves. This aspect of neuromodulatory function provides expansive opportunities for the regulation of excitatory and inhibitory circuitry in the auditory system through a diversity of receptor types (Edeline 2012; Hurley and Sullivan 2012). Receptor diversity allows even single neuromodulators to create sophisticated profiles of effects on auditory circuitry. In the partnership between neuromodulatory release and reception, local events at the level of auditory neurons can translate even broad-scale release into highly specific effects on auditory circuitry.

Although neuromodulatory systems clearly have a profound ability to organize auditory activity on both short- and long-term time scales, an integrated view of their function is very much a work in progress. Therefore, the following sections are organized into two major conceptual divisions. The first of these describes a functional "toolbox" highlighting some prominent features of neuromodulatory function in the auditory system. In some of these sections, work in the auditory systems of songbirds provided useful comparative models that emphasize neuromodulation as it relates to the behavioral salience of natural vocal signals. The concepts developed in this section are then applied to speculation about a broad functional role for neuromodulatory systems with high relevance to auditory research: responses of the central auditory system to hearing loss.

9.3.2 A Toolbox of Neuromodulatory Function

9.3.2.1 How is Neuromodulatory Function Measured?

Auditory neuromodulation can be assessed at multiple points along the pathway from release to reception with different methods providing different types of information (Fig. 9.2). The specific sorts of events or states represented by neuromodulatory systems can be inferred from the electrophysiological or transcriptional responses of central neuromodulatory neurons to different types of behavioral events (Fig. 9.2A) (Bharati and Goodson 2006; Gale and Perkel 2010). Variation in neuromodulatory activity within auditory regions can be captured by comparing the levels of neuromodulators and their products in dissected tissues as indicative of *turnover* (a measure of release and subsequent metabolism) (Cransac et al. 1998). Time courses of neuromodulatory activity can be tracked with repeated measurements in behaving animals with microdialysis followed by neurochemical analysis or by electrochemically forcing neuromodulatory oxidation and measuring the corresponding currents with carbon fiber voltammetry (Fig. 9.2B) (Stark and Scheich 1997; Hall et al. 2010).

On the postsynaptic side, responses of auditory neurons to neuromodulators are often presented as changes in spontaneous or evoked spike rate or timing during application of neuromodulatory agonists or antagonists or during stimulation of neuromodulatory centers (Fig. 9.2C) (Edeline et al. 2011; Salgado et al. 2011). Finally, plasticity in modulatory input to auditory regions can be represented by increases or decreases in the density of projections immunolabeled for neuromodulatory synthetic enzymes or selective transporters (Matragrano et al. 2012a; Papesh and Hurley 2012). These different types of measurements contribute to a portrait of relevant neuromodulatory events occurring on multiple timescales with short-term changes in response to behaviorally salient events superimposed on longer state-dependent or experience-dependent fluctuations.

9.3.2.2 Neuromodulators are Sensitive to Behavioral Context: Internal State and Salient Events

Modeling the neuromodulatory regulation of auditory circuitry requires an understanding of the behavioral conditions evoking neuromodulatory release. Neuromodulatory pathways operating within the auditory system are responsive to many of the factors that define behavioral context, including the nature of external

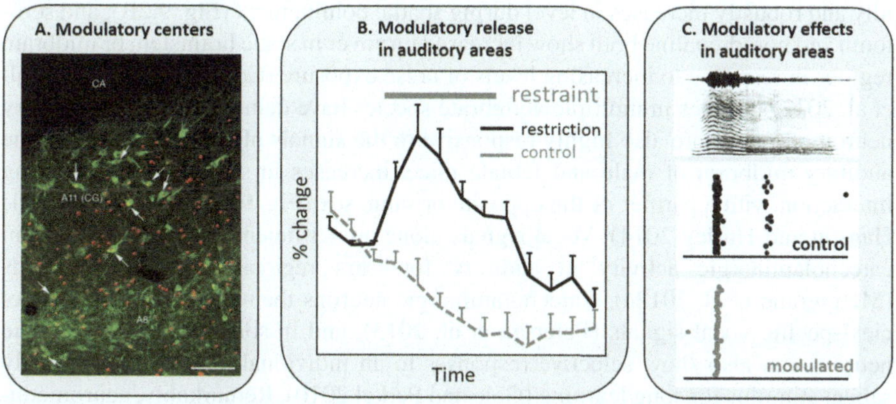

Fig. 9.2 Illustration of methods for measuring neuromodulatory function. **(A)** Co-label of synthetic enzyme for catecholamines (*green*) with a marker of immediate early gene expression (*red*). *Arrows* and *asterisk* indicate double-labeled neurons; *scale bar:* 100 μm. **(B)** Increase in electrochemically measured serotonin in the inferior colliculus during physical restriction. **(C)** Example of modulatory effects on a spike train in response to a vocalization playback. *Top*: oscillogram and spectrogram of a mouse vocalization. *Middle*: Raster plot of the response of a single inferior colliculus neuron to the call in the top panel (*control*). *Bottom panel*: Response of the same neuron to the same call during agonism of serotonin receptors (*modulated*). (*A8*, *A11*, aminergic cell groups; *CA*, cerebral aqueduct; *CG*, central gray of the midbrain) [**A** from Bharati and Goodson (2006), used with permission; **B** adapted from Hall et al. (2012); **C** unpublished data from L. Hurley]

events, internal state, and past experience (Cransac et al. 1998; Hurley and Hall 2011). Neuromodulators are often described as broadly mediating the effects of behavioral arousal or attention but also convey nuanced information on variation within behavioral contexts.

Neurons at many levels of the auditory system respond differently during different phases of the sleep-wake cycle (Velluti 2008). All of the neuromodulatory systems described in this chapter also show activity that is tied to the sleep-wake cycle. Higher levels of firing or different firing modes by neuromodulatory neurons, coupled with greater release of neuromodulators in target areas, typically correspond to waking states but also vary across different phases of the sleep-wake cycle (NA: Berridge and Waterhouse 2003; ACh: Lee et al. 2005; DA: Monti and Monti 2007). Although comparisons of neuromodulatory activity between sleep and waking have been made rarely within the auditory system, the effects of neuromodulators administered to auditory regions may qualitatively or quantitatively depend on the level of arousal (Manunta and Edeline 1999; Cardin and Schmidt 2004). In a similar vein, the levels of at least one neuromodulator, serotonin, rise in the auditory midbrain of mice during recovery from anesthesia, as a relationship to general arousal would predict (Hall et al. 2010).

Neuromodulators in the auditory system respond to behaviorally salient events, from imposed stressors to interaction with conspecifics, a class of behavioral events with special relevance to vocal communication. External stressors quickly increase the activity of multiple neuromodulators in different auditory areas. Serotonin rapidly and robustly increases in level during spatial confinement (Fig. 9.2B), and serotonin and noradrenaline both show increased turnover in some brainstem or midbrain regions in response to increasing levels of noise exposure (Cransac et al. 1998; Hall et al. 2012). Studies in multiple vertebrate species have demonstrated that auditory neuromodulators are also highly responsive to the signals of social partners. In the auditory midbrain of male and female mice, increases in serotonin occur during interaction with a partner of the opposite or same sex (Fig. 9.3A) (Hall et al. 2011; Hanson and Hurley 2014). Vocal signals alone are sufficient to trigger changes in catecholaminergic activity in auditory forebrain regions in some songbirds (Matragrano et al. 2012a). Catecholaminergic neurons themselves respond to species-specific vocal signals (Petersen et al. 2013), and in songbirds, dopaminergic neurons can also show selective responses to an individual's own song, a highly salient stimulus for song learning (Gale and Perkel 2010). Remarkably, neuromodulatory activity may even be selective for behaviorally salient variation within social contexts. Songbirds exposed to more challenging songs of rivals or to "sexier" songs of potential mates have higher levels of catecholaminergic activity in auditory forebrain areas than those exposed to less challenging or less attractive songs (Sockman and Salvante 2008; Sewall et al. 2013. Likewise, elevated serotonin in the inferior colliculus of mice correlates with increased social investigation (Hall et al. 2010).

If a single broad function can be ascribed to neuromodulators in auditory processing, it is that they alter the representation of acoustic stimuli in accordance with salient events, but different neuromodulatory pathways are often described in distinct functional terms. Serotonin has been linked to negative salience and stress as well as to social behavior (Dayan and Huys 2008; Kiser et al. 2012). Acetylcholine has been linked to cue-directed attention and focus (Sarter et al. 2014). Noradrenaline has been linked to arousal and stimulus-directed cognitive shifts (Berridge and Waterhouse 2003; Bouret and Sara 2005) and dopamine has been linked to reward contingencies (Chandler et al. 2014; Pignatelli and Bonci 2015). However, comparison of neuromodulatory responses in the same behavioral paradigms depicts activity in different pathways that, while distinct in some regards, is overlapping in others (Bouret and Sara 2005; Chandler et al. 2014). For instance, multiple neuromodulators represent the behavioral certainty of sensory cues in predicting subsequent events (Sarter et al. 2014; Pignatelli and Bonci 2015). Within the auditory system itself, an example of neuromodulatory overlap is seen in an increase in both serotonergic and dopaminergic metabolites in auditory cortex during associative training sessions (Stark and Scheich 1997). This dual increase was potentially indicative of general stress. However, the dopaminergic metabolite, unlike the serotonergic metabolite, increased most during an initial session for animals that were presented with tones that were predictive of shock, paralleling conditioned behavior. Thus, activity of different neuromodulatory pathways in the auditory system may reflect each other during some behavioral circumstances, but diverge during others in relation to specific behaviors.

In summary, although direct measurement has been rare, neuromodulatory activity within the auditory system is broadly linked to behavioral arousal, and neuro-

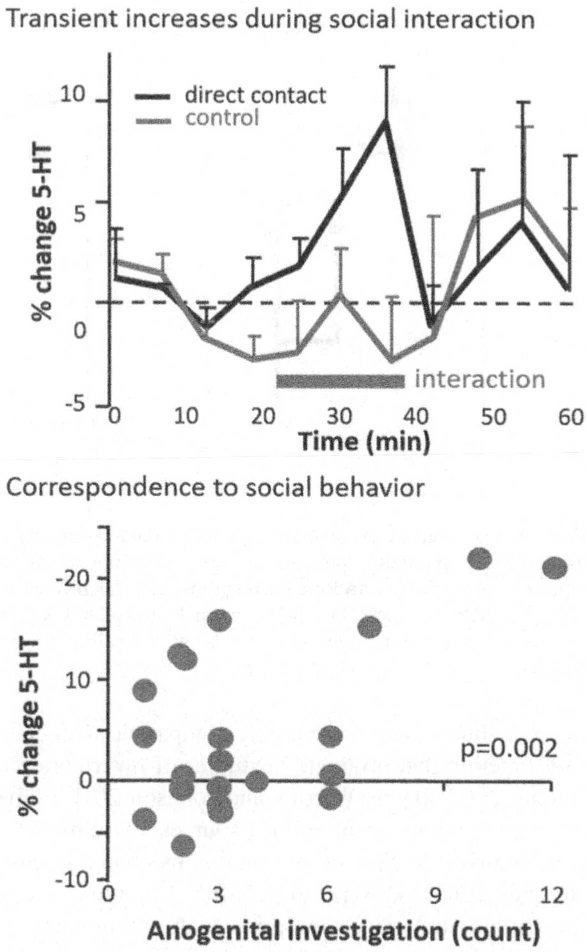

Fig. 9.3 Neuromodulatory activity and effects are related to social context. Serotonin levels (*top*) increase during a social interaction; individual variation in serotonergic change correlates with individual variation in social behavior (*bottom*). P-value taken from multivariate regression. [Modified from Hall et al. (2011)]

modulators also encode information on variation within behavioral contexts like social interaction and associative training. Rather than being entirely separate in functional domains, it is likely that profiles of different neuromodulators signal behaviorally salient events.

9.3.2.3 Neuromodulators Functionally Reconfigure Auditory Circuitry

Approaches to studying the short-term effects of neuromodulators on ongoing auditory processing have included studies in vivo or in brain slice preparations during presentation of sound or electrical stimulation of input pathways and accompanied by stimulation of neuromodulatory centers or application of exogenous agonists and antagonists. Despite this wide range of approaches, most neuromodulatory effects

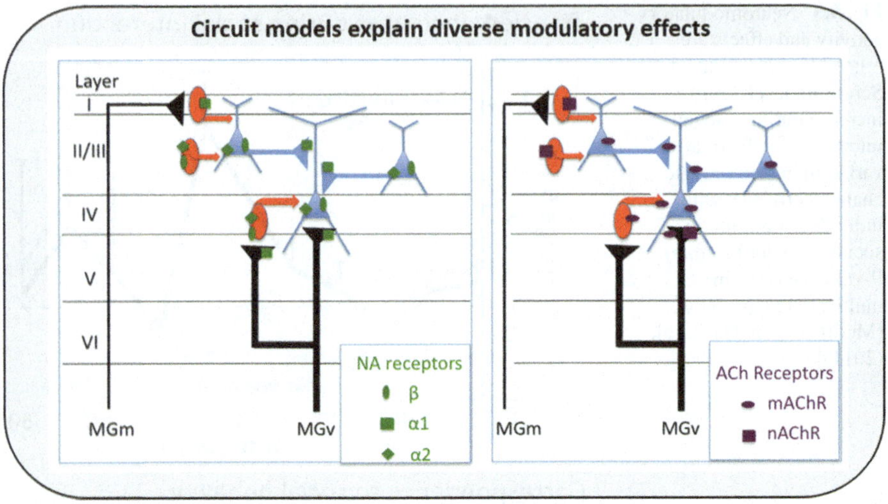

Fig. 9.4 Conceptual representation of neuromodulatory effects in auditory cortex as circuit reconfigurations, emphasizing intracortical versus thalamocortical processing. *Red ovals* and *arrows* represent inhibitory GABAergic interneurons, and *blue triangles and lines* represent glutamatergic neurons. (ACh, acetylcholine; MGm, medial geniculate body, medial subdivision; MGv, medial geniculate body, ventral subdivision; NA, noradrenaline. [From Edeline (2012) courtesy of the author]

in the auditory system are highly compatible with classic models of neuromodulatory function that originate in studies of invertebrate motor circuitry (Marder and Bucher 2007; Harris-Warrick and Johnson 2010). Given a complex neural circuit integrating inputs from multiple sources, these models portray neuromodulators as reconfiguring the flow of information by changing intrinsic properties and synaptic strengths (Fig. 9.4). What this confers on auditory circuitry is the flexibility to favor the configuration most appropriate to the circumstances triggering neuromodulatory release.

Reconfiguration of auditory circuits by neuromodulators occurs at multiple levels of auditory processing. In auditory cortex, several types of neuromodulators alter the balance between thalamocortical and intracortical processing. For example, a cholinergic agonist acting through muscarinic receptors dampens evoked polysynaptic inputs to cortical layer IV neurons and has less of a dampening effect on thalamocortically evoked fast potentials than on intracortically evoked fast potentials. Overall, this combination of effects promotes fast feedthrough processing (Hsieh et al. 2000). In contrast, dopamine acting through D1/D5 receptors prolongs input to the cortex by recruitment of a feedback loop through auditory thalamus, ultimately prolonging horizontal interactions within auditory cortex (Happel et al. 2014). Circuit reconfiguration is also seen at subcortical levels. In the dorsal cochlear nucleus, muscarinic acetylcholine receptors strengthen activity along a polysensory neural pathway by targeting multiple neuron types (fusiform cells: Chen et al. 1998; granule cells: Kőszeghy et al. 2012; cartwheel cells: He et al.

2014). Likewise, by influencing both principal neurons and inhibitory interneurons, serotonin increases the relative weight of polysensory inputs versus auditory inputs (Tang and Trussell 2017).

Acetylcholine, originating in projections from the ventral nucleus of the trapezoid body, also influences sources of auditory input to the cochlear nucleus, dampening cochlear amplification but increasing the responsiveness of T stellate neurons (Fujino and Oertel 2001). This constellation of effects could potentially influence the balance between auditory and polysensory information at the level of projection neurons from the cochlear nucleus (see Trussell and Oertel, Chap. 4).

Effects of dopamine, serotonin, noradrenaline, or acetylcholine have also been variously reported within many nuclei in the ascending auditory system, including the cochlear nucleus (Ebert and Ostwald 1992; Felix et al. 2017), medial nucleus of the trapezoid body (Leão and Von Gersdorff 2002), lateral superior olive (Fitzgerald and Sanes 1999), inferior colliculus (Fig. 9.4A) (Habbicht and Vater 1996; Hurley and Sullivan 2012; Gittelman et al. 2013), and medial geniculate body (Pape and McCormick 1989). Although most of these studies have examined the effects of single neuromodulators at unitary sites in the auditory system, there are two important points to be addressed in establishing the ultimate effects of neuromodulatory release. First, neuromodulators simultaneously acting at multiple auditory sites likely interact (Ma and Suga 2005), although this is a topic that in general has not been well-explored. Second, different neuromodulators commonly converge in their effects on single neuron types, a phenomenon that is exemplified by the effects of dopamine (Bender et al. 2010), acetylcholine (He et al. 2014), and noradrenaline (Kuo and Trussell 2011) on inhibitory cartwheel interneurons in the dorsal cochlear nucleus, but the phenomenon is seen at many sites along the auditory neuraxis. These findings suggest that different neuromodulatory systems do indeed have some of the same auditory targets and may interact in creative ways in line with overlaps in their release patterns.

Precisely how neuromodulators reconfigure particular auditory circuits depends on a range of factors, including which subtype of neuromodulatory receptor is activated, since different receptor types act via different intracellular effectors, ultimately influencing membrane properties in distinct ways (e.g., Ramos and Arnsten 2007; Hannon and Hoyer 2008). If particular receptor types are expressed by excitatory versus inhibitory neurons or in specific subcellular locations, the differences can lead to highly targeted effects on neural circuits and microcircuits. As convincing examples, dopamine modulates calcium influx through T-type channels found exclusively on the axon initial segment (but not on dendrites) of inhibitory cartwheel neurons of the dorsal cochlear nucleus (Bender et al. 2010), and serotonin acting through 5-HT1A receptors alters spike threshold in the axon initial segment of neurons in the medial superior olive (Ko et al. 2016). This results in a selective reduction of the spiking output of these neurons. Receptor type and location may also interact, as occurs for the effects of noradrenaline on the responses of layer II/III pyramidal neurons in auditory cortex (Salgado et al. 2011). Noradrenaline increases the amplitudes of inhibitory currents generated by stimulation of layer II/III inputs via $\alpha 2$ and β adrenergic receptors, but noradrenaline decreases inhibitory currents generated by stimulation of layer I inputs via α1 receptors. The suggested result is an emphasis on the processing of nontonotopic or intracortical inputs.

An emergent property of this wide variety of receptor mechanisms is that neuromodulators often have effects that depend on the spectrotemporal structures of the auditory stimuli presented, extending to different effects on varied species-specific vocalizations (Hurley and Pollak 2005). Such stimulus dependence may be tied to behavioral salience, since manipulating modulatory systems, like the noradrenergic system, influences the ability of auditory neurons to encode relevant stimuli such as vocal signals (Fig. 9.5) (Castelino and Schmidt 2010; Ikeda et al. 2015). Stimulus dependence of neuromodulators fits well with the understanding of the mechanisms described above. Variation along a given stimulus dimension may create variation in the profiles of inputs, which are differentially sensitive to neuromodulation via specific types of receptors. However, this phenomenon raises concerns for interpreting neuromodulatory function in response to natural stimuli based on simpler stimuli like tones, since the two may be very different (Gaucher and Edeline 2015).

In summary, the effects of neuromodulators on auditory processing are prevalent and strong. Even single neuromodulators can reconfigure auditory circuitry through multiple receptor types and in multiple auditory regions, and multiple neuromodulators may converge at the level of single neuron types. This makes the effects of neuromodulators complex but confers the ability to produce a range of behaviorally appropriate outputs from auditory circuitry.

9.3.2.4 Neuromodulators Facilitate Long-Term Plasticity in Adults

Facilitating experience-dependent plasticity in the adult auditory system is a core part of the neuromodulatory portfolio. Within the auditory cortex, a paradigm that has been extensively explored is the facilitation of changes in frequency responsiveness following associative pairing of tones with aversive stimuli. This topic has been reviewed repeatedly from multiple perspectives (e.g., Froemke and Martins 2011; Weinberger 2015), so this subject is only briefly sampled here with special emphasis on the additional aspects of neuromodulatory function detailed in this chapter.

In the associative paradigm, changes in the receptive fields of auditory neurons can be produced by pairing an auditory stimulus with an unconditioned stimulus, like a brief shock, that confers predictive value on the tone (Bakin and Weinberger 1990; Weinberger 2007). Receptive field changes at the level of single neurons produce reorganization of the cortical tonotopic map, such that more neurons are more closely matched with the frequency of the conditioned stimulus (Fig. 9.6A) (Kilgard and Merzenich 1998) in a way that predicts individual variation in behavior (Bieszczad et al. 2013).

Acetylcholine plays an important role in this process. The release of acetylcholine within the auditory cortex tracks conditioning in that increased levels occur following tone–reward pairing, but not control treatment (Butt et al. 2009). Notably, phasic stimulation of the nucleus basalis, a major source of cholinergic input to auditory cortex (Bajo et al. 2014), can substitute for an unconditioned stimulus, so

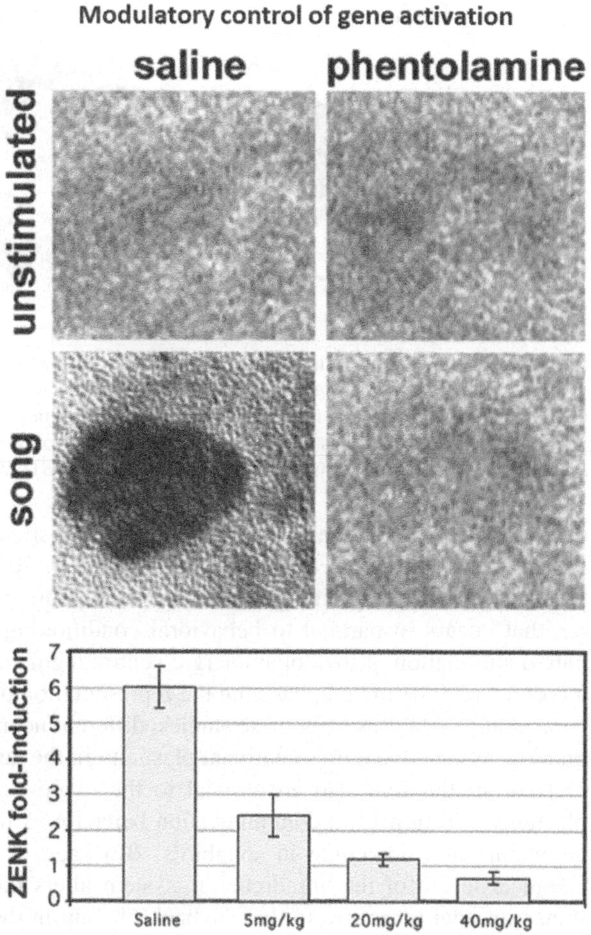

Fig. 9.5 Norepinephrine gates song-triggered gene activation in songbird auditory forebrain. *Top*: Blocking alpha adrenergic receptors with systemically administered phentolamine prevents playback of conspecific song from activating *zenk* transcription in auditory forebrain. *Bottom*: Dose-response relationship for the effects of phentolamine relative to saline injection (as control, shown in both). [Taken from Velho et al. (2012) with permission]

that paired but not unpaired stimulation shifts the best frequencies of many neurons closer to the frequency of the conditioned tone (Bakin and Weinberger 1996; Kilgard and Merzenich 1998). Finally, blocking endogenous sources of acetylcholine from activating muscarinic receptors can prevent many features of the associative changes in frequency tuning (Froemke et al. 2013). A characteristic that this process shares with short-term modulatory plasticity is that facilitation of shifts in tuning by acetylcholine relies on reconfiguration of cortical circuitry. At the level of synaptic inputs that underlie receptive field changes, pairing of a tone with stimulation of nucleus basalis initially causes a reduction of inhibition at the conditioned frequency that is followed by a re-balancing of excitation and inhibition to center around the new best frequency (Froemke and Martins 2011).

Neuromodulators other than acetylcholine also facilitate long-term changes in frequency tuning (Weinberger 2015). Direct application of serotonin or stimulating noradrenergic input causes changes in frequency tuning at the level of single corti-

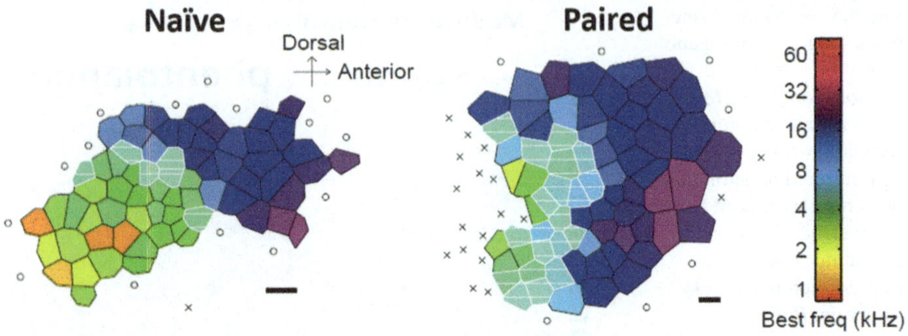

Fig. 9.6 Tonotopic reconfiguration in auditory cortex following pairing of tone presentation with neuromodulatory activation. A 9 kHz tone was paired with stimulation of the cholinergic nucleus basalis. *Light blue polygons* represent regions responding best to the conditioning frequency. *O* and *X* symbols represent sites that did not respond to tones or did not meet criterion values. (*Scale bar:* 200 μm) [Reprinted from Kilgard and Merzenich (1998) with permission from AAAS]

cal neurons, although these may occur in the opposite direction to those facilitated by acetylcholine (Ji and Suga 2007; Edeline et al. 2011). Dopamine also triggers plasticity in frequency tuning. In addition to dopaminergic activity in auditory cortex that occurs in parallel to behavioral conditioning (Stark and Scheich 1997), paired stimulation of the dopaminergic ventral tegmental area with tones produces a cortical remodeling emphasizing the representation of the conditioned frequency (Bao et al. 2001). Based on these studies, different neuromodulatory pathways may underlie associative representational plasticity in the auditory cortex.

Neuromodulators also are crucial to the expression of experience-dependent plasticity during natural communication behavior as illustrated by studies of noradrenaline and dopamine in songbirds. Blockage of noradrenergic receptors or chemical lesion of the noradrenergic system alters the presence or selectivity of transcriptional responses to the playback of song in the auditory forebrain (Lynch and Ball 2008; Velho et al. 2012). An interesting difference of the birdsong paradigm to the associative paradigm in mammals is that the neuromodulatory signal for salience in songbirds is in part triggered by the social stimulus of song itself. This is demonstrated by the responsiveness and selectivity of neuromodulatory neurons to species-specific acoustic signals (Gale and Perkel 2010; Petersen et al. 2013). These types of findings suggest that natural auditory stimuli have intrinsic salience within the context of social behavior that can be further enhanced by factors such as experience or reproductive state (Maney 2013). Such complexity may be typical of the relationships of stimuli to positive or negative salience in the natural world and can inform a view of neuromodulatory function as occurring through mutual instruction with primary sensory systems rather than as a unidirectional relationship.

In summary, neuromodulators help to cement the functional reconfiguration of auditory circuits into lasting changes in stimulus coding. These changes adapt auditory responses in the long term to emphasize stimuli that have occurred during behaviorally salient events like aversive episodes or social interaction.

9.3.2.5 Neuromodulatory Systems are Plastic

In addition to facilitating short- and long-term plasticity in auditory circuits, the portions of neuromodulatory systems localized within auditory regions themselves show a high degree of plasticity in response to changes in both physiological state and peripheral input. This is seen at virtually every level of organization important to modulatory function: in the innervation of auditory regions, the release of neuromodulators, and the expression of receptors. Factors that trigger such plasticity include reproductive state as signaled by gonadal hormones. Priming female songbirds with estradiol increases the density of catecholaminergic and serotonergic fibers, as well as the levels of noradrenaline and of a serotonergic metabolite in the auditory forebrain or midbrain (Matragrano et al. 2011, 2012b). Hearing loss has a significant impact on neuromodulatory projections (Papesh and Hurley 2012), ligand receptor binding (Jin et al. 2006), and receptor expression (Holt et al. 2005; Smith et al. 2014). More subtle changes in peripheral input, such as the makeup of the social environment, can also trigger neuromodulatory plasticity (Sockman and Salvante 2008; Sewall et al. 2013). These types of plasticity allow adaptation of neuromodulatory systems to changes in the average stimulus environment or internal state.

9.3.3 Neuromodulators Help Organize Auditory Responses to Noise and Social Contexts

The wide range of neuromodulatory effects described in an earlier section reflects the aims and approaches of divergent domains of auditory research. In some ways, this diversity makes it difficult to formulate integrated views of the roles of neuromodulators in auditory function. The aim of the following section, therefore, is to gather information on neuromodulation from a spectrum of studies addressing an active area of auditory research: exposure to noise and subsequent hearing loss. Neuromodulatory systems are sensitive to many of the factors related to hearing loss and its related outcomes, including exposure to noise and stress (Knipper et al. 2013). Therefore, it is not especially surprising that many features of neuromodulatory function are consistent with a model of these systems contributing to the central auditory response to noise exposure and hearing loss.

9.3.3.1 Exposure to Noise Recruits Neuromodulatory Systems

Neuromodulators could play a role in the very earliest responses of the auditory system to noise exposure. Increased transcriptional activity by modulatory neurons themselves during exposure to noise, a paradigm for inducing stress, occurs in multiple modulatory systems (Campeau and Watson 1997). Measurements along the auditory neuraxis confirm that neuromodulatory activity responds to noise. Serotonin levels increase rapidly and remain elevated in the inferior colliculus in response to moderate noise levels (Hall et al. 2012). Serotonergic and noradrenergic turnover are also influenced by the presentation of noise, although this may vary among regions or stimulus intensities (Cransac et al. 1998). These findings suggest that, in general, neuromodulators are recruited by noise, but this recruitment may vary across auditory regions or with stimulus characteristics.

9.3.3.2 Olivocochlear Pathways Under Strong Neuromodulatory Control

Because one proposed role for medial olivocochlear projections is to protect cochlear function during exposure to potentially damaging noise (Maison and Liberman 2000; Le Prell et al. 2003), it is interesting that retrogradely labeled olivocochlear neurons are in close proximity to both noradrenergic fibers (Woods and Azeredo 1999; Mulders and Robertson 2000) and serotonergic fibers (Thompson and Thompson 1995; Woods and Azeredo 1999). Moreover, the application of either noradrenaline or serotonin has predominantly excitatory effects on neurons in this region (Wang and Robertson 1997), suggesting that neuromodulation could contribute to dampening cochlear output. Indeed, injecting noradrenaline in the vicinity of medial olivocochlear neurons decreases the amplitude of the compound action potential (Mulders and Robertson 2005b). Taken together, these studies suggest that central noradrenaline or serotonin could facilitate efferent control of cochlear responsiveness. This phenomenon could result in the facilitation of either protection from noise or of additional proposed functions of the medial olivocochlear pathway, such as improving the processing of signals in noise (Elgoyhen and Katz 2012).

9.3.3.3 Auditory and Neuromodulatory Circuitry Changes After Hearing Loss

An important model for the response of the central auditory system to cochlear damage proposes that a reduction in peripheral input triggers changes in the balance between excitation and inhibition, leading to compensatory central hyperactivity or altered tonotopic interactions (Salvi et al. 2000; Eggermont 2003). These types of changes may create auditory processing that is dysfunctional in level or timing, or they may lead to perceptual abnormalities like tinnitus and hyperacusis (Møller 2007; Noreña 2011). To the extent that neuromodulatory systems regulate excitatory/inhibitory balance, they could interact with these processes. This is exemplified by the serotonergic system, which has long been noted to dampen auditory gain (Hegerl et al. 2001; O'Neill et al. 2008) and has been linked to inhibition or suppression at the level of single neurons (DeFelipe et al. 1991; Wang et al. 2008). Another feature of the serotonergic system, which has lent itself especially well to hypotheses on central auditory plasticity, is its own sensitivity to hearing loss. Acoustic trauma alters the density of serotonergic projections to the inferior colliculus

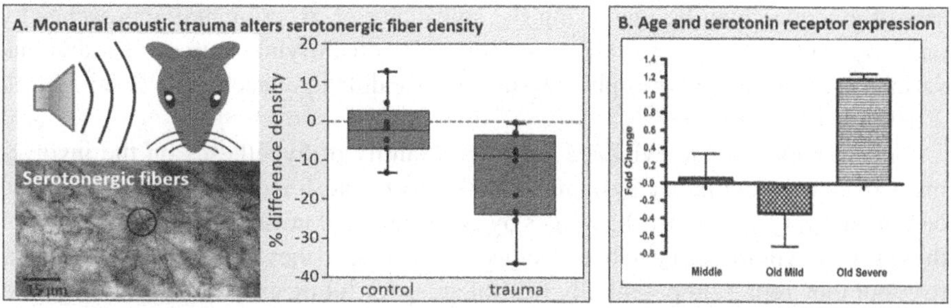

Fig. 9.7 Hearing loss influences the serotonergic system. **(A)** Monaural acoustic trauma decreases the ratio of projections in the contralateral versus ipsilateral inferior colliculus (interaction of side _ treatment, $F_{(1,16)} = 5.90$, Bonferroni post-hoc test $p < 0.05$). Circle in photomicrograph represents "spaceball" sampling technique used to stereologically estimate fiber density. **(B)** In old mice with severe hearing loss, expression of the 5-HT2B receptor is increased. *Middle,* middle-aged mice with good hearing; *Old mild,* old mice with mild hearing loss; *Old severe,* old mice with severe hearing loss. [**A** adapted from Papesh and Hurley (2012); **B** reprinted from Tadros et al. (2007) with permission from Elsevier]

(Fig. 9.7A) (Papesh and Hurley 2012). Multiple independent studies have further documented the upregulation or downregulation of serotonin receptor expression in the same region following hearing loss caused by acoustic trauma (Smith et al. 2014), or cochlear ablation (Holt et al. 2005), or in correspondence with aging (Fig. 9.7B) (Tadros et al. 2007). In a study directly comparing multiple types of serotonin receptors, the 5-HT1B receptor, a type that putatively decreases GABAergic input to inferior colliculus neurons, showed heightened upregulation in response to manipulations, including acoustic trauma (Hurley et al. 2008; Smith et al. 2014). Whether these changes in expression correspond to greater serotonergic control of inhibitory circuitry following hearing loss has not been investigated.

9.3.3.4 Neuromodulators and Auditory Dysfunction Following Hearing Loss

The direct evidence for a link between neuromodulators like serotonin and hearing loss-related disorders, such as tinnitus, is decidedly mixed. Selective serotonin reuptake inhibitors can influence the perception of tinnitus positively, negatively, or not at all, and often in correspondence with the symptoms of depression (Robinson et al. 2007; Baldo et al. 2012). A supporting link between serotonin and tinnitus is provided by salicylate, a drug that produces temporary tinnitus. Acute salicylate injection causes increased immunostaining of immediate early gene products in serotonergic cell groups (Caperton and Thompson 2011) and increases the activity of serotonergic raphe neurons by suppressing inhibitory inputs (Jin et al. 2015). In the inferior colliculus, salicylate triggers a substantial increase in serotonin (Liu et al. 2003) and also diminishes the ability of 5-HT2A receptors to enhance inhibitory postsynaptic currents (Wang et al. 2008). Paths forward through these diverse sets of evidence may be increasing the specificity of experimental approaches by targeting particular receptor types, as well as by quantifying sources of individual variability, such as polymorphisms, in neuromodulatory machinery (Deniz et al. 2010).

In summary, diverse evidence supports a family of hypotheses on the involvement of neuromodulatory systems in both short- and long-term responses of the central auditory system to challenges by environmental noises. At this point, few of these broad hypotheses have been extensively tested, so they remain highly speculative, but they represent potentially fruitful areas for future exploration.

9.4 Summary and Challenges for the Future

Neuromodulators include a variety of signaling molecules and are a part of neural circuits at all levels of the auditory system. The modulators discussed here innervate auditory structures from cochlea to cortex. While much of the innervation arises from modulatory nuclei in the basal forebrain and brainstem, additional projections originate from neurons within auditory nuclei. Together, these modulators play a role in many functions and affect hearing according to internal state, behavioral arousal, and stimulus salience. Mechanisms of modulation are multiple and include the dynamic reconfiguration of circuits that often occurs at multiple levels of the auditory system. Finally, neuromodulators facilitate long-term plasticity. Such plasticity can underlie adaptive changes and may also be implicated in changes, including maladaptive ones, associated with hearing loss or other challenges to the system.

To better characterize circuits (the focal point of this book), it will be essential to identify the relationships between specific cell types in each auditory area and inputs from each modulatory system. Currently, these relationships are largely uncharacterized for the majority of auditory regions. Of course, where multiple sources of a modulator converge on a single target (e.g., for cholinergic inputs to the cochlear nucleus) the source must be related to each target cell type as well. Finally, synapses and receptors must be characterized and related to the physiological effects to understand modulation at a cellular level. To extend this understanding to a systems level, it will be necessary to understand the conditions under which each modulatory system is active and how those inputs affect circuit dynamics and information processing, as well as how these effects in different auditory regions interact with each other. Such information will lead to a more complete understanding of neural modulation of auditory perception and behavior.

Another area where information is incomplete concerns interactions between neuromodulatory cells in different nuclei or even within a single nucleus. For example, cholinergic cells in the PMT project to dopaminergic cells in the ventral tegmental area (Hong and Hikosaka 2014). The LPGi contains cholinergic, serotonergic, and adrenergic cells, each of which has ties to auditory circuits (Van Bockstaele et al. 1993; Bellintani-Guardia et al. 1996). In addition, the LPGi provides a major input to the locus coeruleus, the main source of noradrenergic projections (Luppi et al. 1995). Understanding the interactions between the nuclei will have to be integrated with information on how different modulatory inputs converge on individual cells in auditory nuclei.

As a final reflection, although this chapter has focused on auditory effects of major neuromodulatory pathways, the diffuse projections from neuromodulatory cell groups indicate a broader role in coordinating activity across many neural systems to produce context-appropriate outputs. Ultimately, auditory-driven behaviors are likely to be influenced by parallel neuromodulatory effects on the auditory system and motor or affective systems. Equally broad effects of neuromodulators are likely to be seen during development. The most complete understanding of the influence of neuromodulation on auditory perception may best be achieved by considering these more global interactions.

Acknowledgments The work described here that was completed in the authors' laboratories was supported by grants from the National Institutes of Health (Schofield lab: R01DC004391, F31DC08463, F32DC012450, and F31DC014228; Hurley lab: RO3DC006608, R01DC008963, F32DC000391). The content is solely the responsibility of the authors and does not necessarily represent the official views of the National Institutes of Health. Space limitations prevent citation of many original studies. The authors acknowledge those researchers here with gratitude for their understanding.

Compliance with Ethics Requirements Brett R. Schofield declares that he has no conflict of interest.

Laura Hurley declares that she has no conflict of interest.

References

Altschuler, R. A., & Shore, S. E. (2010). Central auditory neurotransmitters. In A. Rees & A. R. Palmer (Eds.), *The oxford handbook of auditory neuroscience: Vol. 2. The auditory brain* (pp. 65–89). Oxford: Oxford University Press.

Bacci, A., Huguenard, J. R., & Prince, D. A. (2005). Modulation of neocortical interneurons: Extrinsic influences and exercises in self-control. *Trends in Neurosciences, 28*(11), 602–610.

Bajo, V. M., Leach, N. D., Cordery, P. M., Nodal, F. R., & King, A. J. (2014). The cholinergic basal forebrain in the ferret and its inputs to the auditory cortex. *European Journal of Neuroscience, 40*(6), 2922–2940. https://doi.org/10.1111/ejn.12653.

Bakin, J. S., & Weinberger, N. M. (1990). Classical conditioning induces CS-specific receptive field plasticity in the auditory cortex of the guinea pig. *Brain Research, 536*(1-2), 271–286.

Bakin, J. S., & Weinberger, N. M. (1996). Induction of a physiological memory in the cerebral cortex by stimulation of the nucleus basalis. *Proceedings of the National Academy of Sciences of the United States of America, 93*(20), 11219–11224.

Baldo, P., Doree, C., Molin, P., McFerran, D., & Cecco, S. (2012). Antidepressants for patients with tinnitus. *Cochrane Database of Systematic Reviews, 9*, CD003853. https://doi.org/10.1002/14651858.CD003853.pub3.

Bao, S., Chan, V. T., & Merzenich, M. M. (2001). Cortical remodelling induced by activity of ventral tegmental dopamine neurons. *Nature, 412*(6842), 79–83. https://doi.org/10.1038/35083586.

Behrens, E. G., Schofield, B. R., & Thompson, A. M. (2002). Aminergic projections to cochlear nucleus via descending auditory pathways. *Brain Research, 955*(1-2), 34–44.

Bellintani-Guardia, B., Schweizer, M., & Herbert, H. (1996). Analysis of projections from the cochlear nucleus to the lateral paragigantocellular reticular nucleus in the rat. *Cell and Tissue Research, 283*(3), 493–505.

Bender, K. J., Ford, C. P., & Trussell, L. O. (2010). Dopaminergic modulation of axon initial segment calcium channels regulates action potential initiation. *Neuron, 68*(3), 500–511. https://doi.org/10.1016/j.neuron.2010.09.026.

Benson, T. E., & Brown, M. C. (1990). Synapses formed by olivocochlear axon branches in the mouse cochlear nucleus. *The Journal of Comparative Neurology, 295*, 52–70.

Berridge, C. W., & Waterhouse, B. D. (2003). The locus coeruleus-noradrenergic system: Modulation of behavioral state and state-dependent cognitive processes. *Brain Research Reviews, 42*(1), 33–84. doi:S0165017303001437.

Bharati, I. S., & Goodson, J. L. (2006). Fos responses of dopamine neurons to sociosexual stimuli in male zebra finches. *Neuroscience, 143*(3), 661–670. https://doi.org/10.1016/j.neuroscience.2006.08.046.

Bieszczad, K. M., Miasnikov, A. A., & Weinberger, N. M. (2013). Remodeling sensory cortical maps implants specific behavioral memory. *Neuroscience, 246*, 40–51. https://doi.org/10.1016/j.neuroscience.2013.04.038.

Bosch, D., & Schmid, S. (2008). Cholinergic mechanism underlying prepulse inhibition of the startle response in rats. *Neuroscience, 155*(1), 326–335. https://doi.org/10.1016/j.neuroscience.2008.04.018.

Bouret, S., & Sara, S. J. (2005). Network reset: A simplified overarching theory of locus coeruleus noradrenaline function. *Trends in Neurosciences, 28*(11), 574–582. https://doi.org/10.1016/j.tins.2005.09.002.

Brown, M. C. (2011). Anatomy of olivocochlear neurons. In D. K. Ryugo, R. R. Fay, & A. N. Popper (Eds.), *Auditory and vestibular efferents* (pp. 17–37). New York: Springer.

Butt, A. E., Chavez, C. M., Flesher, M. M., Kinney-Hurd, B. L., Araujo, G. C., Miasnikov, A. A., et al. (2009). Association learning-dependent increases in acetylcholine release in the rat auditory cortex during auditory classical conditioning. *Neurobiology of Learning and Memory, 92*(3), 400–409. https://doi.org/10.1016/j.nlm.2009.05.006.

Campbell, M. J., Lewis, D. A., Foote, S. L., & Morrison, J. H. (1987). Distribution of choline acetyltransferase-, serotonin-, dopamine-beta-hydroxylase-, tyrosine hydroxylase-immunoreactive fibers in monkey primary auditory cortex. *The Journal of Comparative Neurology, 261*, 209–220.

Campeau, S., & Watson, S. J. (1997). Neuroendocrine and behavioral responses and brain pattern of c-fos induction associated with audiogenic stress. *Journal of Neuroendocrinology, 9*(8), 577–588.

Caperton, K. K., & Thompson, A. M. (2011). Activation of serotonergic neurons during salicylate-induced tinnitus. *Otology and Neurotology, 32*(2), 301–307. https://doi.org/10.1097/MAO.0b013e3182009d46.

Cardin, J. A., & Schmidt, M. F. (2004). Noradrenergic inputs mediate state dependence of auditory responses in the avian song system. *The Journal of Neuroscience, 24*(35), 7745–7753. https://doi.org/10.1523/JNEUROSCI.1951-04.2004.

Castelino, C. B., & Schmidt, M. F. (2010). What birdsong can teach us about the central noradrenergic system. *Journal of Chemical Neuroanatomy, 39*(2), 96–111. https://doi.org/10.1016/j.jchemneu.2009.08.003.

Challis, C., Boulden, J., Veerakumar, A., Espallergues, J., Vassoler, F. M., Pierce, R. C., et al. (2013). Raphe GABAergic neurons mediate the acquisition of avoidance after social defeat. *The Journal of Neuroscience, 33*(35), 13978–13988. https://doi.org/10.1523/JNEUROSCI.2383-13.2013.

Chandler, D. J., Waterhouse, B. D., & Gao, W. J. (2014). New perspectives on catecholaminergic regulation of executive circuits: Evidence for independent modulation of prefrontal functions by midbrain dopaminergic and noradrenergic neurons. *Frontiers in Neural Circuits, 8*, 53. https://doi.org/10.3389/fncir.2014.00053.

Chen, K., Waller, H. J., & Godfrey, D. A. (1998). Effects of endogenous acetylcholine on spontaneous activity in rat dorsal cochlear nucleus slices. *Brain Research, 783*(2), 219–226.

Commons, K. G. (2015). Two major network domains in the dorsal raphe nucleus. *The Journal of Comparative Neurology, 523*(10), 1488–1504. https://doi.org/10.1002/cne.23748.

Cooper, J. R., Bloom, F. E., & Roth, R. H. (2003). *The biochemical basis of neuropharmacology* (8th ed.). New York: Oxford University Press.

Cransac, H., Cottet-Emard, J. M., Hellstrom, S., & Peyrin, L. (1998). Specific sound-induced noradrenergic and serotonergic activation in central auditory structures. *Hearing Research, 118*(1-2), 151–156.

Darrow, K. N., Simons, E. J., Dodds, L., & Liberman, M. C. (2006). Dopaminergic innervation of the mouse inner ear: Evidence for a separate cytochemical group of cochlear efferent fibers. *The Journal of Comparative Neurology, 498*(3), 403–414.

Dayan, P., & Huys, Q. J. (2008). Serotonin, inhibition, and negative mood. *PLoS Computational Biology, 4*(2), e4. https://doi.org/10.1371/journal.pcbi.0040004.

DeFelipe, J., Hendry, S. H., Hashikawa, T., & Jones, E. G. (1991). Synaptic relationships of serotonin-immunoreactive terminal baskets on GABA neurons in the cat auditory cortex. *Cerebral Cortex, 1*(2), 117–133.

Deniz, M., Bayazit, Y. A., Celenk, F., Karabulut, H., Yilmaz, A., Gunduz, B., et al. (2010). Significance of serotonin transporter gene polymorphism in tinnitus. *Otology and Neurotology, 31*(1), 19–24. https://doi.org/10.1097/MAO.0b013e3181c2dcbc.

Descarries, L., & Mechawar, N. (2008). Structural organization of monoamine and acetylcholine neuron systems in the rat CNS. In A. Lajtha & E. S. Vizi (Eds.), *Handbook of neurochemistry and molecular neurobiology* (3rd ed., pp. 1–20). New York: Springer.

Ebert, U., & Ostwald, J. (1992). Serotonin modulates auditory information processing in the cochlear nucleus of the rat. *Neuroscience Letters, 145*, 51–54.

Edeline, J. M. (2012). Beyond traditional approaches to understanding the functional role of neuromodulators in sensory cortices. *Frontiers in Behavioral Neuroscience, 6*, 45. https://doi.org/10.3389/fnbeh.2012.00045.

Edeline, J. M., Manunta, Y., & Hennevin, E. (2011). Induction of selective plasticity in the frequency tuning of auditory cortex and auditory thalamus neurons by locus coeruleus stimulation. *Hearing Research, 274*(1-2), 75–84. https://doi.org/10.1016/j.heares.2010.08.005.

Eggermont, J. J. (2003). Central tinnitus. *Auris Nasus Larynx, 30*, 7–12.

Elgoyhen, A. B., & Katz, E. (2012). The efferent medial olivocochlear-hair cell synapse. *Journal of Physiology–Paris, 106*(1-2), 47–56. https://doi.org/10.1016/j.jphysparis.2011.06.001.

Felix, R. A., 2nd, Elde, C. J., Nevue, A. A., & Portfors, C. V. (2017). Serotonin modulates response properties of neurons in the dorsal cochlear nucleus of the mouse. *Hearing Research, 344*, 13–23. https://doi.org/10.1016/j.heares.2016.10.017.

Fitzgerald, K. K., & Sanes, D. H. (1999). Serotonergic modulation of synapses in the developing gerbil lateral superior olive. *Journal of Neurophysiology, 81*(6), 2743–2752.

Froemke, R. C., & Martins, A. R. (2011). Spectrotemporal dynamics of auditory cortical synaptic receptive field plasticity. *Hearing Research, 279*(1-2), 149–161. https://doi.org/10.1016/j.heares.2011.03.005.

Froemke, R. C., Carcea, I., Barker, A. J., Yuan, K., Seybold, B. A., Martins, A. R. O., et al. (2013). Long-term modification of cortical synapses improves sensory perception. *Nature Neuroscience, 16*(1), 79–88. https://doi.org/10.1038/nn.3274.

Fujino, K., & Oertel, D. (2001). Cholinergic modulation of stellate cells in the mammalian ventral cochlear nucleus. *The Journal of Neuroscience, 21*(18), 7372–7383.

Fuller, P., Sherman, D., Pedersen, N. P., Saper, C. B., & Lu, J. (2011). Reassessment of the structural basis of the ascending arousal system. *The Journal of Comparative Neurology, 519*(5), 933–956. https://doi.org/10.1002/cne.22559.

Gale, S. D., & Perkel, D. J. (2010). A basal ganglia pathway drives selective auditory responses in songbird dopaminergic neurons via disinhibition. *The Journal of Neuroscience, 30*(3), 1027–1037. https://doi.org/10.1523/JNEUROSCI.3585-09.2010.

Gaucher, Q., & Edeline, J. M. (2015). Stimulus-specific effects of noradrenaline in auditory cortex: Implications for the discrimination of communication sounds. *Journal of Physiology, 593*(4), 1003–1020. https://doi.org/10.1113/jphysiol.2014.282855.

Gittelman, J. X., Perkel, D. J., & Portfors, C. V. (2013). Dopamine modulates auditory responses in the inferior colliculus in a heterogeneous manner. *Journal of the Association for Research in Otolaryngology, 14*(5), 719–729. https://doi.org/10.1007/s10162-013-0405-0.

Gómez-Nieto, R., Horta-Junior, J. A., Castellano, O., Herrero-Turrion, M. J., Rubio, M. E., & López, D. E. (2008). Neurochemistry of the afferents to the rat cochlear root nucleus: Possible synaptic modulation of the acoustic startle. *Neuroscience, 154*(1), 51–64. doi:S0306-4522(08)00246-7. https://doi.org/10.1016/j.neuroscience.2008.01.079.

Habbicht, H., & Vater, M. (1996). A microiontophoretic study of acetylcholine effects in the inferior colliculus of horseshoe bats: Implications for a modulatory role. *Brain Research, 724*(2), 169–179.

Hall, I. C., Rebec, G. V., & Hurley, L. M. (2010). Serotonin in the inferior colliculus fluctuates with behavioral state and environmental stimuli. *Journal of Experimental Biology, 213*(7), 1009–1017. https://doi.org/10.1242/jeb.035956.

Hall, I. C., Sell, G. L., & Hurley, L. M. (2011). Social regulation of serotonin in the auditory midbrain. *Behavioral Neuroscience, 125*(4), 501–511. https://doi.org/10.1037/a0024426.

Hall, I. C., Sell, G. L., Chester, E. M., & Hurley, L. M. (2012). Stress-evoked increases in serotonin in the auditory midbrain do not directly result from elevations in serum corticosterone. *Behavioral Brain Research, 226*(1), 41–49. https://doi.org/10.1016/j.bbr.2011.08.042.

Hannon, J., & Hoyer, D. (2008). Molecular biology of 5-HT receptors. *Behavioral Brain Research, 195*(1), 198–213. https://doi.org/10.1016/j.bbr.2008.03.020.

Hanson, J. L., & Hurley, L. M. (2014). Context-dependent fluctuation of serotonin in the auditory midbrain: The influence of sex, reproductive state and experience. *Journal of Experimental Biology, 217*(4), 526–535. https://doi.org/10.1242/jeb.087627.

Happel, M. F., Deliano, M., Handschuh, J., & Ohl, F. W. (2014). Dopamine-modulated recurrent corticoefferent feedback in primary sensory cortex promotes detection of behaviorally relevant stimuli. *Journal of Neuroscience, 34*(4), 1234–1247. https://doi.org/10.1523/JNEUROSCI.1990-13.2014.

Harris-Warrick, R. M., & Johnson, B. R. (2010). Checks and balances in neuromodulation. *Frontiers in Behavioral Neuroscience, 4*, 47. https://doi.org/10.3389/fnbeh.2010.00047.

He, S., Wang, Y. X., Petralia, R. S., & Brenowitz, S. D. (2014). Cholinergic modulation of large-conductance calcium-activated potassium channels regulates synaptic strength and spine calcium in cartwheel cells of the dorsal cochlear nucleus. *The Journal of Neuroscience, 34*(15), 5261–5272. https://doi.org/10.1523/JNEUROSCI.3728-13.2014.

Hegerl, U., Gallinat, J., & Juckel, G. (2001). Event-related potentials. Do they reflect central serotonergic neurotransmission and do they predict clinical response to serotonin agonists? *Journal of Affective Disorders, 62*(1-2), 93–100.

Holt, A. G., Asako, M., Lomax, C. A., MacDonald, J. W., Tong, L., Lomax, M. I., et al. (2005). Deafness-related plasticity in the inferior colliculus: Gene expression profiling following removal of peripheral activity. *Journal of Neurochemistry, 93*(5), 1069–1086. https://doi.org/10.1111/j.1471-4159.2005.03090.x.

Hong, S., & Hikosaka, O. (2014). Pedunculopontine tegmental nucleus neurons provide reward, sensorimotor, and alerting signals to midbrain dopamine neurons. *Neuroscience, 282*, 139–155.

Hsieh, C. Y., Cruikshank, S. J., & Metherate, R. (2000). Differential modulation of auditory thalamocortical and intracortical synaptic transmission by cholinergic agonist. *Brain Research, 880*(1-2), 51–64.

Hurley, L. M., & Hall, I. C. (2011). Context-dependent modulation of auditory processing by serotonin. *Hearing Research, 279*(1-2), 74–84. https://doi.org/10.1016/j.heares.2010.12.015.

Hurley, L. M., & Pollak, G. D. (2005). Serotonin modulates responses to species-specific vocalizations in the inferior colliculus. *Journal of Comparative Physiology A: Neuroethology, Sensory, Neural, and Behavioral Physiology, 191*(6), 535–546. https://doi.org/10.1007/s00359-005-0623-y.

Hurley, L. M., & Sullivan, M. R. (2012). From behavioral context to receptors: Serotonergic modulatory pathways in the IC. *Frontiers in Neural Circuits, 6*, 58. https://doi.org/10.3389/fncir.2012.00058.

Hurley, L. M., & Thompson, A. M. (2001). Serotonergic innervation of the auditory brainstem of the Mexican free-tailed bat, *Tadarida brasiliensis*. *The Journal of Comparative Neurology, 435*(1), 78–88.

Hurley, L. M., Tracy, J. A., & Bohorquez, A. (2008). Serotonin 1B receptor modulates frequency response curves and spectral integration in the inferior colliculus by reducing GABAergic inhibition. *Journal of Neurophysiology, 100*(3), 1656–1667. https://doi.org/10.1152/jn.90536.2008.

Ikeda, M. Z., Jeon, S. D., Cowell, R. A., & Remage-Healey, L. (2015). Norepinephrine modulates coding of complex vocalizations in the songbird auditory cortex independent of local neuroestrogen synthesis. *The Journal of Neuroscience, 35*(25), 9356–9368. https://doi.org/10.1523/JNEUROSCI.4445-14.2015.

Jacobs, B. L., & Azmitia, E. C. (1992). Structure and function of the brain serotonin system. *Physiological Reviews, 72*(1), 165–229.

Ji, W., & Suga, N. (2007). Serotonergic modulation of plasticity of the auditory cortex elicited by fear conditioning. *The Journal of Neuroscience, 27*(18), 4910–4918. https://doi.org/10.1523/JNEUROSCI.5528-06.2007.

Jin, Y. M., Godfrey, D. A., Wang, J., & Kaltenbach, J. A. (2006). Effects of intense tone exposure on choline acetyltransferase activity in the hamster cochlear nucleus. *Hearing Research, 216*, 168–175. https://doi.org/10.1016/j.heares.2006.02.002.

Jin, Y., Luo, B., Su, Y. Y., Wang, X. X., Chen, L., Wang, M., et al. (2015). Sodium salicylate suppresses GABAergic inhibitory activity in neurons of rodent dorsal raphe nucleus. *PLoS One, 10*(5), e0126956. https://doi.org/10.1371/journal.pone.0126956.

Kamke, M. R., Brown, M., & Irvine, D. R. (2005). Origin and immunolesioning of cholinergic basal forebrain innervation of cat primary auditory cortex. *Hearing Research, 206*(1-2), 89–106. https://doi.org/10.1016/j.heares.2004.12.014.

Kawai, H., Lazar, R., & Metherate, R. (2007). Nicotinic control of axon excitability regulates thalamocortical transmission. *Nature Neuroscience, 10*(9), 1168–1175. https://doi.org/10.1038/nn1956.

Kilgard, M. P., & Merzenich, M. M. (1998). Cortical map reorganization enabled by nucleus basalis activity. *Science, 279*(5357), 1714–1718.

Kiser, D., Steemers, B., Branchi, I., & Homberg, J. R. (2012). The reciprocal interaction between serotonin and social behaviour. *Neuroscience and Biobehavioral Reviews, 36*(2), 786–798. https://doi.org/10.1016/j.neubiorev.2011.12.009.

Klepper, A., & Herbert, H. (1991). Distribution and origin of noradrenergic and serotonergic fibers in the cochlear nucleus and inferior colliculus of the rat. *Brain Research, 557*, 190–201.

Knipper, M., Van Dijk, P., Nunes, I., Ruttiger, L., & Zimmermann, U. (2013). Advances in the neurobiology of hearing disorders: Recent developments regarding the basis of tinnitus and hyperacusis. *Progress in Neurobiology, 111*, 17–33. https://doi.org/10.1016/j.pneurobio.2013.08.002.

Ko, K. W., Rasband, M. N., Mesegeur, V., Kramer, R. H., & Golding, N. L. (2016). Serotonin modulates spike probability in the axon initial segment through HCN channels. *Nature Neuroscience, 19*, 826–834. https://doi.org/10.1038/nn.4293.

Kőszeghy, A., Vincze, J., Rusznak, Z., Fu, Y., Paxinos, G., Csernoch, L., et al. (2012). Activation of muscarinic receptors increases the activity of the granule neurones of the rat dorsal cochlear nucleus—a calcium imaging study. *Pflügers Archives—European Journal of Physiology, 463*(6), 829–844. https://doi.org/10.1007/s00424-012-1103-1.

Kuo, S. P., & Trussell, L. O. (2011). Spontaneous spiking and synaptic depression underlie noradrenergic control of feed-forward inhibition. *Neuron, 71*(2), 306–318. https://doi.org/10.1016/j.neuron.2011.05.039.

Le Prell, C. G., Dolan, D. F., Schacht, J., Miller, J. M., Lomax, M. I., & Altschuler, R. A. (2003). Pathways for protection from noise induced hearing loss. *Noise Health, 5*(20), 1–17.

Le Prell, C. G., Halsey, K., Hughes, L. F., Dolan, D. F., & Bledsoe, S. C., Jr. (2005). Disruption of lateral olivocochlear neurons via a dopaminergic neurotoxin depresses sound-evoked auditory nerve activity. *Journal of the Association for Research in Otolaryngology, 6*(1), 48–62.

Leão, R. M., & Von Gersdorff, H. (2002). Noradrenaline increases high-frequency firing at the calyx of Held synapse during development by inhibiting glutamate release. *Journal of Neurophysiology, 87*(5), 2297–2306.

Lee, C. C., & Sherman, S. M. (2010). Drivers and modulators in the central auditory pathways. *Frontiers in Neuroscience, 4*, 79. https://doi.org/10.3389/neuro.01.014.2010.

Lee, M. G., Hassani, O. K., Alonso, A., & Jones, B. E. (2005). Cholinergic basal forebrain neurons burst with theta during waking and paradoxical sleep. *The Journal of Neuroscience, 25*(17), 4365–4369. https://doi.org/10.1523/JNEUROSCI.0178-05.2005.

Levitt, P., & Moore, R. Y. (1978). Noradrenaline neuron innervation of the neocortex in the rat. *Brain Research, 139*(2), 219–231.

Li, L. Y., Xiong, X. R., Ibrahim, L. A., Yuan, W., Tao, H. W., & Zhang, L. I. (2015). Differential receptive field properties of parvalbumin and somatostatin inhibitory neurons in mouse auditory cortex. *Cerebral Cortex, 25*(7), 1782–1791. https://doi.org/10.1093/cercor/bht417.

Liu, J., Li, X., Wang, L., Dong, Y., Han, H., & Liu, G. (2003). Effects of salicylate on serotonergic activities in rat inferior colliculus and auditory cortex. *Hearing Research, 175*(1-2), 45–53.

Luppi, P. H., Aston-Jones, G., Akaoka, H., Chouvet, G., & Jouvet, M. (1995). Afferent projections to the rat locus coeruleus demonstrated by retrograde and anterograde tracing with cholera-toxin B subunit and *Phaseolus vulgaris* leucoagglutinin. *Neuroscience, 65*(1), 119–160.

Lynch, K. S., & Ball, G. F. (2008). Noradrenergic deficits alter processing of communication signals in female songbirds. *Brain, Behavior and Evolution, 72*(3), 207–214. https://doi.org/10.1159/000157357.

Ma, X., & Suga, N. (2005). Long-term cortical plasticity evoked by electric stimulation and acetylcholine applied to the auditory cortex. *Proceedings of the National Academy of Sciences of the United States of America, 102*(26), 9335–9340. https://doi.org/10.1073/pnas.0503851102.

Maison, S. F., & Liberman, M. C. (2000). Predicting vulnerability to acoustic injury with a noninvasive assay of olivocochlear reflex strength. *The Journal of Neuroscience, 20*(12), 4701–4707.

Maney, D. L. (2013). The incentive salience of courtship vocalizations: Hormone-mediated 'wanting' in the auditory system. *Hearing Research, 305*, 19–30. https://doi.org/10.1016/j.heares.2013.04.011.

Manunta, Y., & Edeline, J. M. (1999). Effects of noradrenaline on frequency tuning of auditory cortex neurons during wakefulness and slow-wave sleep. *European Journal of Neuroscience, 11*(6), 2134–2150.

Marder, E., & Bucher, D. (2007). Understanding circuit dynamics using the stomatogastric nervous system of lobsters and crabs. *Annual Review of Physiology, 69*, 291–316. https://doi.org/10.1146/annurev.physiol.69.031905.161516.

Matragrano, L. L., Sanford, S. E., Salvante, K. G., Sockman, K. W., & Maney, D. L. (2011). Estradiol-dependent catecholaminergic innervation of auditory areas in a seasonally breeding songbird. *European Journal of Neuroscience, 34*(3), 416–425. https://doi.org/10.1111/j.1460-9568.2011.07751.x.

Matragrano, L. L., Beaulieu, M., Phillip, J. O., Rae, A. I., Sanford, S. E., Sockman, K. W., et al. (2012a). Rapid effects of hearing song on catecholaminergic activity in the songbird auditory pathway. *PLoS One, 7*(6), e39388. https://doi.org/10.1371/journal.pone.0039388.

Matragrano, L. L., Sanford, S. E., Salvante, K. G., Beaulieu, M., Sockman, K. W., & Maney, D. L. (2012b). Estradiol-dependent modulation of serotonergic markers in auditory areas of a seasonally breeding songbird. *Behavioral Neuroscience, 126*(1), 110–122. https://doi.org/10.1037/a0025586.

Mellott, J. G., Motts, S. D., & Schofield, B. R. (2011). Multiple origins of cholinergic innervation of the cochlear nucleus. *Neuroscience, 180*, 138–147. https://doi.org/10.1016/j.neuroscience.2011.02.010.

Mesik, L., Ma, W. P., Li, L. Y., Ibrahim, L. A., Huang, Z. J., Zhang, L. I., et al. (2015). Functional response properties of VIP-expressing inhibitory neurons in mouse visual and auditory cortex. *Frontiers in Neural Circuits, 9*, 22. https://doi.org/10.3389/fncir.2015.00022.

Metherate, R. (2011). Functional connectivity and cholinergic modulation in auditory cortex. *Neuroscience and Biobehavioral Reviews, 35*(10), 2058–2063. https://doi.org/10.1016/j.neubiorev.2010.11.010.

Miller, D. J., Lackey, E. P., Hackett, T. A., & Kaas, J. H. (2013). Development of myelination and cholinergic innervation in the central auditory system of a prosimian primate (*Otolemur garnetti*). *The Journal of Comparative Neurology, 521*(16), 3804–3816. https://doi.org/10.1002/cne.23379.

Møller, A. R. (2007). The role of neural plasticity in tinnitus. *Progress in Brain Research, 166*, 37–45. https://doi.org/10.1016/S0079-6123(07)66003-8.

Monti, J. M., & Monti, D. (2007). The involvement of dopamine in the modulation of sleep and waking. *Sleep Medicine Reviews, 11*(2), 113–133. https://doi.org/10.1016/j.smrv.2006.08.003.

Motts, S. D., & Schofield, B. R. (2009). Sources of cholinergic input to the inferior colliculus. *Neuroscience, 160*(1), 103–114. doi:S0306-4522(09)00235-8. https://doi.org/10.1016/j.neuroscience.2009.02.036.

Motts, S. D., & Schofield, B. R. (2010). Cholinergic and non-cholinergic projections from the pedunculopontine and laterodorsal tegmental nuclei to the medial geniculate body in guinea pigs. *Frontiers in Neuroanatomy, 4*, 137. https://doi.org/10.3389/fnana.2010.00137.

Mulders, W. H., & Robertson, D. (2000). Morphological relationships of peptidergic and noradrenergic nerve terminals to olivocochlear neurones in the rat. *Hearing Research, 144*(1-2), 53–64.

Mulders, W. H., & Robertson, D. (2001). Origin of the noradrenergic innervation of the superior olivary complex in the rat. *Journal of Chemical Neuroanatomy, 21*(4), 313–322.

Mulders, W. H., & Robertson, D. (2004). Dopaminergic olivocochlear neurons originate in the high frequency region of the lateral superior olive of guinea pigs. *Hearing Research, 187*(1-2), 122–130. doi:S0378595503003083.

Mulders, W. H., & Robertson, D. (2005a). Catecholaminergic innervation of guinea pig superior olivary complex. *Journal of Chemical Neuroanatomy, 30*(4), 230–242.

Mulders, W. H., & Robertson, D. (2005b). Noradrenergic modulation of brainstem nuclei alters cochlear neural output. *Hearing Research, 204*(1-2), 147–155. https://doi.org/10.1016/j.heares.2005.01.009.

Muñoz, W., & Rudy, B. (2014). Spatiotemporal specificity in cholinergic control of neocortical function. *Current Opinion in Neurobiology, 26*, 149–160. https://doi.org/10.1016/j.conb.2014.02.015.

Noreña, A. J. (2011). An integrative model of tinnitus based on a central gain controlling neural sensitivity. *Neuroscience and Biobehavioral Reviews, 35*(5), 1089–1109. https://doi.org/10.1016/j.neubiorev.2010.11.003.

Niu, X., Tahera, Y., & Canlon, B. (2007). Environmental enrichment to sound activates dopaminergic pathways in the auditory system. *Physiology and Behavior, 92*(1-2), 34–39. https://doi.org/10.1016/j.physbeh.2007.05.020.

Oertel, D., Wright, S., Cao, X. J., Ferragamo, M., & Bal, R. (2011). The multiple functions of T stellate/multipolar/chopper cells in the ventral cochlear nucleus. *Hearing Research, 276*, 61–69. https://doi.org/10.1016/j.heares.2010.10.018.

O'Neill, B. V., Croft, R. J., & Nathan, P. J. (2008). The loudness dependence of the auditory evoked potential (LDAEP) as an in vivo biomarker of central serotonergic function in humans: Rationale, evaluation and review of findings. *Human Psychopharmacology, 23*(5), 355–370. https://doi.org/10.1002/hup.940.

Pape, H. C., & McCormick, D. A. (1989). Noradrenaline and serotonin selectively modulate thalamic burst firing by enhancing a hyperpolarization-activated cation current. *Nature, 340*(6236), 715–718. https://doi.org/10.1038/340715a0.

Papesh, M. A., & Hurley, L. M. (2012). Plasticity of serotonergic innervation of the inferior colliculus in mice following acoustic trauma. *Hearing Research, 283*(1-2), 89–97. https://doi.org/10.1016/j.heares.2011.11.004.

Patel, N. C., & Bickford, M. E. (1997). Synaptic targets of cholinergic terminals in the pulvinar nucleus of the cat. *The Journal of Comparative Neurology, 387*(2), 266–278.

Paxinos, G. (1995). *The rat nervous system* (2nd ed.). San Diego: Academic Press.

Petersen, C. L., Timothy, M., Kim, D. S., Bhandiwad, A. A., Mohr, R. A., Sisneros, J. A., et al. (2013). Exposure to advertisement calls of reproductive competitors activates vocal-acoustic

an
8(
Pigna
sy
ne
Ramo
pr
ph
Robin
Re
Ryug
Ne
Salga
La
21
Salvi,
da
Sarter
hy
Sarter
tic
lat
br
Sarter
of
or
Schof
te
Re
Sewal
in
Or
Shern
co
jn.
Sherri
ec
Smith
lat
he
Sockr
an
De
Stark,
di
lit
Steria
er
m
Storn
ne
an

The manufacturer's authorised representative in the EU is Springer Nature Customer Service Centre GmbH, Europaplatz 3, 69115 Heidelberg, Germany. If you have any concerns regarding our products, please contact ProductSafety@springernature.com

Printed and bound by CPI Group (UK) Ltd, Croydon, CR0 4YY

30/09/2026

02225655-0006